**Direct Current Machines**

# Direct Current Machines

**M G SAY**
PHD, MSC, DIC, FIEE, FRSE

**E OPENSHAW TAYLOR**
BSC, DIC, FIEE, SEN MIEEE, FRSE

*Professors Emeriti of Electrical Engineering*
*Heriot–Watt University Edinburgh*

A HALSTED PRESS BOOK

**JOHN WILEY & SONS**
NEW YORK

First published in Great Britain in 1980
by Pitman Publishing Limited

Published in the U.S.A. by Halsted Press,
a Division of John Wiley & Sons, Inc., New York

© M. G. Say & E. Openshaw Taylor 1980

**Library of Congress Cataloging in Publication Data**

Say, Maurice George.
    Direct current machines.

    "A Pitman international text."
    "A Halsted Press book."
    Bibliography: p.
    Includes index.
    1.  Electric machinery—Direct current.      I.   Taylor,
    Eric Openshaw, joint author.    II.    Title.
TK2611.S29    1980         621.313′2         79–19519
ISBN  0–470–26838–7

Printed and bound in Great Britain
at The Pitman Press, Bath

# Contents

# Preface

Dr A. E. Clayton of Manchester published *The Performance and Design of Direct Current Machines* in 1927. The book, several times revised, was a standard and widely read text. The present authors have undertaken the task of carrying on the tradition.

The half-century since 1927 has seen radical changes in the technology of d.c. machines. When Dr Clayton wrote, d.c. mains supplies were common: now they are virtually extinct. Electronics meant little more than radio: power electronics now dominates the supply and control of d.c. machines. Permanent magnets then were usually very hard metallic alloys: mouldable ferrites are now in common use for the fields of small production-run machines. Run-of-the-mill d.c. motors were heteropolar and cylindrical, with copper-sector commutators and carbon brushes: the range now includes unconventional shapes, brushless commutation, homopolar structure and even superconducting field windings. Simple devices were adequate to provide starting, braking and speed control: automation in engineering production now demands sophisticated and precision control of speed and load, some-times with very high rates of acceleration and reversal, with dynamic performance in a dominant role.

There has been also a change in the methods of machine analysis with the introduction by R. H. Park and Gabriel Kron of the 'generalized theory'. The conventional d.c. machine fits elegantly into this concept because it has, at least in idealized form, all the essential elements including the quasi- (or pseudo-) stationary rotor windings which, although they physically rotate, preserve a fixed m.m.f. axis. Generalized theory expresses machine behaviour, both steady-state and transient, as a matrix of simultaneous differential equations. But a machine is not merely an impedance matrix.

Design by computer has displaced the earlier methods based largely on empiricism and 'know-how'. The ubiquitous digital computer solves numerical equations in design, but numerical results are meaningful only in relation to the underlying physical principles. A machine is not just a digital print-out: it is a highly complex electromagnetic-mechanical-thermal device.

Clearly there has to be a reassessment of the content of a textbook devoted to the d.c. machine. The present authors are at one with Dr Clayton in the belief that elementary machine behaviour can best be grasped through application of flux-current interaction, which is so easily demonstrated experimentally. The treatment therefore starts with this principle, the prototype forms in which it can be realized, and the essential function of the commutator. Ideal cases are followed by progressive departures from the ideal that affect practical performance, and commutation is discussed in both conventional and novel forms. Generator and motor operation on traditional d.c. supply is included for completeness, but considerable attention is given to performance of motors fed from a.c. supplies through converters. Motors for use in control systems are described in some detail, together with the features of a number of 'special' machines and devices. The final chapter deals with typical constructional features and with the basics of design.

As both control engineering and semiconductor technology already have a voluminous literature, the authors have deemed it preferable to deal with machines themselves rather than with the multifarious control systems in which they find application.

All formulae, whether concerned with electrical, magnetic, mechanical or thermal phenomena, are couched consistently in SI base units. Only in numerical examples have decimal multiples and submultiples been employed for convenience, in conformity with IEC recommendations. Guidance, if needed, can be gained from a booklet, 'Symbols and Abbreviations for Electrical and Electronic Engineering' (Institution of Electrical Engineers, 1979).

Our thanks are due to good friends in the industry and in the Universities, abroad as well as in the U.K. We are particularly indebted to a number of anonymous critics who have made valuable suggestions for the improvement of the text. We hope that readers, too, will help by their comments.

*Guestling, Sussex* 1980

M.G. Say
E. Openshaw Taylor

# 1 Introductory

## 1.1 DIRECT CURRENT MACHINES

Michael Faraday enunciated in 1831 the principle of electromagnetic induction. A year later he demonstrated the first *homopolar* electromagnetic generator — a copper disc rotated by a spindle in an axially directed permanent-magnetic field, with sliding contacts ('brushes') at the edge and centre of the disc to pick off the generated e.m.f. The *heteropolar* version, in which a winding is rotated in the magnetic field of successive N–S pole-pairs, soon followed. The first rotating machines embodying Ampere's *commutator* appeared in 1833. These generators were soon found to be reversible and to act equally well as motors. Modern electromagnetic machines, notwithstanding a century and a half of technological progress, still exploit the basic Faraday principle.

D.C. machines may work as generators or motors or brakes. In the *generator mode* the machine is driven by a prime mover and develops electrical power. In the *motor mode,* power supplied to the machine electrically is converted into mechanical power as output. The *brake mode* is a generator action but with the electrical power either regenerated or dissipated within the machine system, developing in consequence a mechanical braking effect.

Almost all land-based electrical power-supply networks are a.c. systems of generation, transformation, transmission and distribution; and as d.c. supplies are readily derived therefrom by rectification, there is little need for large d.c. generators. Further, it is accepted practice in industry to employ a.c. motors wherever they are inherently suitable or can be given appropriate characteristics by means of power-electronic devices. Yet there remain important fields of application in which d.c. machines offer economic and technical advantage, for the outstanding property of the d.c. motor is its versatility. It can be designed for wide ranges of voltage/current or speed/ torque relations for both steady-state and transient operation. The advent of the controlled power rectifier has made it possible to link a d.c. machine to an a.c. mains supply, and speed-control 'packages' are readily obtainable for industrial application.

Power plants isolated from land-based supply networks (as on road vehicles, ships and aircraft) may embody d.c. machinery. Normally a secondary battery provides a central supply and requires a means for charging it; and although the trend is towards a.c. generators and rectifiers for

1

the battery-charging duty, many wholly d.c. equipments are still in wide use. Small back-up and standby generating plants providing useful energy from windmills and mountain streams often use d.c. generators because the fluctuating and intermittent drive would, with an a.c. generator, produce unacceptable variation in the generated frequency.

D.C. series motors, well adapted to the requirements of traction, drive intercity and rapid-transit trains. In some process drives and mine-winding engines it may be found advantageous to employ d.c. machinery for effective speed control. Battery vehicles for fork-lift trucks and delivery vans are likely to become more common. Portable drills, sewing machines and hand tools are well served by the 'universal' series motor on single-phase a.c. supply, but work at higher efficiency on d.c. Miniature d.c. motors working from dry cells operate razors, cameras, tape-recorders and similar small loads.

Automation has brought about a resurgence of interest in precision d.c. machines. High-energy permanent magnets, epoxy-resin winding encapsulation and advanced brush or electronic commutation have made possible a range of control machines that are smaller, cheaper and more reliable than their a.c. rivals, particularly in ratings of a few watts. With power transistors such machines can fulfil basic control functions in feedback systems. At higher ratings the 'crossfield' amplifier machine has been applied. A class exemplified by the 'stepper' motor is operated by d.c. pulses to act as a precision transfer device between the digital information presented by the input driver and the corresponding mechanical motion imparted to the load. While having some resemblance to an a.c. machine, the supply is undeniably unidirectional, so that the question 'Is it a d.c. or an a.c. machine?' has a justifiably equivocal answer.

When a simple conductor (typically a disc) is moved through a permanent-magnetic field, circulating currents flow that develop heat and impose a braking force on the conductor. The effect is put to use in eddy-current brakes and couplings, and particularly for providing electrical-energy meters with a braking torque proportional to the speed of the meter movement.

Most industrial motors have heteropolar field systems and a cylindrical shape. There are however, several other possible forms such as the disc, the linear and the tubular. Further, the homopolar field arrangement has been developed to provide large current at low voltage for electrochemical processes. The homopolar shape lends itself to the use of superconducting field windings, by means of which very high working flux densities can be achieved. But whatever the topology of the machine, and be it homo- or heteropolar, the essential requirement is that a current flow in a direction at right-angles to that of the flux of a permanent or electromagnet in such a way as to utilize most effectively the mechanical force developed by flux-current interaction.

### Classification

D.C. machines are built for ratings from megawatts to milliwatts. For convenient reference they may be roughly classified as follows:

*Industrial* Large generators; and motors for mills, cranes, machine tools and general industrial drives of ratings above a few kilowatts, and usually supplied from a.c. mains through rectifiers.

*Small* Motors for hand-tools, domestic equipments and similar mains-fed applications; and automobile starter motors. Their construction is a simplification of that for conventional industrial machines.

*Traction* Motors for railways and battery-fed road vehicles.

*Miniature* Motors of a few watts output for intermittent loads not requiring precision control.

*Control* Machines associated with open- and closed-loop control systems; they are commonly provided with permanent-magnet fields.

*Special* Linear machines, actuators, eddy-current devices, motors with superconducting field windings, and unconventional designs.

## 1.2 MODELS

Electrical science is concerned, on the microscopic scale, with the properties and phenomena that stem from the sub-atomic electrical nature of matter; and on the macroscopic scale, with the concept of electromagnetic-field theory applied to matter in bulk, as codified in the Maxwell electromagnetic-field equations. For the present purpose we adopt a technological viewpoint. If the magnetic field is taken as associated with a current, and the current itself as a continuous flow rather than as a stream of discrete particles, a more direct approach to machines is achieved on a basis of simple concepts. What we seek is a *model* on which analysis can be based. One model, that of flux–current interaction, gives a physical picture of the mechanism of electro-mechanical energy conversion; a second is the dynamic-circuit representation which gives a concise and powerful analysis of overall machine behaviour under steady-state and, particularly, transient operating conditions. Both models are related to the concept of magnetic flux.

### Magnetic flux

A magnetic field is a region of space in which useful physical effects occur. A pictorial model of the field can be made by drawing closed loops of flux such that their direction and spacing are measures of the direction and concentration of the flux, i.e. the vector flux density. In present context the magnetic circuit is composed partly of ferromagnetic material, partly of an airgap. The magnetic material serves to 'guide' the flux in a desired path; the airgap is necessary to make the useful magnetic-field effects readily accessible.

The lines drawn in a flux plot have no real existence, and must not be regarded as like the bristles on a brush which move bodily with the brush-head. In a given magnetic field the flux density may change direction, become weaker in some regions and stronger in others: but the lines do not 'move' into new positions. All we can say is that, if we map the field at various instants, the patterns differ.

Engineers regard a magnetic flux $\Phi$ as 'produced' by electric current, explicit in the case of an electromagnet, implicit for a permanent magnet. A current $i$ develops, around any path that links it, a magnetomotive force $F = i$ [ampere]. The effect of a current can be multiplied by coiling the electric circuit into $N$ turns so that around a path linking all the turns the m.m.f. is $F = Ni$ [ampere or ampere-turn]. The m.m.f. is distributed along the path to give a path element of length d$x$ the magnetic field strength $H$ [ampere (-turn) per metre]. The summation of $H \cdot \mathrm{d}x$ around a single-loop closed path is $F$. At any point, $H$ gives rise to a flux density $B = \mu H$ [tesla], where $\mu$ is the absolute permeability of the ambient medium. Summation of $B$ over the area available to the flux path gives the total flux $\Phi$ [weber].

The 'law' of the magnetic circuit relates the total flux $\Phi$ to the total m.m.f. $F$ through the expressions

$$\Phi = F/S = F\Lambda \tag{1.1}$$

where $S$ is the total reluctance [ampere (-turn) per weber] and $\Lambda = 1/S$ is the total permeance [weber per ampere (-turn)] of the magnetic circuit. For a part-length $x$ [metre] of material of absolute permeability $\mu$ and having a uniform cross-section $a$ over which the flux density $B$ is everywhere the same, the m.m.f. is $F_x = H \cdot x$. We may therefore write $F_x = \Phi S_x = \Phi/\Lambda_x$ where

$$S_x = F_x/\Phi = Hx/\mu Ha = x/\mu a \quad \text{and} \quad \Lambda_x = \mu a/x$$

for the part-reluctance and part-permeance. For a succession of parts $x, y, z \ldots$ in series, through which the same flux passes,

$$F = F_x + F_y + F_z + \ldots \quad \text{and} \quad S = S_x + S_y + S_z + \ldots$$

For parts in parallel that share the flux,

$$F = F_x = F_y = F_z = \ldots \quad \text{and} \quad \Lambda = \Lambda_x + \Lambda_y + \Lambda_z + \ldots$$

For fields in free space or in air, the absolute permeability is the magnetic constant $\mu = \mu_0 = 4\pi/10^7 \cong 1/800\,000$, which means that $H \cong 800\,000\,B$. For fields in ferromagnetic material, $\mu$ is very much greater, the ratio $\mu/\mu_0 = \mu_r$, the relative permeability, ranging from a few hundreds to a half-million. Consequently $H$ lies between $2B$ and $2000\,B$. But in such materials $\mu_r$ depends on $B$ and varies with the degree of magnetic saturation. Nevertheless, it is sometimes permissible to ignore the requirements of ferromagnetic parts of a magnetic circuit and to assume that the whole m.m.f. is impressed on the airgap: this in effect assumes that the relative permeability of the material is infinite.

*Flux Plots*   In a practical device it is necessary to know the distribution of the magnetic field. As this is three-dimensional, an exact solution is difficult;

but a two-dimensional sketch can in many cases adequately represent the field in the most significant region — the airgap. The plot can be drawn by applying a few simple rules relating to the space between the bounding surfaces of an otherwise ferromagnetic magnetic circuit:

(i) The permeability of the ferromagnetic material ('iron') is assumed to be infinite, so that the gap surfaces become magnetic equipotentials. Intermediate equipotential lines may be added to aid the plotting process.

(ii) The flux is mapped by lines between the surfaces, drawn always orthogonal to the equipotentials and making with them a network of curvilinear squares. The plot can then be used for quantitative assessments.

(iii) Only one flux can occupy the gap, but it may be derived as the vector superposition of component flux densities produced by individual currents.

*Maxwell Stress*   A general principle formulated by Maxwell is that forces on magnetic bodies are transmitted across the gap between them by a system of two stresses: (i) a tensile stress of value $\frac{1}{2}BH$ along the line of action of the flux (i.e. a flux line), and (ii) a compression stress of the same value in all directions perpendicular to the flux lines. In an elementary way, the flux lines can be considered to have an 'elastic thread' property, with the tendency to shorten lengthwise and thicken laterally. With practice (and with due regard to the fictive nature of the flux lines) an immediate impression of magnetic force can be gained from a flux plot.

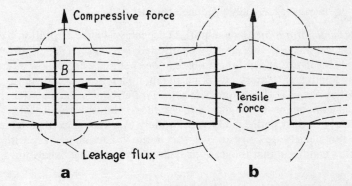

1.1 Maxwell stresses

Fig. 1.1 shows two iron bars forming part of a magnetic circuit. In (*a*) the polar surfaces are close together and the flux is mainly concentrated between the surfaces. The density *B* is large, and so also is *H*, so that $\frac{1}{2}BH$ represents a strong 'tensile' attraction force between the pole-faces. Not all the flux contributes to axial attraction: some, the leakage flux, exists at the flanks of each pole. Flux crossing the boundary between air and a highly permeable ferromagnetic material must enter or leave the boundary almost at right-angles to the surface, so that tensile stress due to leakage flux cannot augment the pole-face attraction. All the 'compressive' stresses balance out by symmetry.

Now consider case (*b*). The greater reluctance of the long airgap reduces the total flux and the pole-face flux density, and there is more leakage because there is little difference in the reluctance of useful and leakage paths. The force of attraction is consequently much less than in case (*a*). The comparison brings out the importance of ascertaining the flux distribution if reliable estimates of mechanical force are to be made. It might be good enough in (*a*) to assume the flux to be uniform over the pole-faces and to pass from face to face directly with negligible leakage. But this will not do for (*b*), which can be dealt with only by aid of a flux plot.

*Magnetic Field Energy*  The Maxwell stress concept implies that energy to the value $\frac{1}{2}BH$ is stored in a unit cube of the space occupied by a magnetic field: thus $\frac{1}{2}BH$ is the magnetic energy density. The complete energy storage in a system is consequently given by $w_f = \frac{1}{2}\Phi F$ [joule]. If an elemental relative displacement d$x$ of parts in a magnetic field results in a change d$w_f$ of the stored energy, the work done in making the displacement is given by

$$dw_f = f_e \cdot dx \quad \text{or} \quad f_e = dw_f/dx \tag{1.2}$$

by the principle of virtual work, [1] where $f_e$ is the force that the magnetic field engenders between the parts. If the elemental movement is angular, d$\theta$, then

$$dw_f = M_e \cdot d\theta \quad \text{or} \quad M_e = dw_f/d\theta \tag{1.3}$$

where $M_e$ is the magnetically produced torque.

*Leakage and Saturation*  Two important aspects of field distribution have practical significance. One, already mentioned, is *leakage*. There is no magnetic insulator, and although flux can be encouraged to follow a ferromagnetic path and to cross a working airgap, it cannot be confined completely thereto. The second is *saturation*. Ferromagnetic materials are essential components of the magnetic circuits of almost all electrical machines, but their permeability varies widely over the range of economically usable flux density. In consequence, flux varies in a nonlinear way with current, a condition that is occasionally crucial to machine behaviour.

## Flux–current interaction model

Important and useful effects occur in magnetic fields in which ferromagnetic parts can move under magneto-mechanical forces by translation or rotation, where mechanical forces are produced on currents, and where electromotive forces are generated by changes of the flux linking an electric circuit. These are set out in Sect. 1.3, using a direct application of the Maxwell stress concept. Although limited to simple cases, the approach gives a direct physical appreciation of the operation of a machine system.

## Dynamic-circuit model

Here the flux $\Phi$ due to a current $i$ is described in terms of an inductance parameter $L$. The flux linked with an electric circuit is then expressed as $N\Phi = \Psi = Li$, and the associated field energy is $w_f = \frac{1}{2}Li^2$. The analysis is thus couched in electric-circuit terms, and what is a highly complex electromagnetic system associated with a defined conducting path can be handled with the aid of network theory. Lumped parameters of resistance and inductance are set up to represent electric, thermal and magnetic-field properties, associated with the circuit properties of voltage, current and e.m.f., to provide a concise shorthand of the system that they model. Account is no longer taken directly of the magnetic field distribution, although flux plotting may sometimes be necessary in establishing the parametric values.

In machines, an essential property is the disturbance of fields by displacement of parts in the process of electrical/mechanical energy conversion, so that the inductances are functions of position. Each electric circuit of the machine has a behaviour expressed by a differential equation, and the device is modelled by as many such equations as there are distinct electric circuits.

As a motor or a generator has mechanical attachments in the form of a load or a driver, both dynamic, the mechanical system reacts on the electrical system to affect acceleration, speed and position. A complete analysis therefore requires consideration of (i) the electric circuit equations, (ii) the energy-conversion process, and (iii) the equation of motion; that is, the whole electromagnetic/mechanical linked system.

## Other models

A machine is as much an electromagnetic-field device as is a wave-guide, and must internally satisfy the Maxwell equations. The field pattern could be expressed in Laplacian or Poissonian terms, and energy transfer across the gap found from application of the Poynting vector. Alternatively, the wave-impedance concept could be applied to unify field and circuit theory. An outline of such models is given by Nasar [2]. Rigorous field theory is elegant and satisfying, but difficult to solve because the topology introduces very complicated boundary conditions. As we have simpler ways, less general but more tractable, it is helpful to employ them.

## 1.3 PRINCIPLES

In common with other electromagnetic/mechanical devices, the operation of the d.c. machine can be explained in terms of basic principles concerned (i) with the development of magneto-mechanical forces and (ii) with the induction of e.m.f.s by the rate of change of flux linkage.

## Alignment force

When a magnetic field is established in an ambient low-permeability medium (such as air), pieces of high-permeability material (such as iron) experience mechanical forces tending to align them with the direction of the field in such a way as to reduce the reluctance of the system.

In the elementary machine of Fig. 1.2($a$), a fixed ferromagnetic member ('stator') is excited by the current in a 'field' winding to produce a flux in the magnetic circuit. A structurally polarized (but unexcited) ferromagnetic 'rotor' member is placed in the gap between the stator poles N and S. A flux plot, drawn in accordance with the rules set out in Sect. 1.2, is shown in ($b$) for the upper gap region (with the radial gap length exaggerated for clarity).

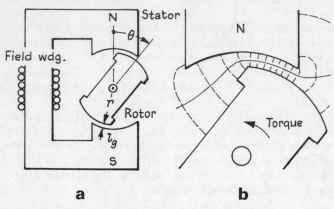

a                              b

1.2 Alignment principle

The stator and rotor facing gap surfaces are taken as equipotentials and one intermediate equipotential has been added. Over the uniform and short gap of length $l_g$ the flux density is high: but it is radial, and from the Maxwell stress concept can result only in a radial attraction between rotor and stator. This force is balanced by an opposite attraction at the lower pole. In the lateral regions of the gap outside the short-gap region, the flux lines terminate on either stator or rotor pole flanks, and the Maxwell tension forces act in a direction to turn the rotor counterclockwise into alignment with the stator poles, increasing the area of the short-gap region and reducing the magnetic-circuit reluctance.

The torque can be deduced from eq. (1.3). If $\theta$, the angle between the stator and rotor magnetic axes, is such as to give a relatively large polar overlap, the flux and gap energy can be taken as wholly contained in the short-gap regions, where the field energy density is $\frac{1}{2}BH = \frac{1}{2}B^2/\mu_0$. With a rotor of axial length $l$, the active gap volume is $2ll_g(r\theta)$, and the gap energy is $w_f = B^2 ll_g r\theta/\mu_0$. An elemental increase $d\theta$ in the displacement angle reduces the gap energy, and equating this to the mechanical work done, $M_f \cdot d\theta$, gives the torque on the rotor

$$M_f = dw_f/d\theta = -B^2 lrl_g/\mu_0 \qquad (1.4)$$

opposing the angular movement and acting to align the magnetic axes.

Now the flux across the short gap is radial and can produce only *radial* force, and eq. (1.4) seems to ignore the fact that *tangential* force is developed by flux entering or leaving the pole flanks, external to the short-gap region. Why? The answer is that an increase in $\theta$ reduces the radial flux, throwing more flux into the torque-developing flanks. The simple analysis above will therefore not be valid if $\theta$ is large enough to minimize the short-gap flux, and the virtual-work principle would then require an integration of the Maxwell tensions for the system as a whole.

## Interaction force

When a current $i_2$ lies in and perpendicular to a magnetic field of density $B_1$ over an active length $l$, a mechanical force

$$f_e = B_1 l i_2 \tag{1.5}$$

is developed on it in a direction perpendicular to both current and field. In $(a)$ of Fig. 1.3 the uniform flux density $B_1$ is established by a field winding else-where on the magnetic circuit, and a linear current sets up a self field super-posed on it. The resultant distribution of flux density is that in $(b)$; and $(c)$ shows that the direction of the mechanical force $f_e$ is unchanged if *both* flux and current directions are reversed. The force acts on the current, which could, for example, be an electron beam; but if it is the flow in a material conductor, the force is transferred to the metal. The direction of the force $f_e$ can be inferred directly from the 'elastic thread' idea.

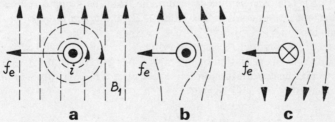

a         b         c

1.3 Interaction principle

To realize this effect in a machine, $B_1$ is set up in the gap between the poles of an electro- or permanent-magnet system, and $i_2$ is conveyed to the conductor from an external source. In Fig. 1.4 the airgap, of length $l_g$, extensive in the plane of the diagram and of depth $l$ measured into the plane, is excited by an m.m.f. $F_1$ to produce as in $(a)$, the flux of uniform density $B_1 = \mu_0 H = \mu_0 F_1/l_g$. A resulting force of attraction of $\frac{1}{2}B_1^2/\mu_0$ per unit area makes it necessary to hold the polar surfaces apart.

Now let the primary field system be unexcited, but let a current $i_2$ be set up as in $(b)$, in a conductor placed centrally in the gap. The m.m.f. $i_2$ circulates a flux that crosses the gap twice. For each gap the m.m.f. $F_2 = \frac{1}{2}i_2$ generates a flux of density $B_2 = \frac{1}{2}i_2\mu_0/l_g$. With both $F_1$ and $F_2$ acting together as in $(c)$, the resultant gap-flux pattern is asymmetric. The lateral force $f_e$ is obtained from the difference of Maxwell compressive stresses on

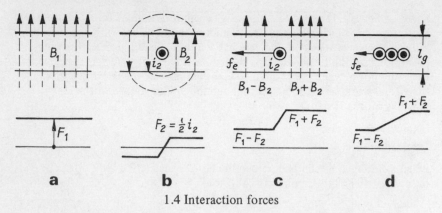

a               b               c               d

1.4 Interaction forces

the two sides of the conductor, or from the change in gap energy that would result from an elemental displacement $dx$ of the conductor. This changes the energy density in the gap volume $l_g l \cdot dx$ from $\frac{1}{2}(B_1 + B_2)^2/\mu_0$ to $\frac{1}{2}(B_1 - B_2)^2/\mu_0$; whence

$$dw_f = \frac{1}{2}[(B_1 + B_2)^2 - (B_1 - B_2)^2]ll_g \cdot dx/\mu_0 = 2B_1 B_2 ll_g \cdot dx/\mu_0$$

But $B_2 = \frac{1}{2}i_2\mu_0/l_g$, so that the force on the current is

$$f_e = dw_f/dx = B_1 l i_2$$

as in eq. (1.5). The force is thus determinable from the undistorted density $B_1$ and the current $i_2$ immersed in it, so long as $B_1$ and $i_2$ are directed mutually at right-angles.

If a belt of $N_2$ conductors, each carrying $i_2$, occupies the gap as in Fig. 1.4(d), there is a gradation of the m.m.f. per gap $F_2$ from $+\frac{1}{2}N_2 i_2$ to $-\frac{1}{2}N_2 i_2$ to give peak corresponding flux densities $\pm B_2$. Writing therefore $N_2 i_2$ in place of $i_2$ in eq. (1.5) gives

$$f_e = B_1 l N_2 i_2 \tag{1.6}$$

as if each conductor made its own contribution. Large interaction forces can be produced: if typically $B_1 = 0.79$ T and the product of $i_2$ and the number of conductors per metre is 32 kA (a value readily obtained in a large machine), the interaction force per square metre of gap surface is 25 kN. But note that the force of attraction between polar surfaces is ten times as great.

*Slotting.*   Conductors placed centrally in the gap must sustain the whole interaction force and be bonded together to counter deformation. Where the conductor current exceeds a few milliamperes, the conductors may be attached to one of the ferromagnetic members as a surface winding, Fig. 1.5(a): the total force is unchanged. With large currents an attachment firm enough to withstand the thrust may be difficult, and a common method of securing conductors is to sink them into slots formed in the gap surface of the iron, Fig. 1.5(b). This has two advantages: (i) the force is now almost

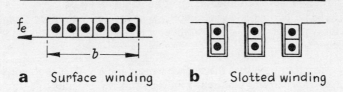

**a** Surface winding     **b**   Slotted winding

1.5 Winding location in airgap

completely transferred to the slot walls and is taken by the strong iron teeth, and (ii) the gap length may, if required, be shortened to reduce the m.m.f. $F_1$ needed to establish the primary field flux density $B_1$. It is a surprising fact that the interaction force is still the same.

One explanation is this. The field pattern in Fig. 1.6($a$) is that of a current $i_2$ of a conductor in a slot set near to a primary polar surface, all ferromagnetic parts having a high permeability. The self flux links the current, part crossing the main gap twice, part crossing the slot-width once.

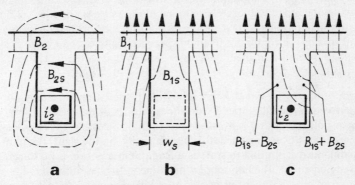

**a**        **b**        **c**

1.6 Magnetic field in slot

The density $B_2 = \frac{1}{2}\mu_0 i_2/l_g$ in the main gap develops attraction between the gap surfaces. The density $B_{2s} = \mu_0 i_2/w_s$ within the slot and above the conductor tends to pull together the sides of the adjacent teeth but has no translational resultant. With $i_2$ zero but with the gap magnetized as in ($b$), part of the flux provides a density $B_{1s}$ at the slot walls. Although none of it passes from wall to wall, we infer from the Maxwell stress concept that there will be some attraction between the teeth. Patterns ($a$) and ($b$) are superimposed in ($c$) for an excited gap and a slot current in combination. Resultant flux densities now appear at the slot walls. The horizontal forces on the right- and left-hand walls are respectively proportional to $(B_{1s} + B_{2s})^2$   and   $(B_{1s} - B_{2s})^2$, and their difference gives a resultant horizontal translational force. The densities naturally vary over the slot depth, but if integrated they yield a total resultant of $f_e = B_1 l i_2$, where $B_1$ is the average gap density due to the gap excitation alone. The field pattern ($c$) shows that the interaction force is concentrated on the sides of the teeth near to the gap, leaving the conductor almost entirely relieved of direct magnetomechanical stress.

It is to be noted that $B_{2s}$ generated by $i_2$ is greater than for an isolated conductor, with the result that a conductor in a slot has an increased self inductance.

*Current Sheet*  Suppose that the surface winding in Fig. 1.5(*a*) comprises $N$ conductors each carrying a current $i$. In the limit this equates to a single broad, thin, sheet-like conductor stuck to the gap surface and carrying a uniformly distributed total current $Ni$. The current-sheet concept simplifies the basic analysis of m.m.f. and flux distributions for an idealized uniformly distributed winding, and applies even to a slotted winding if the secondary effects of the teeth are ignored.

A band of $N = 8$ conductors each carrying $i = 75$ A and evenly spread over a width $b = 0.6$ m on the flat or cylindrical gap surface of a machine gives the product $Ni = 600$ ampere-conductors. Its total m.m.f. is the same as that of a sheet carrying 600 A with uniform linear distribution. The equivalent sheet has a linear current density $A = Ni/b = 1000$ A/m, and it can represent any band of conductors spread uniformly over the same breadth $b$ and having the same product $Ni$. For a machine of disc shape (Sect. 2.2 and 2.3) it is more convenient to write the current density of the equivalent sheet in terms of angle: a conductor band $Ni$ spread radially over an angle $\beta$ is represented by a current sheet of density $A$ such that $A\beta = Ni$. Thus if $N = 8$, $i = 75$ A and $\beta = 1.5$ rad (86°), then $A = Ni/\beta = 400$ A/rad.

*Limitations*  Eq. (1.5) and (1.6) are true only if the currents, acting alone, give rise to equal flux densities $B_2$ on the two sides of the sheet. In Fig. 1.7(*a*) the upper pole of width $b$ is isolated, and the single current $i_2$ is placed off-centre. The density $B_2'$ on the left is less than $B_2''$ on the right, the m.m.f. $i_2$ adjusting its distribution to suit. The force of interaction is no longer given exactly by eq. (1.5). The *heteropolar* machine in (*b*) has, however, a succession of N and S poles, providing an additional magnetic path (of the same reluctance if saturation be neglected) for the flux of $i_2$ in adjacent poles. The densities $B_2'$ and $B_2''$ are equalized, and eq. (1.5) can be applied with confidence.

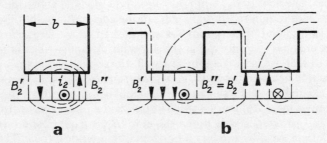

1.7 Interaction force

Interaction force on a rotor current implies that there must be an equal reaction force on the stator structure. For both cases in Fig. 1.7 the reaction arises from the Maxwell tensions at the pole flanks, just as for the alignment-force case in Fig. 1.2. An exception to this rule occurs in the homopolar machine, of which Fig. 1.8 shows typical features. A copper cylinder 'cup' is set to occupy the annular airgap between the inner N and outer S poles of a potmagnet, which provides a uniform radial gap-flux density $B_1$. If a current

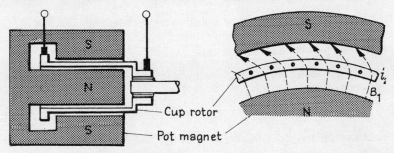

1.8 Homopolar machine

$i_2$ is fed through the cup from brush to brush to give a uniform axial current density, a circumferential m.m.f. $i_2$ is developed around the outside of the cup (but not inside it). The flux crossing the *inner* gap is unaffected: it is still radial and of density $B_1$. But in the *outer* gap the flux results from the combination of $B_1$ (radial) and $B_2$ (circumferential) and does not enter the outer pole radially. As a result there are Maxwell tension and compression forces that sum to a total reaction $B_1 l i_2$.

### Electromotive force

The essential physical system for the induction of an e.m.f. comprises an electric and a magnetic circuit, mutually interlinked. Summation of all the products $\Phi N$ of flux and turns gives the total flux linkage $\Psi$. The e.m.f. exists when the linkage is changing, and its time integral is equal to the change. This is Faraday's law. The direction of the e.m.f. is such as to oppose the change: if the electric circuit were closed, the induced e.m.f. would circulate a current and develop a self flux, the linkage of which tends to restore the level of the original linkage by opposing a rise or supplying a deficiency. This is the Faraday–Lenz law.

Consider a straight conductor (with fixed external connections to form a closed electric circuit) of active length $l$, and set at right-angles to the direction of a flux of uniform density $B_1$. Let the conductor be moved at speed $u$ in a direction at right-angles to both $B_1$ and $l$. Then in unit time the conductor sweeps out an area $lu$ in which there exists a flux $B_1 lu$. This is the rate of change of the circuit linkage, giving the e.m.f. induced by motion as

$$e_r = B_1 lu \tag{1.7}$$

The effect is often described as the result of 'flux-cutting' by the conductor.

The essentials of the system are those of Fig. 1.3. If the conductor carrying $i_2$ is moved in the direction of the force $f_e$ at speed $u$, mechanical work at the rate $p_m = f_e u$ is done. But the motion of the conductor generates an e.m.f. $e_r$ in the circuit in which $i_2$ is flowing, so that an electrical power $p_e = e_r i_2$ has to be supplied. From eq. (1.5) and (1.7)

$$p_m = f_e u = B_1 l\, i_2 u = B_1 l u i_2 = e_r i_2 = p_e \qquad (1.8)$$

This is the essence of electromechanical energy conversion. The rate of conversion at any instant is a simple, direct and intrinsically 'perfect' process, which underlies the behaviour of most d.c. and a.c. machines.

*General Case*   Using the conventions given in Sect. 1.5, the Faraday–Lenz law can be stated as

$$e = + \, \mathrm{d}\Psi/\mathrm{d}t \qquad (1.9)$$

The linkage $\Psi$ that a circuit may have with a flux $\Phi$ can change in a number of ways:

(i) Suppose the flux to be momentarily unvarying, the circuit may move through it.

(ii) Suppose that the circuit is at rest with respect to the flux, the flux may change in magnitude.

(iii) If the circuit moves through a time-varying flux, both of the foregoing occur together and can at any instant be summed.

Eq. (1.9) can be written in partial differentials to show these explicitly. Let the circuit move at linear speed $u = x/t$ in a flux $\Phi$: then

$$e = \left[ \frac{\partial \Psi}{\partial x} \cdot \frac{\mathrm{d}x}{\mathrm{d}t} + \frac{\partial \Psi}{\partial t} \right] = \left[ \frac{\partial \Psi}{\partial x} u + \frac{\partial \Psi}{\partial t} \right] = e_r + e_p \qquad (1.10)$$

Alternatively, if the machine is a rotary one with an angular speed $\omega_r = \theta/t$,

$$e = \left[ \frac{\partial \Psi}{\partial \theta} \cdot \frac{\mathrm{d}\theta}{\mathrm{d}t} + \frac{\partial \Psi}{\partial t} \right] = \left[ \frac{\partial \Psi}{\partial \theta} \omega_r + \frac{\partial \Psi}{\partial t} \right] = e_r + e_p \qquad (1.11)$$

The two components of the e.m.f. $e$ are:

*Motional*   The motional e.m.f. $e_r$ in the circuit resulting from the linear speed $u$ or the angular speed $\omega_r$ is concerned, as we have seen, with electrical/mechanical energy conversion. The terms $\partial \Psi/\partial x$ and $\partial \Psi/\partial \theta$ represent in effect the distribution of the flux density in the active region.

*Pulsational*   The pulsational (or 'transformer') e.m.f. $e_p$ appears in the circuit if the flux linking it changes with time. It provides a means whereby magnetic energy is stored, or transferred from one circuit to another when they are magnetically coupled.

If an electric circuit carries a current $i$, the product $e_r i$ is the rate of electrical/mechanical energy conversion; and $e_p i$ is the rate of energy storage or transfer by magnetic means. The only measurable circuit e.m.f., however is the sum $e = e_r + e_p$ of the motional and pulsational components as given by eq. (1.9).

## 1.4 PRACTICAL FORMS

The essential d.c. machine has two members. One (the 'field') is an electro- or permanent-magnet system providing a working magnetic field in an airgap. The other (the 'armature') is an arrangement of current-carrying conductors so orientated as to develop interaction force and e.m.f., or is so shaped as to give rise to an alignment or reluctance force.

One member, often the field system, is a fixed 'stator'; the other is the moving member, called a 'rotor' if the motion is rotary and a 'runner' if the motion is linear. The machine can be realized in a variety of forms, Fig. 1.9.

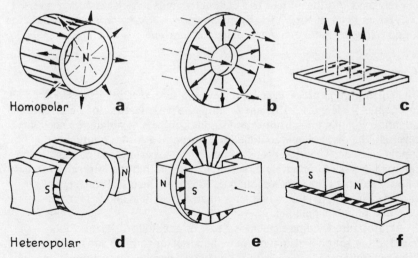

1.9 Practical forms of d.c. machine

Here the direction of the working field flux is indicated by dotted lines and the armature current by thick arrows. All arrangements — cylindrical, disc and linear — preserve the mutual perpendicularity of field, current and motion.

### Homopolar machine

In Fig. 1.9, the *cylindrical* shape (*a*) resembles that in Fig. 1.8, but the cup may be replaced by an assembly of axial conductors joined at each end to a slip-ring for external connection through the sliding contact provided by fixed 'brushes'. The *disc* form (*b*) may also be made up of a number of radial conductors with slip-rings and brushes at the inner and outer radii. The runner in the *linear* machine (*c*) could be an assembly of separate conductors, or a liquid metal as in an electromagnetic pump, or a conducting gas as in a magnetohydrodynamic generator. In all cases the field system is the stator, and slip-ring/brush connections are necessary to lead current to and from the moving armature.

### Heteropolar machine

The full range of ratings can be realized in the *cylindrical* form, Fig. 1.9(*d*). A saving of axial length where space is restricted is achieved by the *disc* form (*e*) but the speed may be limited by centrifugal force at the disc circumference. The flat *linear* form (*f*), though practicable, is uncommon except for limited excursions of the runner.

In the heteropolar machine it is immaterial whether the moving member is the field or the armature, the choice being made on grounds of mechanical convenience. But the current to a conductor must have its direction reversed as it passes from an N-pole to an S-pole region, a *commutation* process that is a prime characteristic of a heteropolar d.c. machine.

### Commutation

The commutation process can be achieved in several ways, discussed in detail in Chapter 5. Here it is sufficient to show the basic features of a conventional armature winding using brushes and sliding contacts to obtain the necessary reversal. The discussion is confined to the case of a rotor armature for a 2-pole machine. The object is to set up a current pattern of two semicircumferential sheets of oppositely directed current, so that the interaction forces of both halves of the armature shall be additive. The conditions sought are those in Fig. 1.3, (*b*) for a conductor in an S-pole field and (*c*) for the same conductor in an N-pole field.

An armature winding element is a pair of conductors, P and Q in Fig. 1.10(*a*), set on a diameter. An end-connector forms P and Q into a turn: if current enters at Q it leaves at P, so giving opposite current directions on the two sides. If, as in (*b*), Q is joined by a front connector to R, near P, the first step has been taken in forming a distributed winding. (The end-connectors lie outside the active gap region and contribute no torque.)

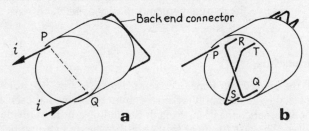

1.10 Formation of commutator-armature winding

If the connection sequence is continued with further turns, the cylindrical surface of the rotor is eventually fully wound, and the final conductor can be connected to P to form a *closed* winding.

When the winding is to be accommodated in a slotted rotor core, it is found difficult to accommodate the end-connectors unless the active conductors are arranged in two layers. Coil-side P is at the top of one slot, its companion Q at the bottom of another slot diametrically opposite, the coil-ends being shaped to suit. In this way a complete winding, symmetrical and compact, is assembled from separate single- or multiturn coils, fitted successively around the rotor with their ends interleaved. Intercoil joints are then made to connect successive coils and close the winding.

The winding layout is more readily appreciated if the cylindrical gap surface is 'unrolled' and drawn 'in the flat'. Fig. 1.11(*b*) is such a diagram, for a 2-pole machine with 12 turns or coils AB, CD . . . PQ . . . XY. The closed winding can be traced in the sequence ABC . . . XYA. Full lines indicate

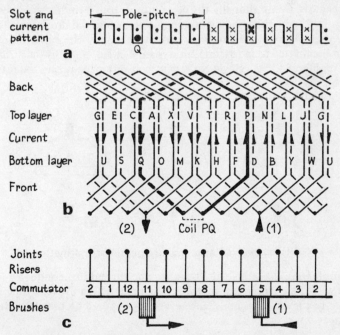

1.11 Layout of two-pole commutator-armature winding

upper conductors, broken lines the lower ones. Successive conductors or coil-sides ABC. . . are spaced by an angular span of approximately $\pi$ rad. To this closed, symmetrical 12-coil winding let two external connections, (1) and (2), be made, with a current fed in at (1) and out at (2). At (1) the current divides between conductors H and J. Each half-current has a separate path through the winding to meet via U and V and to leave at (2). In the active slot-conductors the currents form two belts, each of 12 conductors in 6 slots, to yield the current pattern (*a*) that we seek.

When the winding moves, the current pattern is kept stationary by shifting (1) and (2) backwards by the same angle that the rotor rotates forward. One way of achieving this is to mount on the rotor shaft a *commutator,* a cylindrical assembly of copper sectors (one for each coil), insulated from each other and from the shaft, and connected as in (*c*) to the front end-connector joints in sequence. Current is fed to the winding by a pair of brushes, usually of carbon with additives, held in boxes fixed to the stator frame. In the diagram, current enters the winding through brush (1) and sector 5, and leaves by sector 11 and brush (2). As the whole winding moves to the left, sectors 4 and 10 take the place of 5 and 11, and so on. Thus with respect to stator and brushes, and consequently with respect to the field system, the *current pattern remains at rest* as a result of the switching action of the commutator/brush device.

The physical position of the brushes depends on the angle of the front end-connectors. Conventionally the commutator rotor winding is represented as in Fig. 1.12, with the winding as a simple circle and the brushes set to coincide with the axis of the resultant armature m.m.f. The current directions in (i) can be omitted as in (ii); and for dynamic-circuit analysis the armature can be considered to be a single equivalent coil (iii) magnetizing the rotor along the brush axis.

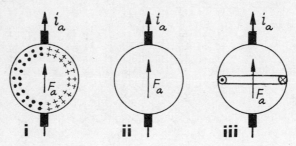

1.12 Conventional diagrams of commutator armature

Commutation is a complicated process. The current in a conductor is reversed while its commutator sector is in contact with a brush, a time that may be only a few milliseconds, and reversal is opposed by the effective coil inductance.

## 1.5 CONVENTIONS

An electromagnetic machine is a device that links an electrical energy system (such as a supply network or a battery) to a mechanical one (such as a prime-mover, a train or a machine-tool), and its behaviour is affected by its terminal energy systems. A generator may trip out if a fault occurs on the network it feeds; starting a large motor may depress the voltage of its power supply; a drive subject to rapid change of load may undergo cyclic fluctuations of speed.

While steady-state performance is informative, the transient performance is more fundamental and has more engineering significance, and it demands a consideration of the energy balance.

### Energy balance

A machine accepts energy in two forms — electrical and mechanical — from its terminal systems. By convention we take energy *input* to be *positive,* so that energy output is regarded as a negative input. As the function of the machine is energy conversion, one of its terminal inputs will normally be negative. Of the total energy input, some is converted, some stored and the rest dissipated in loss. Reckoned from an initial zero-energy condition, the energy balance is

> (electrical energy input) + (mechanical energy input) is equal to
> (stored field energy) + (stored mechanical energy) + (dissipation)

Comparable relations apply to energy changes and to energy rates (l.e. powers):

$$\text{Energy change:} \quad \mathrm{d}w_e + \mathrm{d}w_m = \mathrm{d}w_f + \mathrm{d}w_s + \mathrm{d}w \tag{1.12}$$

$$\text{Energy rate:} \quad p_e + p_m = \mathrm{d}w_f/\mathrm{d}t + \mathrm{d}w_s/\mathrm{d}t + p \tag{1.13}$$

The input electrical and mechanical powers $p_e$ and $p_m$ are together equal to the rates of increase of stored magnetic-field and mechanical energy $\mathrm{d}w_f$ and $\mathrm{d}w_s$ with the addition of the power $p$ dissipated in conductor $I^2 R$ loss, core loss, windage and friction. The storage rates are left as differentials because there are practical limits to energy storage: a magnetic field cannot increase indefinitely in ferromagnetic material, and if the kinetic energy in a rotating mass is continuously raised its speed becomes dangerously high and failure under centrifugal force will occur. In a d.c. machine operating in the *steady* state, neither the field energy nor the stored mechanical energy is subject to change, and eq. (1.13) becomes $p_e + p_m = p$.

### Electrical system

In Fig. 1.13 an electrical power source of voltage $v$ is connected to the winding of a machine. The effective resistance of the conducting path is $r$, and

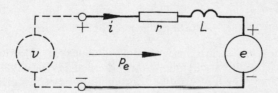

1.13 Electric circuit conventions

the leakage inductance due to non-useful magnetic flux is $L$. The voltage $v$ is opposed by an e.m.f. $e = e_r + e_p$. The direction of the current $i$ is positive when it flows in at the positive source terminal, and the input electrical power is then $p_e = vi$. The applied-voltage and electrical power relations are

$$v = ri + L(di/dt) + e_r + e_p$$
$$p_e = ri^2 + Li(di/dt) + e_r i + e_p i \qquad (1.14)$$

*Operating Mode*  Let steady-state conditions obtain, when both $L(di/dt)$ and $e_p$ are zero. Then

$$V = rI + E_r \quad \text{and} \quad p_e = rI^2 + E_r I$$

If $E_r$ is less than $V$, a positive current $I$ flows and the input power $p_e$ comprises $E_r I$ converted from electrical to mechanical form, together with an $I^2 R$ loss: the machine acts as a *motor*. But if $E_r$ is greater than $V$, the current reverses, i.e. $I$ becomes negative. Conversion is now from mechanical to electrical form, and the terminal electrical input is negative: the machine acts as a *generator* delivering electrical power to the source.

To illustrate, let $V = 10$ V and $r = 1$ $\Omega$. If $E_r = 8$ V, the current is positive and of value $I = (V - E_r)/r = 2$ A. The source provides $p_e = 10 \times 2 = 20$ W, of which $I^2 R$ loss accounts for 4 W and $8 \times 2 = 16$ W is converted by the machine as a motor. Conversely, if $E_r = 12$ V the current is again 2 A but is reversed. The action of the machine as a generator converts $12 \times 2 = 24$ W, of which 4 W is dissipated in internal resistance and 20 W is delivered to the source, the terminal input power $p_e$ now being negative.

Fig. 1.14($a$) shows a full-pitch turn of a 2-pole armature connected to a source of voltage $V$ and rotating counterclockwise (viewed from the source

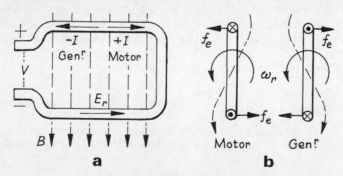

1.14 Motor and generator modes

end) at angular speed $\omega_r$ in a magnetic field of density $B$. The motional e.m.f. $E_r$ opposes $V$. The current directions for motor and generator action are indicated in ($a$) and the forces on the active coil-sides in ($b$). In the motor mode the forces $f_e$ combine as a torque in the direction of rotation to provide mechanical output power. In the generator mode the interaction torque opposes rotation, calling for a mechanical input torque to maintain rotation.

In both modes the directions of rotation and flux are the same, the motor and generator modes being distinguished only by reversal of the current and torque.

As the torque in the motor mode is an output, it is properly regarded as negative. In the generator mode the electrical torque is part of the positive mechanical torque input.

### Diagrams

Conventions for 2-pole cylindrical and linear heteropolar machines, Fig. 1.15, have the following consistent features:

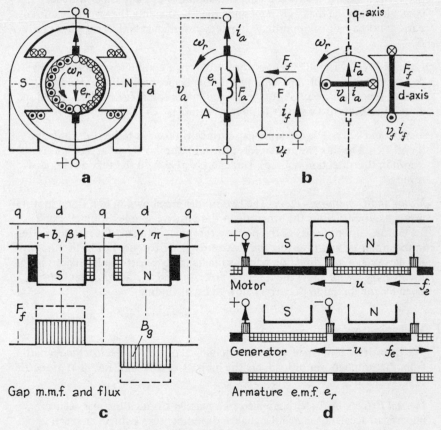

1.15 Diagram conventions

(i) The right-hand field pole has N-polarity, the left hand pole has S-polarity, and the field m.m.f. is $F_f$ per pole.

(ii) The brush axis conventionally indicates the magnetic axis of the rotor m.m.f. $F_a$ per pole.

(iii) For a rotary machine the angular speed $\omega_r$ is counterclockwise. For a linear machine the speed $u$ is from right to left. These apply to both motor and generator modes. The direction of the motional e.m.f. that results is indicated in ($a$) and ($d$).

(iv) The pole-pitch is $Y$ and the pole-arc is $b$. In a rotary machine these can also be represented by the angles $\pi$ and $\beta$. The ratio pole-arc/pole-pitch is $b/Y$ (or $\beta/\pi$) as indicated in ($c$).

*Direct and Quadrature Axes*  The centre-line of the field system, the line of action of $F_f$, is called the *direct axis* (d-axis). The *quadrature axis* (q-axis) is electrically at right angles to the d-axis in a rotary machine, or mid-way between successive d-axes for a linear machine. Normally the brush axis coincides with the q-axis: it represents the line dividing the armature current pattern into a pair of oppositely directed current sheets that together produce $F_a$.

*Current Sheets*  The direction of the current in a conductor is denoted by a dot or a cross, for outward and inward flow respectively. For extended current sheets the array of dots is merged into a solid black block, and the array of crosses into a square hatching, as in ($d$).

*M.M.F. and Flux*  The armature gap surface is taken as reference. Under an S-pole field the flux is directed from the armature to the pole face and is drawn in this direction as in ($c$). Under an N-pole field the flux direction is reversed.

*Motor and Generator Action*  The two modes are shown in ($d$). For a motor, positive current enters the armature at the positive source terminal. For a generator, current leaves at the positive terminal, reversing the current pattern in the armature and reversing the electromagnetic torque. For both modes the motional e.m.f. direction remains unchanged. 'Short-hand' versions in ($b$) for a rotary *motor* show the terminal polarity, and the m.m.f. directions as given by the direction of current flow in each winding.

## Mechanical system

The quantities concerned with the mechanical nature of the machine, apart from the prime-mover or load, are the inertial, elastic and frictional properties of its structure.

*Inertial Effects*  A force $f_m$ applied to a mass $m$ through the mass-centre imparts an acceleration $a$ and a change of momentum (which, at speed $u$, is $mu$). The force determines the rate of change of momentum, $f_e = \mathrm{d}(mu)/\mathrm{d}t$. For a constant mass, then

$$f_m = m(\mathrm{d}u/\mathrm{d}t) = ma$$

A torque $M_m$ applied to rotate a mass having a moment of inertia $J$ about its axis of rotation imparts an angular acceleration $\alpha$ and a change of angular momentum (which, at angular speed $\omega_r$, is $J\omega_r$). If the inertia remains constant, then

$$M_m = J(\mathrm{d}\omega_r/\mathrm{d}t) = J\alpha$$

A mass in motion stores kinetic energy, given by $w_a = \frac{1}{2}mu^2$ or $w_a = \frac{1}{2}J\omega_r^2$ for linear and rotary motion respectively.

*Elastic Effects*   The restoring force or torque of a perfect spring or comparable elastic member (such as a shaft) when deformed is proportional to the linear displacement $x$ or the angular displacement $\theta$:

$$f_m = (1/k)x \qquad M_m = (1/k)\theta$$

The compliance $k$, the reciprocal of stiffness, is defined in terms of distance/force or angle/torque. If an elastic member is initially unstrained and has one end fixed, and if the other end is displaced at speed $u$ (or twisted at angular speed $\omega_r$), the deforming force or torque is

$$f_m = \frac{1}{k}\int u\cdot\mathrm{d}t \qquad M_m = \frac{1}{k}\int \omega_r\cdot\mathrm{d}t$$

An elastic member under strain stores the potential energy

$$w_s = \frac{1}{2}x^2/k = \frac{1}{2}kf_m{}^2 \qquad w_s = \frac{1}{2}\theta^2/k = \frac{1}{2}kM_m{}^2$$

'Hard' springs, and elastic members deformed beyond the limit of proportionality as expressed by the Hooke law, do not behave as simply as this.

*Friction Effects*   These dissipate energy in heat, usually irrecoverable, in bearings and commutator/brush members and in the fanning of ambient air due to motion ('windage'). The effects are complex and nonlinear, but may sometimes be idealized into analytic functions. Coulomb friction assumes a force or torque of constant value, and viscous friction is taken as proportional to speed.

*Combined Effects*   The action of energy storing and dissipating elements, considered to be separable, in a *limited-motion* linear mechanism can be inferred from $(a)$ in Fig. 1.16. The force $f_m$ imparts the same speed $u$ to the mass $m$, to one end of the spring of compliance $k$, and to the piston of the dashpot of viscous damping coefficient $d$, with respect to a 'frame'. The limited rotary system $(b)$ is analogous. Then

$$f_m = ma + du + \frac{1}{k}x = m\frac{\mathrm{d}u}{\mathrm{d}t} + du + \frac{1}{k}\int u\cdot\mathrm{d}t$$

$$M_m = J\alpha + d\omega_r + \frac{1}{k}\theta = J\frac{\mathrm{d}\omega_r}{\mathrm{d}t} + d\omega_r + \frac{1}{k}\int \omega_r\cdot\mathrm{d}t$$

In a practical *rotating* system the end of the shaft in $(b)$ is not, of course, fixed, nor are the speeds concerned with the various parameters the same; the terms must then be considered individually.

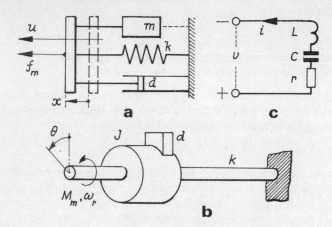

1.16 Mechanical systems

The electric circuit (*c*) is an analogue of (*a*) and (*b*), for its equation of behaviour is $v = L(\mathrm{d}i/\mathrm{d}t) + ri + q/C$. The counterpart of the kinetic energy $\frac{1}{2}mu^2$ or $\frac{1}{2}J\omega_r^2$ is the magnetic-field energy $\frac{1}{2}Li^2$, and similarly for the potential energies. Each system has two sites of energy storage and a damping component, and has a comparable pattern of response. The inference is that an electromechanical system, whether of limited motion or rotating, may exhibit damped oscillations and resonances.

## Electromechanical system

The single-circuit machine in Fig. 1.17 has, in general, positive power inputs $p_e$ and $p_m$, and positive input torques $M_e$ and $M_m$ in the direction of rotation.

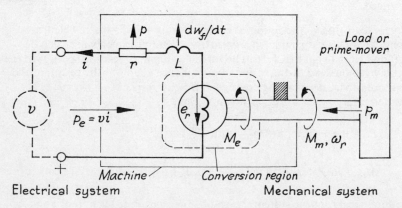

1.17 Electro-mechanical machine system

*Motor Mode*  For this condition, $p_e$ is positive and $p_m$ is negative. The applied voltage $v$ drives a positive current against the e.m.f. to provide a gross input power $p_e = vi$ and a net power $e_r i$ which is converted to a corresponding input mechanical power $M_e \omega_r$, driving the shaft. Deduction of internal mechanical loss and storage gives the net useful output $M_m \omega_r$, i.e. a mechanical *input* of $(-M_m)\omega_r$.

*Generator*  Here $p_e$ is negative (implying current reversal) and $p_m$ is positive. The shaft is driven by the gross mechanical input torque $M_m$ to give $p_m = M_m \omega_r$, and a net converted power $M_e \omega_r = e_r(-i)$. Deduction of the electrical loss and storage results in a terminal *input* $v(-i)$, i.e. an electrical output of $vi = p_e$.

*Conversion Region*  A practical machine has obvious points of attachment to its energy systems (the source-terminals and the shaft). But it is useful to concentrate attention on the *conversion region* within the dotted enclosure in Fig. 1.17, because this is concerned only with the basic quantities $e_r$ and $i$ for the electrical end, and $\omega_r$ and $M_e$ (or $u$ and $f_e$) for the energy conversion process. Outside this region we can account separately for electrical loss and leakage inductance, friction loss and mechanical energy storage. Then if, at any instant, the motional component of the e.m.f. is $e_r$ and the current is $i$, we have

$$e_r i = M_e \omega_r \quad \text{or} \quad e_r i = f_e u \tag{1.15}$$

for the essential electrical/mechanical energy-conversion process.

The machine is now in an analysable parametric form. Its action involves applied voltage and input current, torque and speed, conversion, storage and loss. Evaluation is based on:

| | | |
|---|---|---|
| *Electrical:* | voltage, current | Faraday–Lenz and Kirchhoff laws |
| *Conversion:* | e.m.f., current, magnetic force, displacement | Alignment, induction and interaction laws |
| *Mechanical:* | force, displacement, speed | Newton laws |

## Torque balance

The instantaneous electromagnetic torque is related to the other significant torque components by

$$M_e + M_l + M_a + M_s + M_f + M_d = 0 \tag{1.16}$$

*Electromagnetic Torque $M_e$* is a function of current and flux as controlled by applied voltage, current, rotational e.m.f., and speed.

*Load or Prime-mover Torque $M_l$* for the motor mode is determined by the characteristics of the driven load and is usually speed-dependent. In the generator mode it is replaced by $M_m$ for the prime-mover torque characteristic.

*Inertial Torque* $M_a$ accounts for kinetic energy storage resulting from a change of speed.

*Strain Torque* $M_s$ is associated with elastic deformation of shaft and frames resulting from impulsive or fluctuating shaft torques, flexible couplings, and belt or chain or wire-rope drives.

*Friction and Damping Torques* $M_f$ and $M_d$ account for loss torques in friction and windage, and in eddy currents, etc. When both can be taken as proportional to speed, then $(M_f + M_d) = k_d \omega_r$, where $k_d$ is an appropriate 'damping' coefficient.


### 1.6 PERFORMANCE

Generators and motors cannot be considered in isolation. Their performance is related to the characteristics of the electrical and mechanical systems to which they are connected, as well as to their own constructional features. Performance duties may require a wide control of speed, frequent starting and stopping, braking and reversal, operation on fluctuating or impulsive loads, and conformity with specified conditions of ambient temperature and internal temperature-rise. A specified performance must be built into the machine by appropriate design.

For the purposes of analysis, operating conditions are simplified to those of immediate interest. Sometimes, for example, the electrical system can be taken as an *infinite busbar* of constant voltage and capable of delivering or absorbing energy without limit. Sometimes the speed can be considered as constant so that inertial effects vanish, or be subject to small cyclic variations about a mean as when a generator is driven by an internal combustion engine. Where such simplifications are inadmissible, iterative computer solutions have to be sought.

Basic steady-state performance is discussed in Chapter 6, and simple methods for dealing with a number of transient problems in Chapter 7.

# 2 Prototype Machines

## 2.1 PROTOTYPE D.C. MACHINES

Each of the homo- or heteropolar configurations in Fig. 1.9 comprises two members: one, the stator, is fixed; the other moves with respect to it as a rotor or runner. Here the stator is taken as providing the field system; this is essential for the homopolar machine, but in the heteropolar type the field and armature functions are sometimes interchanged.

A basic 2-pole arrangement is assumed for the heteropolar machine, because although 4, 6 or 8 poles are more usual in practice, such structures can be regarded simply as assemblies of successive 2-pole units.

*Specific Loadings* The gap flux density $B_g$ produced by the field member is taken as constant over the area of each pole-face, and to average $B_1$ over the area of a whole pole-pitch, as in Fig. 2.1. As the armature conductors move,

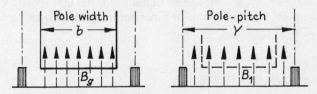

2.1 Pole-face and mean gap flux densities

their individual motional e.m.f.s are related by eq. (1.7) to the flux density in which they lie, but as all the conductors within a pole-pitch between brushes are in series, their sum is the same as if each moved in the *average* flux density $B_1$. The total gap flux per pole can be expressed as $\Phi_g = B_1 Yl$, where $Y$ is the peripheral length of a pole-pitch and $l$ is the active length of a conductor. In a similar way, the interaction force is the same as if each conductor lay in the mean flux density $B_1$, with the mean force per conductor given by eq. (1.5). The assembly of currents in the armature conductors gives current-sheet patterns of density $A_2$. For cylindrical and linear machines $A_2$ is a linear current density [A/m of periphery]; for disc machines it is an angular density [A/rad].

The densities $B_1$ and $A_2$ are respectively the *specific magnetic* and *electric loadings* which, together with the physical dimensions and the speed, determine the conversion power. Saturation of the ferromagnetic members limits practical values of $B_1$ to rather less than 1 T (unless superconducting

27

field systems are employed). The value chosen for $A_2$ depends on the $I^2 R$ loss in the armature and the efficacy of the cooling.

EXAMPLE 2.1: A cylindrical machine has an armature of diameter $D = 0.35$ m and active length $l = 0.28$ m. The mean gap flux density is $B_1 = 0.65$ T and the armature electric loading is $A_2 = 23800$ A/m. Calculate the conversion power at a speed of 860 r/min ($\omega_r = 90$ rad/s, $n = \omega_r/2\pi = 14.3$ r/s).

The peripheral force per unit area of gap surface is $B_1 A_2$; the gap surface area is $\pi Dl$; the total force is $f_e = \pi Dl B_1 A_2$; and the torque is $M_e = \frac{1}{2} D f_e$. Thus the conversion power is

$$P = M_e \omega_r = \frac{1}{2} \pi D^2 l B_1 A_2 \omega_r = \pi^2 D^2 l B_1 A_2 n \tag{2.1}$$

$$= 75\ 000 \text{ W} = 75 \text{ kW}$$

The result applies to both homo- and heteropolar machines. Electrically it is the product $E_r I_2$ of the armature rotational e.m.f. and input current, the individual voltage and current values being determined by the number of armature turns and poles.

*Prototypes* We first consider idealized homopolar and 2-pole heteropolar machines to establish basic e.m.f., current, torque and speed relations in the conversion region of Fig. 1.17. In every case $E_r I_2 = M_e \omega_r$ (or $f_e u$); and further $E_r/\omega_r = M_e/I_2$, so that the motional e.m.f. per unit speed [V per rad/s] is identical with the torque or force per unit armature current [N-m per A, or N per A].

## 2.2 HOMOPOLAR MACHINES

Although it is still an unusual machine, improvements in current collection and the application of superconducting field windings have stimulated an interest in the homopolar cylindrical, linear and disc forms.

### Cylindrical

For the construction in Fig. 1.8, the rotor is a conducting cylinder of diameter $D$ and active length $l$, spinning at angular speed $\omega_r$ in a radial field of uniform flux density $B_1$. The motional e.m.f., and the torque for a rotor current $I_2$ assumed to flow axially, are

$$E_r = \frac{1}{2} Dl B_1 \omega_r \quad \text{and} \quad M_e = \frac{1}{2} Dl B_1 I_2 \tag{2.2}$$

In terms of the armature current-sheet density $A_2 = I_2/\pi D$, the conversion power $P = E_r I_2 = M_e \omega_r$ is given by eq. (2.1). The rotor could alternatively have $Z$ equi-spaced axial conductors each carrying $I_2/Z$ and connected to individual slip-rings. External series connection of the armature conductors would give a total e.m.f. $ZE_r$ and a current $I_2/Z$ for the same converted power.

## Linear

A useful linear machine is the liquid-metal pump, Fig. 2.2, in which a current $I_2$ is passed through the metal where it flows in a magnetic field of uniform

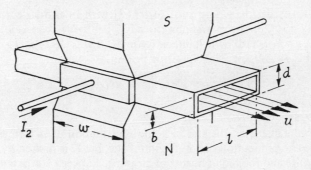

2.2 Homopolar machine: linear

flux density $B_1$. Liquid is moved at speed $u$ against a hydraulic pressure-difference $\Delta p$ at a volume rate $Q = blu$ across the field. The e.m.f., force and conversion power are

$$E_r = B_1 lu \quad f_e = B_1 lI_2 = bl\Delta p \quad P = E_r I_2 = f_e u \tag{2.3}$$

## Disc

The disc armature, Fig. 2.3, of effective outer and inner diameters $D$ and $d$, carries a uniform annular distribution of the radial current $I_2$ in an axial field of uniform density $B_1$ and spins at angular speed $\omega_r$. Consider an elemental

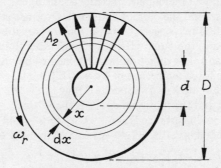

2.3 Homopolar machine: disc

annulus of radius $x$ and radial length $dx$. It moves at peripheral speed $\omega_r x$ and its motional e.m.f. is $de_r = B_1 \omega_r x \cdot dx$. The total disc e.m.f. between diameters $d$ and $D$ is therefore

$$E_r = B_1 \omega_r \int_{\frac{1}{2}d}^{\frac{1}{2}D} x \cdot \mathrm{d}x = \tfrac{1}{8}(D^2 - d^2)B_1\omega_r \tag{2.4}$$

The annulus carries the whole rotor current $I_2$ with an angular current density $A_2 = I_2/2\pi$. The interaction force is $\mathrm{d}f_e = B_1 I_2 \cdot \mathrm{d}x$, developing a torque $\mathrm{d}M_e = B_1 I_2 x \cdot \mathrm{d}x$ which integrates to $M_e = \tfrac{1}{8}(D^2 - d^2)B_1 I_2$, showing again that $E_r I_2 = M_e \omega_r$. The converted power is

$$\begin{aligned}
P &= \tfrac{1}{8}(D^2 - d^2)\,B_1 I_2 \omega_r = \tfrac{1}{4}\pi(D^2 - d^2)B_1 A_2 \omega_r \\
&= \tfrac{1}{2}\pi^2(D^2 - d^2)B_1 A_2 n
\end{aligned} \tag{2.5}$$

The larger the diameter, the greater is the contribution of an elemental annulus to the production of e.m.f. and torque, but $D$ is limited by rotational stress, flexure and the requirements of clearance between stator and rotor members. The low inherent e.m.f. can be increased by splitting the disc into a number of sectors connected externally in series, and by using a very strong magnetic field.

EXAMPLE 2.2: A conducting disc of diameter 0.30 m, with a central hole of diameter 0.10 m, is mounted on a shaft and rotated at 1000 r/min in a uniform axial field of density 0.76 T. Determine the e.m.f. between contacts on the outer and inner rims of the disc.
The angular speed is $\omega_r = 105$ rad/s, and applying eq. (2.4) gives

$$E_r = \tfrac{1}{8}\pi(0.30^2 - 0.10^2)\,0.76 \times 105 = 2.50 \text{ V}$$

EXAMPLE 2.3: An electromagnetic pump, Fig. 2.2, has a magnetic field of mean density $B_1 = 0.50$ T and channel dimensions $b = 20$ mm, $l = 200$ mm, $w = 500$ mm. The liquid-metal path presents a resistance $r = 0.36$ mΩ between electrodes. Find the current and applied voltage required to develop a fluid flow $Q = 0.005$ m$^3$/s against an inlet/outlet pressure-difference of $\Delta p = 100$ kN/m$^2$, and the pump efficiency, neglecting fluid friction.
The field, current and flow directions are assumed to be mutually perpendicular. The force required is $bl\Delta p = B_1 lI_2 = 0.02 \times 0.20 \times 100 \times 10^3$ = 400 N, whence $I_2 = 4.0$ kA. The speed of flow is $u = Q/bl = 0.005/0.02$ × 0.20 = 1.25 m/s, and the motional e.m.f. is therefore $E_r = B_1 lu = 0.125$ V. The volt-drop in liquid resistance is $I_2 r = 4.0 \times 0.36 = 1.44$ V. Thus the applied voltage needed is 0.125 + 1.44 = 1.565 V, and the efficiency is

$$E_r I_2/VI_2 = E_r/V = 0.125/1.565 = 0.08 \text{ p.u.}$$

## 2.3 HETEROPOLAR MACHINES

The relation between the cylindrical, disc and linear forms is brought out in Fig. 2.4 in terms of the conversion region, i.e. the gap between the fixed and moving members. If the cylindrical gap region (*a*), shown with its d- and

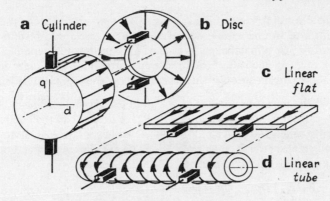

**a** Cylinder     **b** Disc

**c** Linear *flat*

**d** Linear *tube*

2.4 Forms of heteropolar machine

q-axes and armature current directions, has its front radius contracted and its rear radius expanded, it becomes the disc form (*b*); and if, instead, it is cut axially and 'unrolled' the result is the flat linear form (*c*); and finally if (*c*) is 're-rolled' into a tube, the linear tube shape (*d*) is obtained. The cylindrical machine has been in common use for well over a century and its technology is well established. The disc configuration is more recent: it suits applications in which the axial length must be short. Linear machines are usually a.c. induction motors in which only one member need be fed with a conducted supply; a linear d.c. machine requires both members to be so fed (unless the field is a permanent magnet) and its linear excursion is practically limited.

### Cylindrical

Fig. 2.5 shows a stator field system with excited N and S poles centred on the d-axis, and a rotor armature of diameter $D$ and axial length $l$ carrying

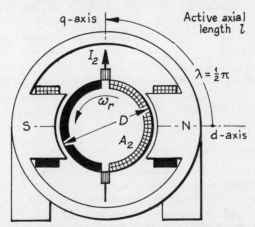

2.5 Heteropolar machine: cylindrical

symmetrical current sheets of linear density $A_2$ magnetizing along the q-axis. Let the rotor be wound with a total of $N_2$ turns. These are formed by the brushes into a pair of parallel paths each of $\frac{1}{2}N_2$ turns (or $N_2$ conductors), each path carrying one-half of the total input current $I_2$. The linear current density of the current sheets is therefore $A_2 = (\frac{1}{2}I_2N_2)/(\frac{1}{2}\pi D) = I_2N_2/\pi D$.

*E.M.F.*  When the rotor rotates at $\omega_r = 2\pi n$ in the gap field of mean density $B_1$, the mean e.m.f. per conductor is $e_r = B_1 l \frac{1}{2}D\omega_r$. As all the conductors of a path are in series, the path e.m.f. between brushes is $N_2 e_r$. An alternative expression employs the flux per pole $\Phi = B_1 l \frac{1}{2}\pi D$, the speed $n$ and the total number of active armature conductors $Z_2 = 2N_2$. Then

$$E_r = B_1 l N_2 \tfrac{1}{2} D\omega_r = \Phi N_2 \omega_r/\pi = \Phi Z_2 n \tag{2.6}$$

As the two armature paths are in parallel, eq. (2.6) represents the e.m.f. between brushes. As a result of the commutation process, $E_r$ is, in the steady state, a constant direct e.m.f.

*Torque*  Considering the $2N_2$ conductors, each carrying $\frac{1}{2} I_2$ and lying in a mean gap flux density $B_1$, the torque is

$$M_e = B_1 l \left(2N_2 \times \tfrac{1}{2}I_2\right) \tfrac{1}{2} D = B_1 l N_2 I_2 \tfrac{1}{2} D = \Phi N_2 I_2/\pi \tag{2.7}$$

Commutation makes the torque contributions of each side of the armature additive.

*Conversion Power*  The two foregoing equations give $E_r I_2 = M_e \omega_r$ for the converted power $P$, which can be expressed as in eq. (2.1):

$$P = \tfrac{1}{2}\pi D^2 l B_1 A_2 \omega_r = \pi^2 D^2 l B_1 A_2 n \tag{2.8}$$

*Torque Angle*  With a d-axis field and a q-axis armature m.m.f., the angle between these m.m.f. axes is $\lambda = \frac{1}{2} \pi$ rad, as indicated in Fig. 2.5. The torque is maximized because each armature current sheet lies wholly in a field of one polarity. Were the brush axis turned into coincidence with the d-axis field, making $\lambda = 0$, there would under each pole be equal numbers of conductors carrying oppositely directed currents, developing opposing torques with zero resultant. Thus torque is a function of $\lambda$, called the *torque angle*. In some a.c. machines both the gap flux density and the armature current-sheet pattern can be taken as sinusoidal, in which case the function is $\sin \lambda$; but this is not appropriate for a d.c. machine.

## Linear

For the flat form in Fig. 2.6, the current sheets of linear density $A_2$ lie in a mean gap flux density $B_1$ extending over each pole-pitch $Y$. The runner moves at speed $u$, the stationary brushes ensuring that the current-sheet pattern remains fixed. Let the runner have an active width $l$, turns $N_2$ per pole-pair and an input current $I_2$: then the motional e.m.f. between brushes, and the translational force and converted power per pole-pair, are

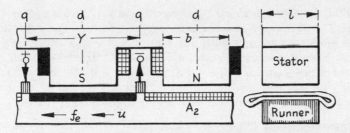

2.6 Heteropolar machine: linear (flat)

$$E_r = B_1 l u N_2 \quad f_e = B_1 l I_2 N_2 \quad P = 2Y l B_1 A_2 u \tag{2.9}$$

In the tubular construction, the whole of each armature turn is active. The e.m.f., force and power are given by eq. (2.9) with $l$ replaced by $\pi D$, where $D$ is the runner diameter.

## Disc

For the 2-pole arrangement of Fig. 2.7 and an average gap flux density $B_1$, the e.m.f. in a single rotor conductor is identical with that in the homopolar

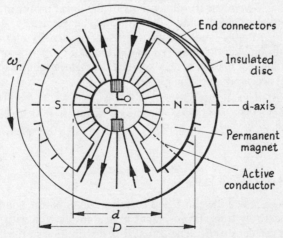

2.7 Heteropolar machine: disc

machine, eq. (2.4). Let the armature have $N_2$ turns and an input current $I_2$: then there are $N_2$ conductors in series between brushes and the angular current density is $A_2 = N_2 I_2 / 2\pi$. Consequently

$$E_r = \tfrac{1}{8}(D^2 - d^2)N_2 B_1 \omega_r \quad M_e = \tfrac{1}{8}(D^2 - d^2)N_2 B_1 I_2$$

and $\quad P = \tfrac{1}{4}\pi(D^2 - d^2)B_1 A_2 \omega_r = \tfrac{1}{2}\pi^2(D^2 - d^2)B_1 A_2 n$ \quad (2.10)

The geometry of a disc machine is more suitable for 6, 8 or more poles.

EXAMPLE 2.4: The cylindrical 2-pole machine in Fig. 2.5 has a rotor diameter $D$ = 80 mm and an active axial length $l$ = 60 mm. The stator field winding produces a uniform radial flux density $B_g$ = 0.50 T under the pole-faces, which each subtend the angle $2\pi/3$ rad ($120°$). The rotor has $N_2$ = 126 turns and its input current is $I_2$ = 4.0 A. Calculate (i) the force of attraction between each pole and the rotor, (ii) the tangential force and the torque, and (iii) the mechanical power developed when the rotor spins at 100 rad/s.

(i) The radial force per unit area is $\frac{1}{2}B_g^2/\mu_0$ = 100 kN/m$^2$. The area of each pole-face is $\frac{1}{3}\pi Dl$ = 0.005 m$^2$ so that the total radial force is 500 N, which resolves to 410 N on the d-axis. The attractive forces at the N and S poles have zero resultant provided that the machine has complete symmetry: if not, the resultant gives rise to an 'unbalanced magnetic pull'.

(ii) The current $\frac{1}{2}I_2$ in the $2N_2$ conductors gives an electric loading $A_2 = N_2 I_2/\pi D$ = 2000 A/m. For the parts of the current sheet within the pole-arcs the tangential force is $f_e = \frac{2}{3}B_g l\pi DA_2$ = 10.0 N; alternatively the whole current sheet lying in the mean gap density $B_1 = \frac{2}{3}B_g$ gives the same result. The torque is $f_e\frac{1}{2}D$ = 0.40 N-m.

(iii) The converted power is $M_e\omega_r$ = 40 W. Thus $E_r$ should be 10 V, a value that is confirmed by application of eq. (2.6).

EXAMPLE 2.5: Given that the dimensions of the linear motor in Fig. 2.6 are $Y$ = 0.25 m, $b$ = 0.15 m and $l$ = 0.10 m, that the gap density under the pole-faces is $B_g$ = 0.40 T, and that the linear current density on the runner is $A_2$ = 10 kA/m, find per pole-pair (i) the net force of attraction on the suspended runner (of mass $m$ = 11.3 kg), (ii) the translational force and (iii) the initial acceleration of the runner.

(i) The force of magnetic attraction is $2(\frac{1}{2}B_g^2/\mu_0)bl$ = 1.92 kN. The gravitational force is $11.3 \times 9.81$ = 110 N. The net force is $1920 - 110$ = 1810 N.

(ii) $f_e = 2B_g lbA_2$ = 120 N.    (iii) $f_e/m$ = 120/11.3 = 10.6 m/s$^2$.


## 2.4 MAGNETOMOTIVE FORCE

So far we have assessed the e.m.f. and torque of an armature winding as resulting respectively from its movement in, and interaction with, a magnetic field of mean gap density $B_1$. The actual flux distribution in the gap in fact is produced by the field and armature m.m.f.s in combination, since each has a magnetizing effect on the magnetic circuit. We now consider the field and armature m.m.f.s both separately and in combination, related for simplicity to the common cylindrical structure. The assumptions are: (i) the airgap accounts for the whole of the magnetic-circuit reluctance; (ii) the radial gap length under the pole-faces is uniform; (iii) leakage and fringing can be ignored; and (iv) the excited armature can be represented by a pair of current sheets without slotting effects.

## Field

To produce a gap density $B_g$ over the pole-face area requires a magnetizing force $H_g = B_g/\mu_0 \cong 800\,000\,B_g$ per unit radial length of gap. Thus for $B_g = 0.8$ T and a gap length $l_g = 5$ mm, the field m.m.f. required is $F_1 = 800\,000 \times 0.8 \times 0.005 = 3200$ A-t/pole regardless of the physical size of the machine. The distribution over the pole-arc of the m.m.f. $F_1$ and the gap flux density $B_g$ produced by it are shown in Fig. 2.8(*a*). In a practical machine the space waveform is modified by leakage and fringing.

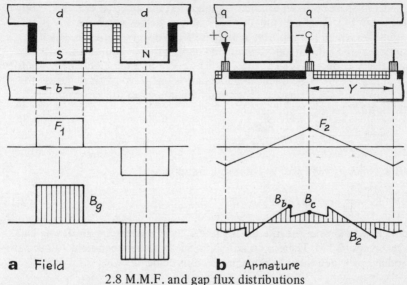

**a** Field              **b** Armature

2.8 M.M.F. and gap flux distributions

*Inductance* The m.m.f. per pole $F_1 = N_1 I_1$ is produced by a field current $I_1$ in a pole winding of $N_1$ turns. The pole-face flux density is $B_g = N_1 I_1 \mu_0/l_g$, the flux per pole is $\Phi_1 = \frac{1}{2}\pi D\,l(b/Y)B_g$. The inductance $L_f$ of a pole winding is $N_1\Phi_1/I_1$, whence

$$L_f = \tfrac{1}{2}\pi N_1^2 D\,l(b/Y)\,\mu_0/l_g \tag{2.11}$$

The pole-arc/pole-pitch ratio $b/Y$ is normally about 2/3. The total inductance of a 2-pole field system depends upon whether the windings are series- or parallel-connected. It introduces a time delay in the response of the field flux to transient changes in field voltage.

## Armature

The current-sheet pattern for a 2-pole armature winding of $N_2$ turns and a brush current $I_2$ produces an m.m.f. distribution of triangular space waveform with respect to the gap surface, Fig. 2.8(*b*). The peak m.m.f. per pole is $F_2 = \frac{1}{4}N_2 I_2$ at the q-axis positions along which the armature magnetizes. The gap reluctance is low under the poles, but high elsewhere.

The distribution of the resulting gap density $B_2$ is in an actual machine modified by fringing, but in any case there is a density $B_c$ on the brush axis. This is the commutation region; each turn, short-circuited by a brush, is moving through a flux of density $B_c$ and has the motional e.m.f. $e_{rc} = 2 B_c l \frac{1}{2} D \omega_r$ induced. The current that circulates through the coil via the brush surface is superimposed on the current being commutated, impairing the process of reversal.

*Inductance*  The energy approach is applied to find the self inductance of a 2-pole armature between brushes. Assume that the gap flux density — shown hatched in Fig. 2.8($b$) — extends *only* over the pole-width $b$. At any point within this region the gap flux density is inferred from the appropriate fraction of the total armature m.m.f. $F_2$ acting there. Thus at the pole edge the density is $B_b = \frac{1}{4} N_2 I_2 (b/Y) \mu_0/l_g$. The corresponding field-energy density is $\frac{1}{2} B_b^2/\mu_0$. The average under the pole-face is one-third of this because the waveform of $F_2$ is triangular. The total field energy in the gap volume $(\pi D \, l \, l_g)(b/Y)$ is

$$w_f = \frac{1}{96} \; \pi N_2^2 I_2^2 D l \, (b/Y)^3 \, \mu_0/l_g$$

But $w_f = \frac{1}{2} L_a I_2^2$, whence the rotor self inductance is

$$L_a = \frac{1}{48} \; \pi N_2^2 D l \, (b/Y)^3 \, \mu_0/l_g \tag{2.12}$$

neglecting linkage in the interpolar gaps. $L_a$ will as a rule be much smaller than $L_f$, eq. (2.11). There is no mutual inductance between field and armature windings provided that the rotor m.m.f. axis coincides with the q-axis.

### Field-armature combination

Superposition of the m.m.f. distributions of Fig. 2.8 gives the resultants in Fig. 2.9, ($a$) for the motor and ($b$) for the generator mode, and corresponding gap flux density distributions $B_g$.

*Armature Reaction*  With zero armature current, the gap m.m.f. is $F_1$ alone, centred on the d-axis. When armature current flows, adding the q-axis m.m.f. $F_2$, the 'distortion' is described as arising from *armature reaction*; it is, of course, essential to the production of the torque (as discussed in Sect. 1.3, Fig. 1.4). In an ideal machine that is all; but in practice there are secondary but important consequences. One, the appearance of $B_c$ in the commutating zone, has already been noted; further effects are the local saturation of the field poles and the greater rotational e.m.f. in armature coils moving in the regions of high gap density which contribute to possible commutator 'flashover'.

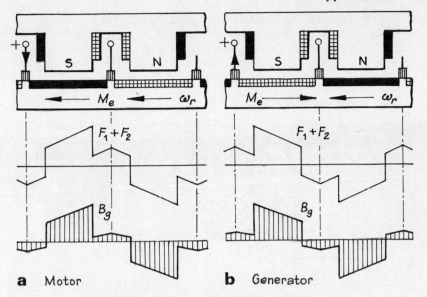

**a** Motor   **b** Generator

2.9 Resultant m.m.f. and gap flux distributions

## 2.5 STEADY-STATE PERFORMANCE

We consider a d.c. machine connected to constant-voltage busbars. Its performance depends on the way in which the stator and rotor members are connected. In the basic electromechanical system, Fig. 2.10, it is convenient to label the field and armature respectively as F and A, and to use subscripts $f$ and $a$ to distinguish between their voltages and currents.

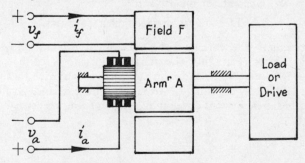

2.10 Basic electromechanical machine system

### Ideal machine

With negligible armature-winding resistance and mechanical loss, the rotational e.m.f. $E_r$ is equal always to the applied armature voltage $V_a$, and the electrical and mechanical output or input torques balance.

*Permanent-magnet Excitation* The field flux $\Phi_f$ is constant so that $E_r$ is directly proportional to the speed $\omega_r$ and can be expressed as $E_r = k_e\omega_r$, where $k_e$ is a constant depending on the number of armature turns. With a constant armature source voltage $V_a$ the speed is invariable. The electrical torque, $M_e = k_t I_a$, is proportional to the armature current, the basic relation $E_r I_a = M_e \omega_r$ showing that $k_t = k_e$. The equations of behaviour are

$$V_a = E_r = k_e\omega_r \qquad M_e = k_t I_a \qquad (2.13)$$

If $I_a$ is positive the machine is a motor. With a mechanical drive the current reverses and the machine generates. Fig. 2.11(*a*) shows the combined motor and generator characteristics relating speed and torque to armature current.

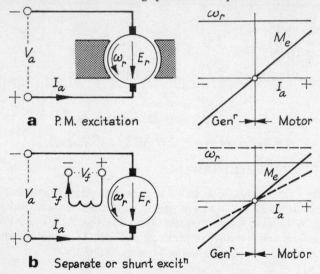

**a**   P.M. excitation

**b**   Separate or shunt excit$^n$

2.11 Ideal 'constant-speed' characteristics

*Separate Excitation*   The working flux is established by a current $I_f$ in pole windings. For a given $I_f$ and flux, the characteristics are as for the permanent-magnet machine, but can be modified, as indicated by the dotted lines in Fig. 2.11(*b*), by adjusting $I_f$. Armature voltage balance results in a lower (or higher) field current developing a higher (or lower) speed. The equations are

$$V_a = E_r = GI_f\omega_r \qquad M_e = GI_f I_a \qquad (2.14)$$

where $G$ is a constant [V per A and per rad/s] embodying the gap dimensions and the number of armature turns. The product $GI_f$ replaces $k_e$ and $k_t$ in eq. (2.13).

*Shunt Excitation*   The field and armature windings are connected in parallel to a common source so that $V_f = V_a = V$. The characteristics are like those for the separately excited machine, a change of field current being obtained by introducing a rheostat into the field circuit.

*Series Excitation* The field and armature windings are connected in series to the source to give $I_f = I_a = I$, Fig. 2.12. The equations of behaviour are

$$V = E_r = GI\omega_r \quad M_e = GI^2 \tag{2.15}$$

The product $I\omega_r$ is a constant, and the speed is therefore inversely proportional to the load current. In the motor mode, a low torque demand means a small current and flux, and consequently a high speed; in fact, for the ideal machine considered the zero-torque condition implies an infinite speed. In passing from the motor to the generator mode, the current is reversed, and as this would also reverse the field flux it is necessary to interchange the field connections.

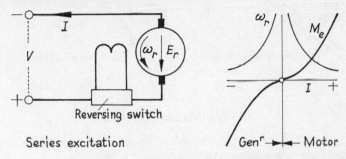

Series excitation                    Gen$^r$ —►◄— Motor

2.12 Ideal 'inverse-speed' characteristics

*Compound Excitation* Characteristics intermediate between those for series and shunt excitation are obtained with two field windings, one shunt or separate and the other is series with the armature. If the windings 'assist' in one mode they will 'oppose' in the other. The functions are called *cumulative* and *differential* compounding. A shunt winding may be incorporated in a series motor to limit the rise of speed on no load.

### Practical machine

Real machines have winding resistance, core and mechanical losses, saturation effects and several phenomena associated with commutation. Armature resistance drop $I_a r_a$ causes $E_r$ to differ from $V_a$, particularly in small machines; magnetic-circuit saturation, augmented by armature reaction, means that the field flux is no longer proportional to the field current; core and friction losses reduce the shaft output of a motor and increases the mechanical torque demand of a generator; and circulating-current loss occurs in the commutation process. Typical modifications of the 'ideal' characteristics are shown in Fig. 2.13 for 'constant-speed' and 'inverse-speed' machines.

*Constant-speed* With permanent-magnet, separate and shunt excitation the variation in armature resistance drop $I_a r_a$ is accommodated by a minor change only in speed.

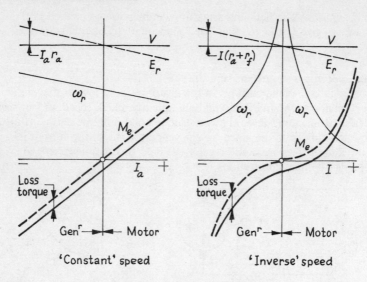

'Constant' speed          'Inverse' speed

2.13 Practical characteristics

*Inverse-speed* This corresponds to a series characteristic in which the torque is, roughly, inversely proportional to speed.

*Adjustable-speed* The term refers to operation at a number of 'constant' speeds by change of field current or of armature voltage.

EXAMPLE 2.6: A 2-pole series motor with 306 armature turns generates an open-circuit armature voltage of 100 V when driven at 50 rad/s with a field current of 5 A. Obtain the speed, gross torque and converted power characteristics for operation on a 200 V supply (i) as an 'ideal' machine, (ii) as a practical machine with a terminal resistance $(r_a + r_f) = 1\Omega$ and a flux $\Phi$ related to the motor current $I$ by

| Current $I$ (A): | 1 | 3 | 5 | 7.5 | 10 | 15 |
|---|---|---|---|---|---|---|
| Flux $\Phi$(mWb): | 5 | 14 | 20 | 24 | 26 | 30 |

(iii), as (ii) but with $\Phi$ (in mWb) reduced by $0.2I$ as a result of armature reaction.

(i) For 5 A and 50 rad/s, $E_r = V = 100 = GI\omega_r$, whence $G = 0.40$. Then for a motor current $I$, $\omega_r = 200/0.4\,I$ and $M_e = 0.4\,I^2$.

(ii) $E_r = V - 1.0\,I = 200 - I$. From eq. (2.6), $E_r = \Phi N_2 \omega_r/\pi = 97.4\,\Phi\omega_r$. The gross torque is $M_e = E_r I/\omega_r$.

(iii) As for (ii), but with the flux $\Phi$ reduced.
The results are tabulated below for comparison.

In (ii) and (iii) the armature $I^2 R$ loss reduces the available conversion power. The effect of saturation is to raise the speed and reduce the torque at the higher current levels, results that are intensified by armature reaction.

| Current (A) | | 1 | 3 | 5 | 7.5 | 10 | 15 |
|---|---|---|---|---|---|---|---|
| E.M.F. (V) | (i) | 200 | 200 | 200 | 200 | 200 | 200 |
| | (ii and iii) | 199 | 197 | 195 | 192.5 | 190 | 185 |
| Speed (rad/s) | (i) | 500 | 167 | 100 | 67 | 50 | 33 |
| | (ii) | 409 | 144 | 100 | 82 | 75 | 63 |
| | (iii) | 425 | 151 | 105 | 88 | 81 | 70 |
| Torque (N-m) | (i) | 0.40 | 3.6 | 10 | 22.5 | 40 | 90 |
| | (ii) | 0.49 | 4.1 | 9.8 | 17.6 | 25 | 44 |
| | (iii) | 0.47 | 3.9 | 9.3 | 16.4 | 24 | 40 |
| Power (kW) | (i) | 0.20 | 0.60 | 1.00 | 1.50 | 2.0 | 3.0 |
| | (ii and iii) | 0.20 | 0.59 | 0.98 | 1.44 | 1.9 | 2.8 |

## Operating point and stability

The steady-state operating point is that for which there is an electrical/ mechanical torque balance at a given speed. For stability, a small change of torque should result in only a minor change of speed to restore balance. In the following, $M_e$ and $M_l$ are the gross electromagnetic and mechanical-load torques, so that the operating point is defined by that at which the two torque/speed curves intersect.

*Motor Mode* In Fig. 2.14($a$), the point P is the intersection of typical torque/speed characteristics. For most mechanical loads the torque rises with speed, i.e. the slope $(dM_l/d\omega_r)$ is positive. P is most closely defined if $(dM_e/d\omega_r)$ is negative and crosses the mechanical characteristic at right-angles.

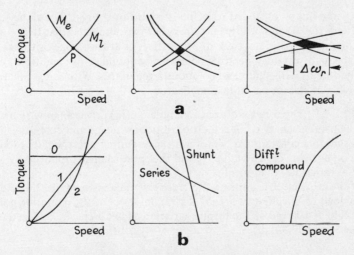

2.14 Motor mode: operating point

In practice, small variations in supply voltage, temperature etc. are inevitable, making P a small region rather than a point. A 'sensitive' condition will occur if the angle of intersection is acute, for then the speed change $\Delta \omega_r$ becomes excessive. Should the characteristics be tangential, or should $dM_e/d\omega_r$ be more positive than $dM_l/d\omega_r$, complete instability ensues and the motor will race. In Fig. 2.14($b$), three idealized mechanical load-torque/speed characteristics are shown: (0) constant torque $M_l = k_0$; (1) speed-proportional torque $M_l = k_1 \omega_r$ typical of a machine-tool; and (2) a fan or centrifugal pump characteristic $M_l = k_2 \omega_r^2$. Comparison with the torque/speed curves of shunt and series motors indicates that operation is stable for all three loads. But instability might occur with a type (2) load driven by a differential-compound motor.

*Generator Mode*   Much depends on the speed of the prime-mover, which on road vehicles and aircraft may vary widely. Where the speed is governed, the generator voltage $V$ can be adjusted by field control, but its current $I$ depends on the voltage/current relation $v/i$ of the load, for which three are given in Fig. 2.15; (0) infinite busbar of constant voltage $V_0$, (1) a secondary battery,

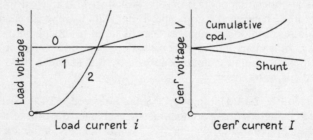

2.15 Generator mode: operating point

and (2) a non-linear resistor of resistance proportional to $i^2$. The slope $dv/di$ is zero for (0) and positive for the other two. Provided that the generator characteristic slope $dV/dI$ is negative, or at least less positive than that of the load, a stable operating point can be found. This condition generally obtains with shunt and compound generators. With series generators the problem of stability is more complicated.

EXAMPLE 2.7: The series motor in Example 2.6(iii) drives a fan with a torque demand given by $\omega_r^2/1250$. The friction, windage and core-loss torques sum to a constant 2.5 N-m. Find the operating point with the motor running on normal voltage of 200 V, and the speed change produced by a ± 2.0 N-m variation in fan torque.
Plots of the gross electrical torque $M_e$ from the table in Example 2.6, and of the total load torque $M_l = (2.5 + \omega_r^2/1250)$, intersect at the operating point of 10.5 N-m at 100 rad/s. The torque variation causes a change of speed over the range 95 − 107.5 rad/s.

## 2.6 TRANSIENT PERFORMANCE

Machines seldom run in a sustained steady state. When changes of voltage or speed or load occur, the new condition is reached from the initial state through a transient period during which the stored field, inertial and elastic energies are redistributed, a process that cannot be instantaneous. To grapple with several electromechanical effects at once, we must re-model the machine more suitably.

Mechanical quantities were discussed in Sect. 1.5. Magnetic fluxes are expressed in terms of the inductance and current of the electric circuits that excite them, using $L_f$ and $L_a$ from eqs. (2.11) and (2.12) or from test. There remains the motional e.m.f. $e_r = k_e \omega_r$ or $e_r = G i_f \omega_r$, a function of the speed and field. With the armature brushes in the q-axis there can be no mutual inductance between field and armature: we require some other parameter by means of which $e_r$ can be related to the field flux.

*Motional Inductance*   Consider a 2-pole cylindrical machine of armature diameter $D$ and active axial length $l$, with $N_f$ field turns/pole, $N_a$ armature turns, q-axis brushes, a pole-arc/pole-pitch ratio $b/Y$, and a flux density $B_g = B_1(Y/b)$ under the pole-faces. From eq. (2.6) the rotational e.m.f. for an angular speed $\omega_r$ is

$$e_r = \tfrac{1}{2} B_g (b/Y) N_a D l \omega_r$$

Writing $B_g = N_f i_f \mu_0 / l_g$ and rearranging gives

$$e_r = [\tfrac{1}{2} N_f N_a D l (b/Y)(\mu_0/l_g)] \; i_f \omega_r = G i_f \omega_r$$

The square-bracketted term gives $G$ explicitly, and inspection of eq. (2.11) for the field inductance $L_f$ shows that

$$G = (1/\pi) \, (N_a/N_f) L_f \tag{2.16}$$

$G$ can be called the *motional inductance*.

### Behaviour equations

It is now possible to model a machine as the electromechanical system of Fig. 2.16 and to express its behaviour by the differential equations

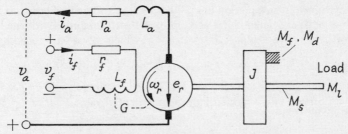

2.16 Electromechanical system model

Field:       $v_f = r_f i_f + L_f(di_f/dt)$

Armature:    $v_a = r_a i_a + L_a(di_a/dt) + Gi_f\omega_r$          (2.17)

Torque:      $M_e = Gi_a i_f = M_a + M_s + M_f + M_d + M_l$

The mechanical torques are those stated in eq. (1.16). Voltages, currents, torques and speeds are all instantaneous values. The electric-circuit equations are more concisely expressed by the matrix

$$
\begin{array}{c} \\ F \\ A \end{array}
\overset{\begin{array}{cc} F & \qquad A \end{array}}{
\begin{bmatrix} v_f \\ v_a \end{bmatrix} = \begin{bmatrix} r_f + L_f p & - \\ G\omega_r & r_a + L_a p \end{bmatrix} \cdot \begin{bmatrix} i_f \\ i_a \end{bmatrix}}
\qquad (2.18)
$$

in which for brevity the operator p is used in place of $d/dt$.

Any prototype heteropolar d.c. machine can now be modelled so long as the inductance and resistance coefficients can be considered to be constant (i.e. if saturation can be disregarded). For a permanent-magnet machine, row F is omitted and $Gi_f$ replaced by $k_e$ or $k_t$. For a shunt machine it is usual to have $v_f = v_a$, while for a series machine $i_f = i_a$. For steady-state conditions the operator p is zero so that

$$V_f = r_f I_f \qquad V_a = r_a I_a + GI_f\omega_r \qquad M_e = GI_f I_a \qquad (2.19)$$

giving the characteristics of Fig. 2.13.

Dynamic circuit analysis deals with the machine in terms of 'lumped parameters'. The physical model is not entirely lost, for magnetic flux-distribution patterns must still be set up to establish the self and motional inductances. In assessing the behaviour of real machines the skill lies in devising parameters adequate for the purpose in an essentially linear system which does not take direct account of leakage, saturation, slotting, commutation and eddy-current reactions.

EXAMPLE 2.8: A small motor with constant separate excitation has an armature of resistance $r_a = 30\Omega$, inductance $L_a = 0$ and a torque constant $k_t = 0.30$ N-m/A. The inertia of the motor and its load is $J = 0.006$ kg-m$^2$ but all other load torques are zero. A constant voltage $V_a = 100$ V is suddenly applied to the rotor at rest. Find the time for the speed to reach 48 r/s (300 rad/s).

The armature voltage equation is $V_a = r_a i_a + k_e\omega_r$. The electrical torque is $M_e = k_t i_a = J \cdot p\omega_r$, whence $i_a = (J/k_t)p\omega_r$. Substituting in the voltage equation gives $V_a/k_e = (J/k_t k_e)r_a \cdot p\omega_r + \omega_r$, which solves to

$$\omega_r = (V_a/k_e)[1 - \exp(-k_e k_t/Jr_a)t] = 333[1 - \exp(-0.5t)]$$

For $\omega_r = 300$ rad/s the time is $t = 4.6$ s. The speed rises exponentially from zero. The current falls exponentially from an initial value 100/30 = 3.33 A.

## 2.7 MULTIPOLAR MACHINES

Practical heteropolar machines may have any integral number $2p$ of poles. Small machines normally have two poles ($p = 1$), as this suits both permanent-magnet and current-excited field systems. Disc machines have 6 or 8, occasionally more, poles because of the shape of the rotor. Large industrial machines may have up to 10 poles, but rarely more and often fewer.

Taking each pole-pitch to extend over $\pi$ rad, the electrical and physical angles in a 2-pole machine are the same. A multipolar machine, considered as a set of 2-pole units each covering $2\pi$ *electrical* radians, has a physical circumferential angle of $2p\pi$ elec. rad but a physical circumference of only $2\pi$ mech. rad. The most direct effect of this concerns the torque, a mechanical quantity, and the angular speed. As will be seen, however, it is easily possible in a machine to associate the torque with the physical speed, and in any case it is always true that the conversion power is given by $E_r I_a = M_e \omega_r$.

## 2.8 VARIABLE-RELUCTANCE MACHINES

Some control machines, in particular the stepping motor described in Chapter 11, rely solely on the alignment principle (Sect. 1.3) exemplified in Fig. 1.2. The alignment torque arises as a result of the rate of change of gap-field energy $w_f$ with rotor displacement, eq. (1.3).

The prototype of such a machine, Fig. 2.17($a$), comprises a current-excited stator pole developing an m.m.f. $F = Ni$ and producing a gap flux $\Phi$ which enters a pole on a structually polarized but unexcited rotor. The gap

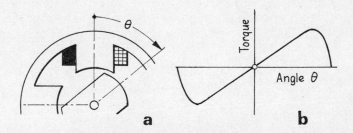

2.17 Variable-reluctance machine

flux is directly proportional to the gap permeance $\Lambda$ or inversely proportional to the gap reluctance $S$, which change with the angle of polar overlap. When the rotor axis is aligned with that of the stator, the position is one for which there is no rate of change of reluctance with position and so the torque vanishes. But if displaced, the rotor experiences a restoring torque, which increases in rough proportion to the displacement angle $\theta$ until the rotor pole is removed from the influence of the stator field, giving the typical torque/displacement relation ($b$). The operating process is to establish the stator field to bring the rotor into alignment, and then to de-energize it. By energizing an adjacent stator pole, the rotor is attracted into alignment with it, producing an angular 'step'.

The variable-reluctance machine is essentially a dynamic device, with a pulsed excitation, a position-dependent torque, and no condition of steady state other than quiescence. It operates under complex electric, magnetic and mechanical conditions of transience.

# 3 Magnetic Circuits

## 3.1 METHODS OF EXCITATION

The field member of a machine produces in the airgap the main or working flux. As magnetic flux is a solenoidal quantity (it follows a closed path), a ferromagnetic structure is required to complete the magnetic circuit. At the poles, a source of m.m.f. is introduced to establish the flux against the circuit reluctance, much of which lies in the airgap. Ferromagnetic materials exhibit a non-linear relation between flux density $B$ and magnetizing force $H$ because of the saturation phenomenon, so that 'average' graphical or tabular $B/H$ relations are required for each.

In most industrial machines the poles are excited by field currents in concentrated coils. In fractional-kilowatt, miniature and control machines the use of permanent-magnet excitation is usually preferred to avoid the large space and field $I^2 R$ loss that current excitation involves, permitting reduction of physical size and a marked improvement in efficiency.

## 3.2 PERMANENT-MAGNET EXCITATION

A permanent magnet (p.m.) is a ferro- or ferri-magnetic material that can sustain a high inherent magnetization at normal ambient temperatures, and act as a passive source of m.m.f. and flux. Its basic properties, as indicated by its hysteresis loop, Fig. 3.1($a$), are (i) *remanence* (or remanent flux density) $B_r$, and (ii) *coercivity* (or coercive field strength) $H_c$. The magnet is operated in the upper left-hand quadrant, shown shaded in ($a$), at some point P on the *demagnetization* characteristic. For the simple circuit ($b$) the operating point ($B, H$) for a p.m. of length $l_m$ and cross-section $a_m$ supplying a flux $\Phi_g$ to an airgap of length $l_g$ and pole area $a_p$ is obtained from

$$\Phi_m = B_m a_m = k_1 \Phi_g = k_1 B_g a_p$$

$$-H_m l_m = k_2 H_g l_g = k_2 (B_g/\mu_0) l_g$$

where $k_1$ and $k_2$, both greater than unity, account respectively for leakage flux and for loss of m.m.f. arising from the fact that the actual $B$ and $H$ values within the magnet are not uniform throughout its volume. The ratio $B/H$ at the operating point is

$$B/H = -(k_1/k_2)(a_p/a_m)(l_m/l_g)\mu_0 = -\lambda\mu_0$$

47

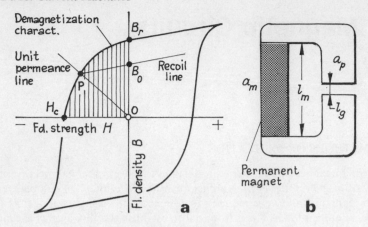

3.1 Basic permanent-magnet system

where $\lambda$ is a permeance coefficient. The operating point P in Fig. 3.1(*a*) is determined by the intersection of the unit permeance line, of slope $-\lambda\mu_0$, and the demagnetization characteristic.

EXAMPLE 3.1: A hard ferrite with a $B/H$ characteristic approximating to a straight line between $B_r = 0.40$ T and $H_c = 200$ kA/m, is to establish a flux density $B_g = 0.25$ T in an airgap of length 5 mm and area 30 mm $\times$ 30 mm, using two ferrite blocks and flux-concentrating pole-shoes. Taking $k_1 = 1.8$ and $k_2 = 1.3$, estimate the volume of each ferrite block.
The optimum magnet working point is $B_m = \frac{1}{2}B_r = 0.20$ T, $H_m = \frac{1}{2}H_c = 100$ kA/m. The pole-shoe area is $a_p = 30 \times 30 = 900$ mm$^2$, the gap flux is $\Phi_g = B_g a_p$, the magnet flux is $\Phi_m = k_1 \Phi_g$, and the magnet cross-section is

$$a_m = \Phi_m/B_m = 2025 \text{ mm}^2$$

Taking $H_m l_m = k_2 H_g l_g = k_2 (B_g/\mu_0) l_g$, then

$$l_m = 1.3(0.25 \times 800\,000)5/100 \times 10^3 = 13 \text{ mm}$$

The volume of each magnet is therefore 26 300 mm$^3$ = 26.3 cm$^3$.

In a machine, armature reaction m.m.f. partially opposes the field m.m.f., shifting the p.m. operating point P along the recoil line, Fig. 3.1(*a*), towards a density $B_o$. The recoil line is assumed to be straight, and parallel to the demagnetization characteristic at $B_r$, with a slope $\mu_c\mu_0$ where $\mu_c$ is the *recoil relative permeability*. In some p.m. materials the demagnetization characteristic is almost rectilinear so that the recoil line coincides with it.

*Equivalent Magnetic Circuit*  In Fig. 3.2(*a*) the various leakages are lumped into an equivalent flux $\Phi_l$ in a path of permeance $\Lambda_l$. The useful flux $\Phi_g$ in the gap permeance and with an armature-reaction m.m.f. $F_a$, occupies a parallel path. The operating point P is on a recoil line and can be taken as an

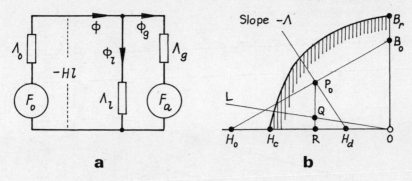

**a**            **b**

3.2 Permanent-magnet system: equivalent circuit

'open-circuit' m.m.f. $F_o$ acting through a permeance $\Lambda_o = \mu_c \mu_0 \, (a_m/l_m)$.
Extending the recoil line back to the $H$-axis at $H_o$ gives $F_o = H_o l_m$ as in (b).
Then $\Phi_m = \Phi_g + \Phi_l$, and equating m.m.f.s

$$-H_m l_m = \Phi_l/\Lambda_l = H_l l_l \quad \text{and} \quad -H_m l_m - F_a = \Phi_g/\Lambda_g = H_g l_g$$

gives $B_m = -\Lambda \, (H_m + H_d)$, where

$$\Lambda = (\Lambda_g + \Lambda_l) \, (l_m/a_m) \quad \text{and} \quad H_d = (F_a/l_m) \cdot \Lambda_g/(\Lambda_g + \Lambda_l).$$

The operating point $P_0$ in Fig. 3.2(b) is located where a line of slope $-\Lambda$,
raised from $H_d$, intersects the recoil line between $B_o$ and $H_o$. If the leakage
unit permeance line OL is added, then $P_0 Q$ and QR correspond respectively
to the gap and leakage fluxes.

## Permanent-magnet materials

The properties of a p.m. material are $B_r, H_c$, recoil permeability, $(BH)$max,
Curie point, ageing characteristic, mechanical working, cost and availability.
The operating point should approach $(BH)$max. The Curie temperature is that
at which inherent magnetization fails, and an associated property concerns
the change of $B_r$ with temperature. The mechanical properties are hardness,
stress limit, elastic modulus, and magnetic stability as a function of time and
stressing. Electrically, a high resistivity reduces eddy-current loss under
changing load conditions, and a high retentivity stabilizes the p.m. over its
working life.

Lynch [3] classifies p.m. materials into four groups according to their
demagnetization characteristics, Fig. 3.3. The optimum $B_r, H_c$ and $(BH)$max
properties are not obtained in any one group, but some deficiencies can be
countered by good magnetic-circuit design. The high coercivities of ceramic
and rare-earth materials withstand armature reaction effects and make
possible magnets of large area and very short length, contrasting with metallic
p.m.s. The four groups have the properties listed: $\theta_c$ is the Curie point, $\delta$ the
density and $\rho$ the resistivity.

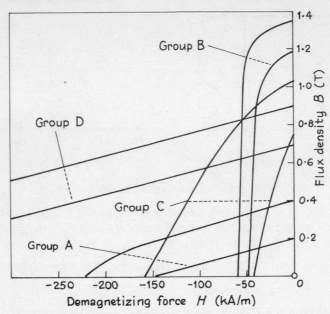

3.3 Permanent-magnet materials: demagnetization characteristics

Permanent-magnet Materials

| Group and Material | $B_r$ T | $H_c$ kA/m | $(BH)$m kJ/m³ | $\mu_c$ – | $\theta_c$ °C | $\delta$ kg/m³ | $\rho$ $\mu\Omega$-m |
|---|---|---|---|---|---|---|---|
| A Ceramic: ferrite Ba–Sr | 0.38 | 200 | 30 | 1.1 | 450 | 4800 | high |
| B Metallic: Alcomax III | 1.26 | 55 | 43 | 3.1 | 850 | 7300 | 500 |
| C Metallic: Hycomax III | 0.8 | 100 | 34 | 2.5 | 850 | 7000 | 500 |
| D Rare-earth: Sm–Co | 0.75 | 600 | 130 | 1.0 | 700 | 8000 | 60 |

A:    For quantity-production machines; finished p.m. is cheap but bulky
B:    Magnet must be long, and may require pole-shoes; operating point must allow reserve against armature reaction
C:    High coercivity gives greater reserve; $(BH)$ not sensitive to operating point; steep recoil line results in greater reduction of flux density with demagnetizing fields; torque/armature-current less linear
D:    As A, but higher energy levels; costly; low remanence may make pole-shoes necessary; length/area relation dictated by $B$ and $H$ requirements

### Magnet arrangements

Some of the possible ways of setting up p.m. field systems for miniature and small machines are illustrated in Fig. 3.4.

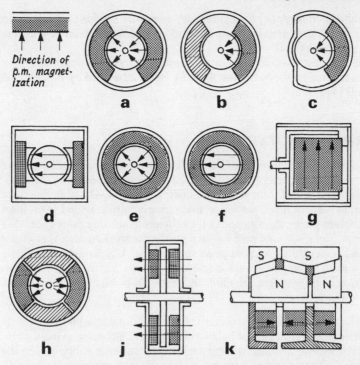

Direction of
p.m. magnet-
ization

a b c

d e f g

h j k

3.4 Small permanent-magnet machines

*Two-pole Fields* These normally have two p.m. segments given the magnetic orientation indicated by arrows. Radial orientation is most advantageous when a segment subtends an angle of about 120°. The segments are fixed by epoxy-resin adhesive or by interpolar spacers within a steel housing which also provides the yokes. The dimensional tolerance in ferrite p.m. profiles is such that the fit may be imperfect, introducing small gaps that increase the magnetic-circuit reluctance. To obtain adequate working flux the axial length of the p.m. segments may be greater than that of the armature. The two-segment p.m. construction (*a*) may sometimes be cheapened by substituting a steel block (*b*) for one magnet, or by modifying the tube profile (*c*). Method (*d*) permits the use of flat magnets and pole-shoes. Annular magnets (*e*) and (*f*) can be cross- or radially magnetized to develop a 2-pole field. In (*g*), the magnet is a cylinder fixed centrally to provide an annular gap for a cup-shaped ('moving-coil') armature.

*Multipolar Fields* For the same armature electric loading the armature reaction per pole is reduced, but the rotor and commutator structures are less simple. The 4-pole machine (*h*), shown as a consequent-pole structure with two p.m.s and two steel blocks, may alternatively have four p.m.s. The poles on one side of the disc machine (*j*) are sometimes omitted and replaced by a steel disc, stationary or carried on the rotor, to complete the magnetic circuit.

The imbricated rotor ($k$) for a stationary-armature machine is built up from axially magnetized annular p.m.s interlinked with steel plates to which pole-shoes are attached.

The field distribution in a p.m. machine is analytically complex. Computational methods are described by Binns *et al*, and by Reichert and Freundl [4].

### 3.3 CURRENT EXCITATION

Fig. 3.5 shows typical forms of current-excited magnetic circuits. In (*a*) for a small machine, the complete magnetic circuit is stamped from core-steel sheet, the stator and rotor stacks then being clamped. The poles and yokes in (*b*) are, in modern machines for rectifier supply, also stamped from sheet. The diagram shows also the position of the commutating poles (Sect. 5.4). The m.m.f. of each main pole has to provide the working flux that crosses the gap, and passes through a pole-pitch span of the yoke, armature teeth and core. It must be larger than the armature-reaction m.m.f. to prevent undue distortion of the gap flux distribution, particularly when speed is controlled by field-weakening.

*Split Field*  Small control machines fed with a constant armature current may be constructed with a pair of balanced field windings on each main pole. Under 'quiescent' conditions these carry equal currents in opposition: the effective excitation is then zero and the armature develops no torque. A control signal to a d.c. amplifier unbalances the field currents to produce a torque proportional to the unbalance and in a direction corresponding to the

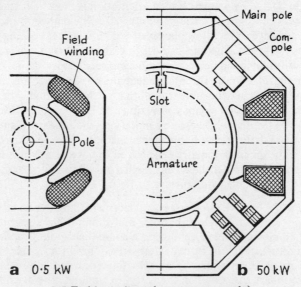

3.5 Field windings (not to same scale)

resultant excitation. The opposing field currents increase the $I^2 R$ loss and heating, but the current level is usually low.

## 3.4 ARMATURE

A laminated construction of steel plates electrically insulated from each other by thin paper, kaolin or other surface treatment is essential to limit eddy-current core loss and to permit the varying flux to penetrate the core. The interplate insulation reduces the effective 'iron' length of a stack of core-plates by a small fraction. The *stacking factor,* the ratio (useful/overall) length of a stack, is typically in the range 0.90–0.97.

The common *cylindrical* armature comprises a stack of silicon-steel stampings each carrying an insulating film on one side. For small machines, Fig. 3.5($a$), the rotor stampings, slotted in the punching process, are assembled on a shaft with their teeth in register. Larger machines, in which annular or sector plates are first punched and then slotted in a second process, employ a shaft-mounted 'spider' to carry the core assembly. Axial-flux *disc* armatures may be built without steel. Runners for *linear* machines are assembled from slotted sheets, which are then stacked and clamped.

## 3.5 MAGNETIZATION CHARACTERISTICS

The $B/H$ characteristics of typical ferromagnetic materials are given for comparison in Fig.3.6, and more detailed curves in Figs.3.7 and 3.8. For

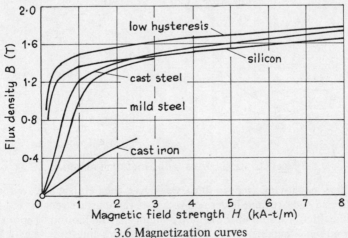

3.6 Magnetization curves

digital computation some analytic relation $B = f(H)$ may be needed. Froehlich's classic formula is $f(H) = H/(a + bH)$, which is adequate at higher saturation levels. Macfadyen *et al* [5] propose

$$B = k_1 [1 - \exp(-k_2 H)] + k_3 [1 - \exp(-k_4 H)] + \ldots + \mu_0 H$$

to give a better fit and to include the 'airgap line'.

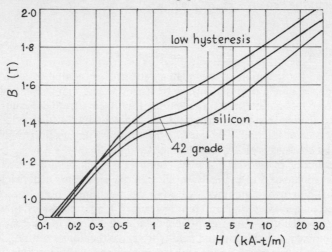

3.7 Magnetization curves: core materials

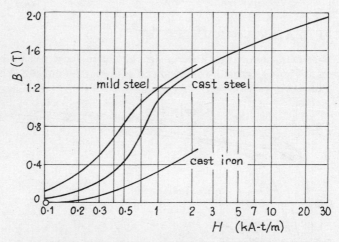

3.8 Magnetization curves: constructional materials

Silicon as an alloying agent in sheet steel raises the permeability, reduces hysteresis and eddy-current losses and increases the tensile strength. But it impairs the ductility, so that the silicon content may be limited to about 3% to avoid punching difficulties. Miniature machines and those (e.g. in aircraft) that require high flux densities to save mass, may have cobalt-iron alloys such as 49Co/49Fe/2V with which saturation densities up to 2.3 T can be reached.

Magnetic sheet material as received from the steelmakers has surface asperities, waviness and variations in thickness. Core stacks have therefore to be kept tight by clamping. A loose core vibrates under magnetostrictive and repulsion forces, but excessive pressure distorts the stack and impairs the magnetic properties.

## Core loss

At any point within the armature teeth and core, the flux density is subject to cyclic reversal as the rotor rotates at speed $n$ in a $2p$-pole field, the basic frequency being $f = pn$ (with $n$ in r/s). In a core plate the changing flux induces an e.m.f. that circulates an 'eddy' current, the result of which is a loss appearing as heat. A further loss occurs because of the cycling of the material through its hysteresis loop. Traditionally the two losses are considered separately, and for a sinusoidal time-variation of flux density of peak $B_m$ and r.m.s. value $B$ the specific loss components [W/kg] are

$$\text{Hysteresis: } p_h = k_h f B_m^x \quad \text{Eddy-current: } p_e = k_e d^2 f^2 B^2 / \rho$$

The factor $k_h$ depends upon the molecular structure of the material, the exponent $x$ lying between 0.8 and 2.3 and for normal flux densities approximating to 2. The thickness $d$ of the sheet takes account of the restriction on eddy-current flow obtained by lamination, and the resistivity $\rho$ shows that the material should, as a conductor, have a high electrical resistance.

The flux variation is not in fact sinusoidal, nor is it profitable (except for the steelmaker) to consider separately the two loss components. The loss is better regarded as a mass effect, for the actual loss in a built-up core is increased by random contact between plate edges, hardening and burring in the punching process, nonhomogeneity of the material and similar unpredictable anomalies. A designer employs empirical curves such as those in Fig.3.9 for the specific core loss in a constructed armature, verified by test results on completed machines.

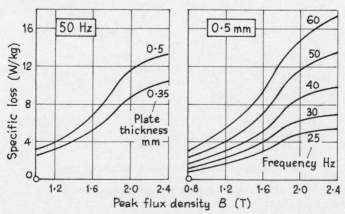

3.9 Core loss

### 3.6 MAGNETIC-CIRCUIT DESIGN

Having available the $B/H$ curves for the materials employed, the magnetic circuit is divided into convenient parts in series, the flux densities are estimated for a specified working flux, and the m.m.f. per unit length found and multiplied by the length of the part. Summation yields the total magnetic-circuit m.m.f. Intricacy of the shapes of the various parts (particularly of the airgap boundaries) makes determination difficult.

The total flux developed by the poles is a combination of the *useful* working flux crossing the gap and the *leakage* flux in the interpolar space external to the armature. When the latter is also excited, armature reaction affects the distribution of the working flux, and also generates slot and end-winding leakage flux components. All the flux components are separately assessed, and superposed on the assumption that the presence of one does not affect another in the same region. This is strictly untrue because superposition may change the saturation conditions. In the magnetic-circuit design of heteropolar machines the initial calculation is done for pole excitation alone, with armature-current effects added thereafter.

### 3.7 HETEROPOLAR CYLINDRICAL MACHINE

The 4-pole structure in Fig. 3.10 is considered. The part-circuit bounded by one pole-pitch has a main path through the pole, airgap and teeth, dividing through the flanking yokes and the armature core. A rough flux pattern is

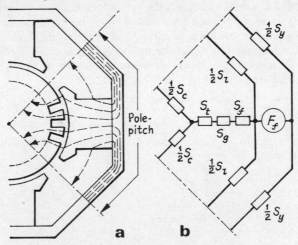

3.10 Magnetic circuit of heteropolar machine

indicated in Fig. 3.10(*a*) and an arbitrary equivalent circuit in (*b*). The m.m.f. per pole $F_f$ works through an array of part-reluctances $S$ for the yokes ($y$), the pole itself ($f$), the airgap ($g$), the teeth ($t$) and the armature core ($c$). The pole leakage flux path is represented by the branches $S_l$.

## Airgap

The gap is bounded by a smooth polar surface on one side, and a slotted and ducted surface on the other. The principal quantities concerned are

$l$ total axial length of core          $y_s$ slot pitch
$l_i$ net iron length of core          $w_o$ slot opening
$l_g$ radial length of gap          $w_d$ duct width

The gap reluctance (or permeance) is solved by use of the analytical work of Carter [6]. In Fig. 3.11, (a) shows a possible gap flux distribution in the

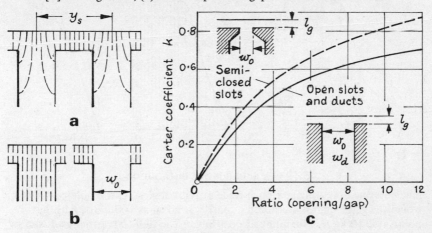

3.11 Gap magnetization

presence of openings in the gap boundary. The gap reluctance is greater than if the opening were ignored, but less than if it were deemed to carry no flux at all (b). The effective 'contracted' slot-pitch is

$$y_s' = y_s - k_o w_o$$

where $k_o$ is a function of the ratio $(w_o/l_g)$ from Fig. 3.11 (c), a distinction being made as to the form, open or semi-closed, of the slots. Radial ventilating ducts are treated in a similar way by contracting the axial length $l$ of the gap surface to $l'$, where

$$l' = l - k_d n_d w_d$$

for $n_d$ ducts of width $w_d$, with $k_d$ a function of $(w_d/l_g)$ from Fig. 3.11(c). Then for an airgap between a pole-face and a dentated rotor the gap m.m.f. is

$$F_g = 800\,000\,K_g B_g l_g \qquad K_g = (l/l')\,(y_s/y_s')$$

where $l_g$ is the gap length at the pole-centre where the gap flux density is $B_g$, and $K_g$ takes account of slot and duct openings.

*Gap-flux Distribution*  Carter [6] has given 'fringing' curves for the flux distribution from a pole-face, including the effect of chamfer, Fig. 3.12(*a*). A direct method is to draw a scaled outline of a half pole-pitch divided into a number of equal intervals, say 6. From each division mark is sketched a flux

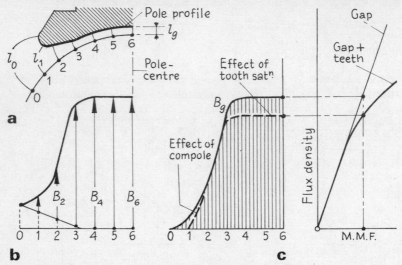

3.12 Gap flux density distribution

line, entering or leaving an iron surface at right-angles. At the pole-centre (6), where the gap length is $l_g$, the flux density is taken as 100 units. At other points it is $100(l_g/l_x)$ for a line at position $x$. The plot (*b*) of the result shows a positive value at position (O): this should in fact be zero, so a straight line is drawn from the ordinate of $B$ at (O) down to the baseline at a point where the chamfer begins, in effect excluding interpolar leakage. The actual rotor surface flux density is taken as the intercept between the straight line and the $B$-plot, to give (*c*). Further adjustments may be made, such as a further exclusion of compole leakage and an adjustment of the flux density near the pole-centre as a result of saturation in the armature teeth in that region, as determined from a characteristic relating gap-flux density to the m.m.f. of the gap plus teeth. These are indicated in (*c*).

### Teeth

Small machines with wire windings have parallel-sided teeth and tapered slots, but in large machines with windings formed from rectangular conductors it is necessary to employ parallel-sided slots with tapered teeth. High tooth saturation densities may make necessary a careful estimate of the required tooth m.m.f.

*High Density*  For highly saturated teeth, the m.m.f. is sufficient to drive appreciable flux through the slots and ducts in parallel with the teeth. Consequently an estimate of the *apparent* tooth density $B_t'$ based on tooth area alone is erroneous. The *true* density $B_t$ is related to the apparent density by

$$B_t' = B_t + \mu_0 H (K - 1)$$

the last term being the flux density in the nonmagnetic parallel paths $K$ is the ratio (gross area of iron and air/net iron area) for each tooth section taken; it must include the stacking factor, and may have values between 1.5 for a machine without ducts and 3 for a large machine with several radial cooling ducts. A graphical estimate of the tooth m.m.f. can be based on a plot of the true density for a number of sections from tooth tip to root, using the $B/H$ curve of the core material to obtain $H$. For moderate taper the three-ordinate method applies Simpson's rule to three equidistant sections to give the mean $H = (H_1 + 4H_2 + H_3)/6$.

*Low Density* For slight taper it is sufficient to find the density $B_{t\frac{1}{3}}$ at a section one-third of the tooth length from the narrower end and to take this as applying over the whole tooth length.

### Cores and yokes

The m.m.f. requirements of cores, yokes and similar sections are based on the choice of suitable regions for the calculation of area and the estimate of flux density.

### Poles

For salient poles, account must be taken of the change in flux that results from inter-polar leakage, complicated by the presence in some cases of compoles in the inter-polar gaps. It may be adequate to assume an empirical leakage coefficient (e.g. 1.2) to relate the pole flux at the yoke end to that facing the gap. If not, then an estimate must be made of the several reluctance paths shown in Fig. 3.10. Appropriate formulae are derived in Sect. 4.12.

EXAMPLE 3.2: The main dimensions of the magnetic circuit of a 4-pole 100 kW 360 V 1400 r/min d.c. shunt motor with 49 rotor slots are given in Fig. 3.13. The full-load gap flux per pole is $\Phi_g = 25.2$ mWb, the pole leakage coefficient is 1.15 and the allowance for armature-reaction effects is 0.13. Estimate the full-load excitation per pole and draw the flux/field-current characteristic for 0.85, 1.0, 1.1 and 1.2 p.u. of normal flux on open circuit.

### Magnetic circuit

The calculation is made for one pole-pitch centred on the polar axis. The flux carried by the pole divides between the yokes and armature core. The ferro-magnetic materials have the low-hysteresis magnetization characteristic in Fig. 3.7, and have an equivalent axial length $l_i = 0.90\, l = 0.90 \times 0.17 = 0.153$ m.

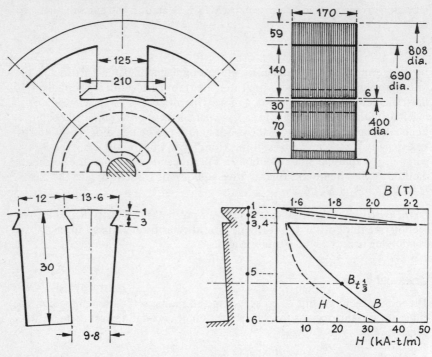

3.13 Magnetic circuit: Example 3.2 (dimensions in mm)

*Armature Core*  Depth below slots, 0.07 m; area, $a_c$ = 0.153 × 0.07 = 0.0107 m$^2$; flux, $\Phi_c = \frac{1}{2}\Phi_g$ = 12.6 mWb; density, $B_c$ = 12.6/0.0107 × 10$^3$ = 1.18 T.

*Teeth*  Slot-pitch, $y_s$ = 25.6 mm; pole-arc, $b$ = 210 mm; teeth per pole-arc, $b/y_s$ = 8.2, say 8; flux per tooth, 25.2/8 = 3.16 mWb. Applying *Simpson's rule* separately to the wedge and slot portions, Fig. 3.13, the tooth widths and areas for the six sections are evaluated and the flux densities found. If the density exceeds about 2 T it is necessary to calculate the ratio $K$:

| Section | Tooth width mm | Tooth area mm$^2$ | Flux density T | Ratio $K$ — | Magnetizing force kA-t/m | kA-t/m |
|---------|------|------|------|-----|------|-----|
| 1 | 13.6 | 2050 | 1.52 | — | 1.2 | — |
| 2 | 11.4 | 1740 | 1.82 | — | 10.0 | — |
| 3 | 9.2 | 1410 | 2.24 | 3 | — | 50 |
| 4 | 13.2 | 2020 | 1.56 | — | 1.6 | — |
| 5 | 11.5 | 1760 | 1.79 | — | 9.0 | — |
| 6 | 9.8 | 1500 | 2.10 | 2.5 | — | 32 |

The appropriate values of $H$ are then

Wedge: $\quad H = \frac{1}{6}[1.2 + 4 \times 10 + 50] = 15.2 \text{ kA-t/m}$
Slot: $\qquad H = \frac{1}{6}[1.6 + 4 \times 9 + 32] \ \ = 11.6 \text{ kA-t/m}$

The total tooth m.m.f. is $(15.2 \times 4 + 11.6 \times 26)/10^3 = 360$ A-t.
Plotting points from the Table gives the *graph* of $B_t$ and $H$. The height of the $H$ curve has a mean of about 13 kA-t, so that the tooth m.m.f. is $13 \times 30 \times 10^{-3} = 390$ A-t. At *one-third* of the tooth length from the root, $B_{t\frac{1}{3}} = 1.86$ T and a corresponding $H = 13$ kA-t/m, whence the total m.m.f. is 390 A-t. For this degree of saturation all three methods give approximately the same figure of 390 A-t.

*Airgap* Gap length, $l_g = 6.0$ mm; slot opening, $w_o = 12$ mm; ratio $(w_o/l_g) = 2.0$, whence $k_o = 0.29$ from Fig. 3.11(c). The effective pole-arc is

$$b' = b - 8k_o w_o = 210 - (8 \times 0.29 \times 12) = 182 \text{ mm} = 0.182 \text{ m}$$

The gap area is $a_g = 0.182 \times 0.17 = 0.031 \text{ m}^2$ and the gap flux density is $B_g = 25.2 \times 10^{-3}/0.031 = 0.813$ T.

*Pole* Area $a_p = 0.125 \times 0.153 = 0.0191 \text{ m}^2$. The flux is assumed to be $\Phi_p = 25.2 \times 1.15 = 29.0$ mWb and the density is $B_p = 1.52$ T.

*Yoke* The area of each branch is $0.01 \text{ m}^2$ and the density is $B_y = 14.5 \times 10^{-3}/0.01 = 1.45$ T.

**Total m.m.f.**

For the normal flux of $\Phi_g = 25.2$ mWb, the results above are collected for summation:

| | Flux $\Phi$ mWb | Area $a$ m$^2$ | Density $B$ T | Mag. fce. $H$ kA-t/m | Length $l_x$ m | M.M.F. $F$ kA-t |
|---|---|---|---|---|---|---|
| Armature core | 12.6 | 0.0107 | 1.18 | 0.29 | 0.10 | 0.03 |
| Teeth | 25.2 | 0.0136 | 1.86 | 13.0 | 0.03 | 0.39 |
| Airgap | 25.2 | 0.0310 | 0.813 | 650 | 0.006 | 3.90 |
| Pole | 29.0 | 0.0191 | 1.52 | 1.1 | 0.14 | 0.15 |
| Yoke | 14.5 | 0.010 | 1.45 | 0.8 | 0.25 | 0.20 |
| | | | | | | 4.67 |

Pole excitation with 0.13 p.u. addition for armature reaction      5.30

**Flux/current relation**

On open circuit there is no addition required for armature reaction.

Repeating the magnetic-circuit calculations for the required p.u. flux values gives:

| Flux per pole | (p.u): | 0.85 | 1.0 | 1.1 | 1.2 |
|---|---|---|---|---|---|
| | (mWb): | 21.4 | 25.2 | 27.7 | 30.2 |
| Field excitation | (kA-t): | 3.60 | 4.67 | 6.44 | 10.09 |
| Airgap excitation | (kA-t): | 3.31 | 3.90 | 4.29 | 4.68 |

This is plotted in Fig. 3.14. At low flux levels the gap m.m.f. predominates, as indicated by the *airgap line*. At 1.0 p.u. flux the gap takes 84% of the excitation: at higher fluxes the saturation of the ferromagnetic parts of the

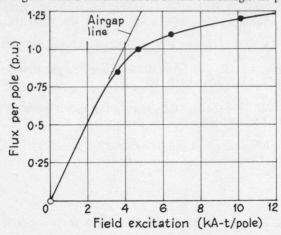

3.14 Flux/excitation characteristic: Example 3.2

magnetic circuit (particularly the teeth) makes heavy demands on the total m.m.f.


## 3.8 HETEROPOLAR DISC MACHINE

Disc machines are usually small, and almost invariably have permanent-magnet excitation with 6, 8 or 10 poles, using sintered Group A materials (Sect. 3.2) which can be magnetized before assembly, or metal p.m.s. from Groups B or C which must have a few turns around each magnet for impulse magnetization after assembly. The p.m.s are arranged on a fixed stator disc. The return flux may be by similarly mounted magnets on the other side of the armature gap, or by a steel disc, usually fixed but occasionally as a backing to the armature.

Group A magnets are light in weight and cheap. They have a high coercivity, but although their remanence is only moderate, their recoil lines lie on the *B/H* characteristic and do not involve minor hysteresis loops.

The total axial length of the airgap determines the length of the magnets: typically the magnet length is about 20 times the gap length, so that a thin armature disc (consistent with mechanical strength and freedom from flexure) is necessary.

The magnets are circular or sectorial, the latter providing a greater proportion of the working flux at the outer diameter. For a given gap flux density the useful flux is increased if inter-magnet leakage is minimized by adequate spacing, but this conflicts with the need for maximum working flux. Fig. 3.15 shows typical flux-density distributions, radial and

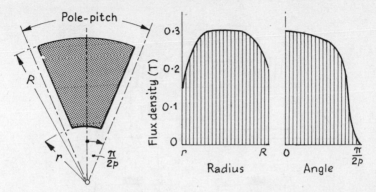

3.15 Gap flux distribution for disc machine

circumferential, for a disc machine. It was shown in eq. (2.10) that the flux is more effective where the peripheral speed is greater. With an increase in the ratio $R/r$ there is more active space for the armature conductors; but there is rather more flux lost from the gap, and the end-connectors of the armature winding must be lengthened. A useful criterion is to maximize the armature power, on which basis Campbell [7] concludes that the optimum diameter ratio is $D/d = \sqrt{3}$.

## 3.9 HETEROPOLAR LINEAR MACHINE

Most d.c. linear motors are designed for special duties, such as pumping, door-closing, cross-chart pen-recorder positioning, actuators and thrustors. The need to wind the runner over its full working length and to provide brushgear and sliding contacts are the main disadvantages, although if the runner excursion is limited it is possible to use fixed taps on the runner winding and to feed them through flexible cables.

Fig. 3.16 shows two realizations of the tubular linear machine. The *short-stroke thrustor* (*a*) has a fixed stator with d.c. excitation. The mild-steel runner carries a single-layer helix winding with flexible central and end connectors to a d.c. supply to form a pair of half-windings having oppositely directed currents. Interaction thrust is developed to left or to right according to the runner supply polarity. *The long-stroke transverser* (*b*) inverts these

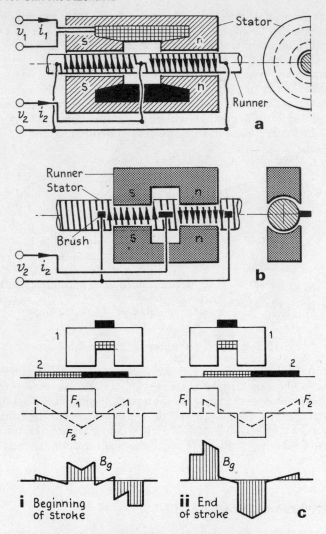

3.16 Heteropolar linear machines

functions: the full-length helix winding is fixed and the permanent-magnet system is the runner. The long stroke is achieved by feeding the active part only of the helix by brushes, carried on the runner and bearing on a stator contact track formed by removing an axial strip of the wire insulation.

For both machines the magnetic circuit is geometrically simple, but the combination of stator and runner m.m.f.s in the two machines is significantly different. In the *traverser* it is substantially that of Fig. 2.9(*a*). In the *thrustor,* however, the working flux is affected by the runner position as indicated in Fig. 3.16(*c*). At the beginning of a stroke (i) the armature m.m.f.

opposes that of the stator field; at the end of a stroke (ii) the reverse is the case, so that the working flux is greater. Green and Paul [8] describe a method of evaluating the gap-flux distribution $B_g$ for any armature position, and show that for a particular thrust/displacement requirement the condition can be met by suitably grading the runner winding.

Analysis of the behaviour of the thrustor and traverser machines is discussed in Sect. 12.3.

### 3.10 HOMOPOLAR MACHINES

#### Linear pump

For the d.c. conduction pump, Fig. 2.2, both gap-flux and channel-current densities must be high to secure an adequate pumping pressure. To avoid duplicating the high-current low-voltage source (e.g. a homopolar generator or a rectifier) it is normal to connect the field winding and the channel path in series. Some important effects impair the efficiency: (i) the channel-current reaction distorts the distribution of the working flux; (ii) the channel current between electrodes strays out of the pole-face region; (iii) the inter-action force disturbs the liquid flow pattern; and (iv) there is considerable pole-flux leakage. Fig. 3.17 shows typical distributions of flux and current densities in the channel as a consequence of these effects. Methods of ameliorating the effects are described in Sect. 12.3.

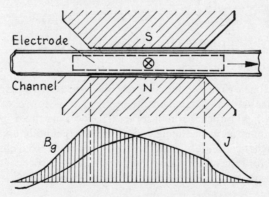

3.17 Flux and current densities in linear pump

#### Disc machine

The Faraday disc, Fig. 2.3, has been applied to low-voltage generators for electrochemical processes. Machines of more normal voltage and of several megawatts rating have been made possible by the production of intense magnetic fields by use of *superconducting* field systems. Superconductors (Sect. 12.6) are elements and alloys that, when cooled below a critical transition temperature (unique for each material but never far above absolute zero) lose all resistivity, so that very high magnetizing currents can be sustained by a d.c. supply of a volt or two. Flux densities of 5 T or more can be developed. At such densities, ferromagnetic materials can play only a secondary role, and the field therefore exists in an 'open' magnetic circuit. The field distribution can be evaluated by use of advanced three-dimensional field theory; and as there can be no effective confining path, there is a wide spread of the flux around the field winding.

An elementary disc motor could take the basic form shown diagramatically in Fig. 3.18. The field winding, in a vessel containing liquid helium and connected to a refrigerator, is geometrically simplified to an annular coil surrounding a disc which can rotate in air on a shaft.

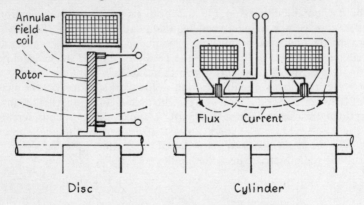

3.18 Elementary homopolar machine

**Cylindrical machine**

The small machine of Fig. 1.8 has an encapsulated rotor winding rotating in a simple annular gap. The large machine in Fig. 3.18 is equivalent to a pair of Faraday discs of cylindrical form in series. The working flux is provided by a pair of circumferential magnetizing coils, and the steel body not only forms part of the magnetic circuit but also serves as a rotating axial conductor between collector rings. Such a machine, driven at high speed, can generate very large currents at terminal voltages of the order of 100 V.

## 3.11 VARIABLE-RELUCTANCE MACHINES

The magnetic conditions in a variable-reluctance machine, Fig. 3.19($a$), change with the relative position of the stator and rotor poles (or teeth) from in-register coincidence (position $i$) to symmetrical out-of-register (position $o$), for which the effective magnetic-circuit permeances are respectively $\Lambda_i$ and $\Lambda_o$. It is the aim to make $\Lambda_i$ as high and $\Lambda_o$ as low as is technologically feasible because the torque depends on the rate of change of permeance (or its counterpart, reluctance) with position. The gap geometry (i.e. the relative proportions of tooth-width $w_t$, slot-width $w_s$, tooth-pitch $Y = w_t + w_s$ and the gap-length $l_g$) is therefore a fundamental design parameter. Fig. 3.19 illustrates the gap geometry for the common rectangular configuration (*a*), and indicates in (*b*) two other possible shapes, the trapezoidal and the sinusoidal.

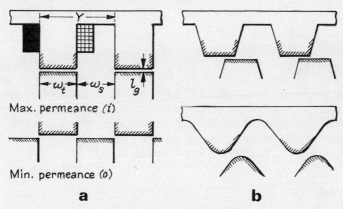

Max. permeance (*i*)

Min. permeance (*o*)

**a**                    **b**

3.19 Variable-reluctance machine

To evaluate the torque it is necessary to obtain the variation of permeance with position. Mukherjee and Neville [11] have derived permeance coefficients $\lambda = \Lambda/\mu_0$ for configuration (*a*) and various gap geometries on the linear assumption that the gap boundary surfaces are magnetic equipotentials. Then the mean torque (the average over the displacement of the rotor from position *i* to position *o*) for a machine with a stator m.m.f. $F$ that provides a gap flux density $B$ in the overlap region is

$$\overline{M}_e = kB^2 [(\lambda_i - \lambda_0)\,(l_g/Y)^2]$$

where $k$ is a function of the rotor diameter and active length. For a given machine, the term in square brackets is to be maximized.

In a practical machine the gap length is very short, and the stator current gives an m.m.f. that drives the overlap regions of gap boundary into a considerable degree of saturation, so that $B$ approximates to the saturation density $B_s$, typically about 2 T. This suggests that the mean torque should become constant: but test shows that the torque continues to increase with current, and also that it is related for certain levels of saturation to the number of teeth. The reason is that local saturation causes the flux to spread to the flanks of the stator and rotor teeth where, as discussed in Sect. 1.3, it is effective in increasing the lateral torque. The evaluation of the magnetic conditions in a variable-reluctance machine is a very difficult problem; some methods of tackling it are given in Ref. [12].

# 4 Windings

## 4.1 MATERIALS

Windings are formed from insulated conductors. For armatures the use of copper is almost unavoidable because of the restriction on winding space in slots. With large industrial machines, aluminium may be an acceptable alternative for field windings. Superconducting windings require conductors highly specialized in both material and formation.

### Conductors

The loss per unit cube in a conductor of resistivity $\rho$ and volume density $\delta$, carrying a current of density $J$, is $J^2\rho$; and the loss per unit mass (the specific loss) is $p = J^2\rho/\delta$. The loss must be dissipated at a rate to limit the temperature-rise of the conductor, chiefly because of the deleterious effect of excessive temperatures on the insulation.

*Copper* The International Annealed Copper Standard (IACS) has at $20^\circ$C a resistivity $\rho = 0.017\ 241\ \mu\Omega$-m, a resistance-temperature coefficient $\alpha = 0.003\ 93$ per $^\circ$C, and a tensile strength $220 - 250$ MN/m$^2$. Hot and cold working (such as wire-drawing) raises the mechanical strength at a small sacrifice in conductivity. Most machine windings are of annealed high-conductivity copper.

*Aluminium* Aluminium conductors to BS 3242 have at $20^\circ$C a resistivity of $0.0325\ \mu\Omega$-m (i.e. 55% of IACS conductivity).

Typical Characteristics at $20^\circ$C

|  | $\rho(\mu\Omega$-m) | $\alpha$(per $^\circ$C) | $\delta$(kg/m$^3$) |
|---|---|---|---|
| Copper: annealed | 0.0173 | 0.003 93 | 8 900 |
| hard-drawn | 0.0178 | 0.003 90 | 8 900 |
| Aluminium: hard-drawn | 0.0325 | 0.003 90 | 2 700 |

Copper has the higher conductivity and mechanical strength. Aluminium is lighter and cheaper, but its contact property, which is impaired by the formation of an oxide coating and by the relaxation of clamped joints by plastic flow, presents constructional problems.

## Insulants

Insulating materials may be organic or inorganic, uniform or heterogeneous in composition, natural or synthetic in origin. An ideal insulant would have (i) high electric strength, maintained at elevated temperatures; (ii) good thermal conductivity to transfer conductor $I^2R$ loss to the surrounding structure and coolant; (iii) mechanical permanence, ease of working and application, and resistance to failure by moisture, vibration, abrasion and bending. All usable materials are subject to thermal limits, none has all the desirable properties; but marked advances in synthetic materials have made them usually preferable to the traditional natural forms.

The insulants are classified as follows, each class being assigned a maximum operating temperature:

Y: Cotton, silk, paper, wood, cellulose, fibre etc., not impregnated nor oil-immersed

A: Class Y materials impregnated with natural resins, cellulose esters, insulating oils etc., also laminated wood, varnished paper, cellulose-acetate film etc.

E: Synthetic-resin enamels, cotton and paper laminates with formaldehyde bonding, etc.

B: Mica, glass fibre, asbestos etc. with suitable bonding substances, built-up mica, glass-fibre and asbestos laminates

F: Class B materials with thermally resistant bonding materials

H: Glass-fibre, asbestos and built-up mica with silicone resin bonds

C: Mica, ceramic, glass, quartz and asbestos, unbonded or with silicone resins of superior thermal stability

### Maximum Permitted Operating Temperature

| Insulation Class: | Y | A | E | B | F | H | C |
|---|---|---|---|---|---|---|---|
| Temperature ($^\circ$C): | 90 | 105 | 120 | 130 | 155 | 180 | >180 |

The figures are based on a 20-year working life under 'average' conditions. Of importance to the life of an insulant is the 'hot-spot' temperature of the winding that it covers. The following notes give an indication of the properties and applications of representative insulating materials.

*Mica* is used by bonding splittings into sheets with shellac, bitumen or synthetic polyester.

*Micafolium* is a wrap of mica splittings bonded to paper. It may be wound on to conductors, then rolled and compressed between heated plates to exclude air and to consolidate the material.

*Glass Fibre* is made from material free from alkali-metal oxides. Tapes and cloths woven from continuous-filament yarns have high resistivity, thermal conductivity and tensile strength in Class B. Thin glass-silk coverings are used for field-coil conductors. Varnishing is necessary to resist abrasion.

*Cotton Fibre* tapes woven from acetylated cotton are much less hygroscopic than ordinary cotton materials.

*Polyamides* such as nylon tapes are thin and strong. Nylon film is one of the new plastics that have adequate thermal stability.

*Synthetic-resin Enamels* of the vinyl-acetate or nylon types have a smooth finish that facilitates wire winding, and are useful bonds.

*Slot Liners* for high-voltage machines may be mica-composites. For small and low-voltage machines use is made of pressboard or varnished cotton cloth.

*Silicones* are semi-organic materials with a basic structure of alternate silicon and oxygen atoms; they are resistant to heat and can be used as binders in Class H. Their water-repellant and anti-corrosive properties are valuable in mill and traction motors, and in small machines in aircraft that must operate over the temperature range 200 to −40°C. Heat transfer is good because of their high thermal conductivity.

*Epoxy Resins* are thermo-setting. They are employed in casting, winding encapsulation, laminate bonding and varnishing.


## 4.2 WINDINGS

Windings with specific functions are required for the stator and rotor members of a machine. For heteropolar machines the field winding is most usually on the stator, with the armature winding and its commutating device on the rotor or runner. Control machines may have several field windings to modify the performance in response to signals derived from speed, torque, current, direction of rotation etc.


### Field

*Shunt* windings for small machines employ coils of a few hundred turns of small-section enamel-insulated wire. In large machines the field current may be large enough to permit the use of rectangular conductor sections wound on metal or pressboard formers and protected at the flanks by insulating washers. The winding may be sectionalized to improve cooling.

*Series* and *Commutating-pole* windings have comparatively few turns. A single-layer coil, strip-on-edge wound, may be used, with thin pressboard between turns (or air-spaced if the coil is rigid). For large machines a conductor of two or more thin strips in parallel may be adopted to ease bending. Small machines employ series windings of circular-section enamelled wire.

*Compensating* (neutralizing) windings for large machines with severe loading conditions are of copper strap, insulated and inserted into slots in the pole-shoes.

### Armature

For heteropolar machines the winding is distributed, generally in slots, and each coil is designed so that its active sides lie under poles of opposite polarity, their span approximating to a pole-pitch.

## 4.3 FIELD WINDINGS

Shunt and separately excited field windings are designed on a basis of their applied voltage, series windings for a known current.

### Shunt

A winding supplied at voltage $V$ to produce an m.m.f. $F = NI$ in $N$ turns each of cross-section $a$ has a resistance $r = \rho L_{mt} N/a = V/I$, whence

$$a = \rho L_{mt} F/V \tag{4.1}$$

where $\rho$ is the conductor resistivity at working temperature, and $L_{mt}$ is the mean length of turn. The field current is then $I = Ja$, the current density $J$ being chosen so that the $I^2 R$ loss $p = VI$ can be dissipated from the winding without the specified temperature-rise being exceeded. The thermal dissipation is a function of the exposed coil surface, which in turn depends on the space factor $k_s$ = (active conductor area/gross coil cross-section) for the coil.

The field time-constant $\tau = L/r$ is roughly settled by the coil cross-section $(Na/k_s)$. Expressing the inductance as $L = qN^2$, then

$$\tau = qN^2 a/\rho L_{mt} N = (Na/k_s) q k_s/\rho L_{mt}$$

which is proportional to the gross coil section; $\tau$ is substantially independent of $r$ because smaller conductors imply more turns.

### Series

For a current $I$, an m.m.f. $F$ and a permissible loss $p$, the winding resistance is $r = p/I^2 = \rho L_{mt} N/a$, whence

$$a = \rho L_{mt} NI^2/p = \rho L_{mt} IF/p \tag{4.2}$$

An initial trial is needed to estimate the loss dissipation.

EXAMPLE 4.1: Obtain the main dimensions of the field coils for a separately excited 4-pole motor to produce 5.30 kA-t/pole. The coils are in series with a field regulator so that the voltage per coil is 55 V. The pole dimensions are given in Fig. 3.13. The working conditions are: mean conductor temperature, 75°C; coil surface temperature-rise, $\theta = 45°C$; cooling coefficient (Sect. 10.6), $c = 0.028°C/W$, per m$^2$ outside exposed coil surface.
A first estimate of the mean length of turn is $L_{mt} = 0.70$ m. From eq. (4.1) with a copper resistivity $\rho = 0.021$ $\mu\Omega$-m at 75°C

$$a = 0.021 \times 10^{-6} \times 0.70 \times 5.3 \times 10^3/55 = 1.42 \times 10^{-6} m^2 = 1.42 \ mm^2$$

The corresponding conductor diameter is $d = 1.35$ mm, insulated with synthetic varnish to $d_i = 1.45$ mm. The space factor is estimated, e.g. from $k_s = 0.75(d/d_i)^2$, to be 0.65.
Assume a current density $J = 1.70$ A/mm$^2$: then $I = Ja = 2.40$ A, $N = F/I = 2210$, $p = VI = 132$ W. The total conductor cross-section is $Na = 3140$ mm$^2$ and the gross coil section is $Na/k_s = 4830$ mm$^2$. The available coil length is 105 mm, whence its thickness is 46 mm. A check on the mean length of turn gives

$$L_{mt} = (125 + 170) + (171 + 216) = 295 + 387 \ mm = 0.68 \ m$$

The exposed cooling surface is $S = 105(2 \times 387) = 81\ 300$ mm$^2$ = 0.0813 m$^2$. The surface temperature-rise is

$$\theta = cp/S = 0.028 \times 132/0.0813 = 45.5°C$$

## 4.4 CYLINDRICAL–ARMATURE WINDINGS

The essentials of a heteropolar armature winding were discussed in Sect. 1.4. It is a two-layer arrangement of identical coils each spanning approximately a pole-pitch. Typical coils for a slotted armature are illustrated in Fig. 4.1.

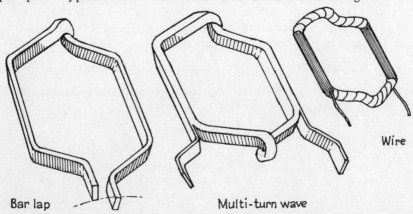

Bar lap            Multi-turn wave            Wire

4.1 Coils for heteropolar slotted armature

For a bar winding, the bar is cut to length and formed by press into a closed ('lap') or open ('wave') diamond shape with one upper and one lower coil-side, and then insulated. A 2- or 3-turn coil is formed from pre-insulated copper strap. Coils for small machines are wound on jigs and subsequently bent to shape.

## Nomenclature

The top and bottom coil-sides are given odd and even numbering respectively. In Fig. 4.2, slot (1) has four coil-sides, 1 and 3 at the top being joined by the

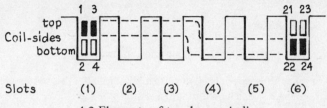

4.2 Elements of two-layer winding

coil-overhang to 22 and 24 at the bottom of slot (6); the lower coil-sides 2 and 4 of slot (1) are part of another coil whose top coil-sides (not shown) occupy a slot to the left. The pitch of a coil in terms of coil-sides must be an odd number (here $22 - 1 = 24 - 3 = 21$). In general, we write

$S$   number of slots in a uniformly slotted armature
$m$   number of coil-sides per slot
$C$   number of coils: $C = \frac{1}{2}mS$
$N$   number of armature turns: $N = Cz$
$z$   number of conductors per coil or coil-side: $z = N/C$
$p$   number of pole-pairs
$a$   number of pairs of parallel armature paths
$y_b$   back pitch (or coil-span) in coil-sides
$y_f$   front pitch in coil-sides
$y_r$   resultant pitch in coil-sides: $y_r = y_b \pm y_f = 2y_c$
$y_c$   commutator pitch in coils or commutator bars between the starts of electrically successive coils.

For symmetry, all the $2a$ parallel paths into which the winding is divided by the brushes must be identical. For this condition, both $S/a$ and $p/a$ must be integral.

## Simple lap winding

Fig. 4.3 shows diagrammatically the appearance of two successive lap coils (the second 'laps' over the first). The start of each single- or multi-turn coil is connected to a commutator sector, with a commutator pitch $y_c = 1$. In a multipolar machine, the set of coils of which the top sides occupy the first double pole-pitch complete a 2-pole 2-path system; and when the whole

Coils

Top coil-sides

Coil pitches

Bottom coil-sides

Commutator

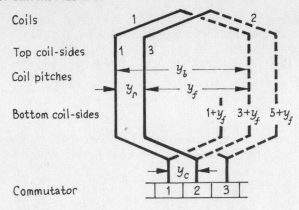

4.3 Elements of simple lap winding

winding is in place there are $p$ such systems, all parallelled by the brush connections, giving $a = p$. The winding rules are therefore

$$y_b - y_f = y_r = 2; y_c = 1; a = p \tag{4.3}$$

*Equalizers* If magnetic asymmetry (due e.g. to an eccentric shaft) causes the e.m.f.s in the $2a$ supposedly identical paths to differ, currents will circulate between paths via the brush contact surfaces. To mitigate this effect, permanent equalizer connections are made at the back of the winding to join points whose e.m.f. should at every instant be the same. Such points are spaced at intervals $C/p$, and a number of equalizer rings, each with $p$ joints so spaced, are fitted. The circulating currents then superpose a partially correcting magnetic field; but where they are appreciable, the actual armature $I^2R$ loss cannot be accurately predicted from external tests.

EXAMPLE 4.2: Draw the developed and sequence diagrams for a 4-pole lap armature winding with 22 single-turn coils. Show the position of the brushes (covering $1\frac{1}{2}$ sector widths) and include two equalizers. With $m = 2$ giving $S = C = 22$, the pole-pitch in coil-sides is $44/4 = 11$. Hence make $y_b = 11$, so that the coil-span is equal to the pole-pitch, and $y_f = 11 - 2 = 9$. The equalizer pitch is $C/a = 11$. The upper diagram in Fig. 4.4 shows the *developed* winding, with equalizers joining coils 1 and 12, and joining 6 and 17 respectively. The *sequence* diagram below is drawn to indicate the succession of coil-sides, and clearly indicates that the brushes divide the winding into four parallel paths ($2a = 2p = 4$). One coil (with coil-sides 19 and 30) is shown by heavy lines for each diagram.

### Simple wave winding

Successive coils in a multipolar wave winding, Fig. 4.5, are located in successive double pole-pitches. It is not feasible to have full-pitch coils as the winding would close on itself after only $p$ coils had been fitted (i.e. after a

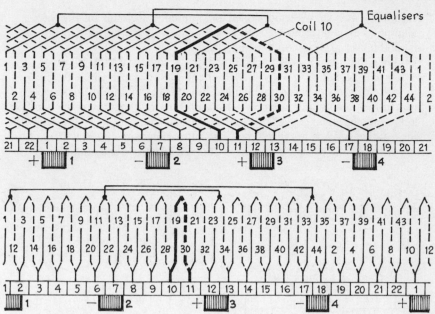

4.4 Four-pole lap winding: Example 4.2

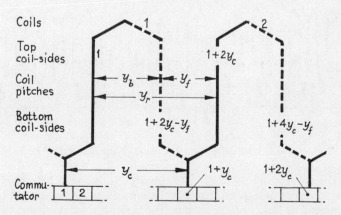

4.5 Elements of simple wave winding

single tour of the armature). The commutator pitch is therefore made $y_c = (C \pm 1)/p$ so that after one tour the last coil-side is one sector ahead or astern of sector 1 at which the first coil started. Applying the conditions of symmetry, $a$ must not exceed $p$, for $p/a$ must be a whole number. In this simple wave winding we have $a = 1$. The winding rules are:

$$y_c = (C \pm 1)/p \,; y_r = 2y_c = y_b + y_f \,; a = 1 \tag{4.4}$$

EXAMPLE 4.3: Draw the developed and sequence diagrams for the wave winding of a 4-pole armature with 21 coils, showing the position of the brushes (covering $1\frac{1}{2}$ sector pitches).

Applying the rules: $y_c = (21 \pm 1)/2 = 10$ or $11$, say $10$; $y_r = 2y_c = y_b + y_f = 20$, say $y_b = 11$, $y_f = 9$. The layout is shown in Fig. 4.6. The sequence diagram explains how the brushes, despite their uniform spacing around the commutator, are associated with only *two* regions of the winding and give a single pair ($a = 1$) of parallel circuits. It is possible to dispense with one positive and one negative brush without making any essential difference, but of course the current per brush is doubled and there is a minor asymmetry.

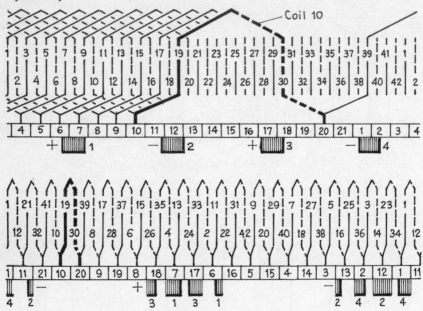

4.6 Four-pole wave winding: Example 4.3

## Limitations

The two foregoing examples show that the 4-pole lap winding has 22 coils but the wave winding has 21, as dictated by the conditions of symmetry. This means that the slotting, commutator and winding pitches differ, and the same armature construction cannot be used for both lap and wave windings at choice. In a *lap* winding the number of coils or commutator sectors $C$ is not restricted except that it must conform to the number of slots $S$ and of coil-sides per slot $m$ in accordance with $C = \frac{1}{2}mS$; and if the equalizers are to be equispaced, then $S$ must be a multiple of $p$. In a *wave* winding, $C$ must be one more or one less than a multiple of $p$: thus for a 4- or 8-pole armature $C$ is odd, and for a 6-pole case $C$ must not be a multiple of 3. The further limitation $S = 2C/m$ means that $S$ is odd and $m$ can be 2, 6 or 10. For a 6-pole winding, $S$ may not be a multiple of 3 nor $m$ a multiple of 6.

*Multiplex Windings* If a simple lap $(a = p)$ or a simple wave winding $(a = 1)$ does not give a usable number of parallel paths, duplex windings with $a = 2p$ and $a = 2$ respectively may be employed: they comprise two identical interleaved simple lap or wave windings. A triplex winding may occasionally be needed. These multiplex windings are rare and will not be discussed in detail.

### Armature resistance

In an armature winding of $N_a$ total turns and $2a$ parallel paths, the total input current $I_a$ divides to give $I_a/2a$ per path. Let each turn have a mean length $L_{mt}$ and a cross-section $s$: then the resistance per path is $(\rho L_{mt}/s)(N_a/2a)$, where $\rho$ is the resistivity of the conductor material at working temperature. The resistance of the $2a$ paths in parallel is therefore

$$r_a = (\rho L_{mt}/s) \, N_a/(2a)^2 \tag{4.5}$$

The resistance between the armature positive and negative terminals must include the resistance of the brushes themselves and of the brush/commutator contact surfaces. The latter is nonlinear, and is usually regarded as equivalent to a constant volt drop (e.g., 1 V) per brush-set.

## 4.5 DISC-ARMATURE WINDINGS

Fig. 4.7 shows the winding of Fig. 4.6 redrawn with the conductors radial instead of axial. The result is a 4-pole disc wave winding, electrically identical but different mechanically in that the usual disc armature has no iron and no slots. The winding is formed by use of a jig. After the radial

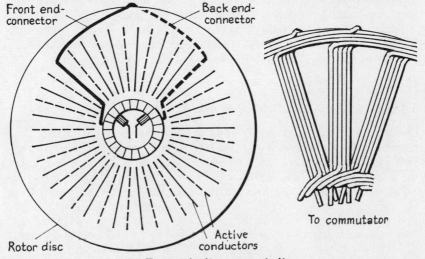

4.7 Four-pole disc wave winding

commutator has been connected, the jig is removed and the winding encapsulated. A second method originated with the windings electrodeposited on an insulated disc (the so-called 'printed-circuit' technique), but punched copper or aluminium sheet is preferred. The front and back half-windings are placed on the faces of an epoxy impregnated glass-fibre disc, and the resin then cured under pressure. The centre hole and the sheet edges are cut away, and the halves of the end-connectors welded together to form the complete winding. Up to three such windings could be mounted on the shaft and connected in series.

The wire-wound construction is provided with a radial or face commutator. With the punched form the brushes bear directly on the conductors at the inner diameter, the disadvantage being that the life of the machine is limited by wear. In each case it is usual to employ a wave winding to exploit the facility of a single brush-pair.

The winding in Fig. 4.7 is for a 4-pole machine, in order to demonstrate its essential electrical similarity with that for a cylindrical machine. However, for the better utilization of the conductors (in particular a reduction of the length of the 'idle' overhang) a 6- or 8- or 10-pole construction is usual. Typical limits of size are diameters of 250–300 mm and thicknesses up to 10 mm. The ironless rotor has no hysteresis, saturation or slot-cogging effects, but there may be appreciable eddy currents in the conductors with the punched-sheet construction.

## 4.6 MOVING-COIL ARMATURE WINDING

Micromachines such as control motors and tachogenerators, for which the rotor inertia may have to be very small, can be built with windings basically identical with those of Sect. 4.4 except that their cup-shaped construction and absence of iron obviate the winding restrictions imposed by slotting. The windings are formed from skeins of very fine wire of hard-drawn copper or aluminium. With the help of fibre glass and casting resins, the mechanical strength of such a winding is satisfactory.

## 4.7 POLARIZED ARMATURE WINDINGS

Cheap but effective miniature d.c. motors for battery-operated duty in record-players, windscreen wipers and models must be capable of starting even with low battery voltage. A permanent-magnet field is employed (unless the machine has also to work on a.c.) and the armature is reduced to a 3- or 5-slot form. The p.m. and current-excited constructions are shown at (*a*) and (*b*) in Fig. 4.8. The p.m. motor has a 2-pole ceramic ring magnet with radial or transverse magnetization on the stator, enclosed in a mild-steel cylinder. The 3-slot rotor has a 3-sector commutator and a pair of diametral brushes. The winding connections are indicated in (*a*) and the brush position in (*b*).

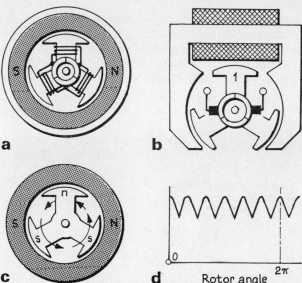

**a**      **b**

**c**      **d**    Rotor angle   $2\pi$

4.8 Polarized-armature machines

The motor operates on the alignment principle. With the rotor in position (*a*) or (*b*), the winding of rotor tooth 1 is connected directly across the positive and negative brushes; teeth 2 and 3 are in series and carry about one-half of the current of tooth 1. The tooth polarities (*c*) show that magnetic attraction and repulsion forces are additive to produce a counterclockwise torque. The tooth polarities interchange as the armature rotates, maintaining a unidirectional but pulsating torque that has six ripples per revolution as in (*d*), a feature termed *pole-sensitivity*. The effect can be mitigated if the stator p.m. is magnetized with a 3rd-order space harmonic.

In motors of rating up to 10 W the brush and bearing friction absorbs appreciable torque, and a short gap length must be used to obtain the highest practicable gap-flux density. For higher ratings it is feasible to employ ten or more teeth and the pole-sensitivity is much less prominent.

## 4.8 BRUSHLESS WINDINGS

Electromagnetically, the stator and rotor functions of a machine can be inverted, putting the field system on the rotor. There is no advantage if conventional commutation is retained, for the commutator sectors are now fixed and the brushgear must rotate at the same speed as that of the rotor field. Solid-state switching by transistors or thyristors, triggered by position sensors can, however, replace the brushgear by fully electronic commutation, endowing small machines with a valuable control facility. Speed-controlled flywheels and torque-controlled reaction wheels for the attitude control of aircraft, for example, require small, reliable and efficient d.c. machines that are well suited to brushless electronic commutation. Machines for rapid start/ stop and high acceleration/retardation duty are also suitable for inversion, the only mechanical elements then subject to wear being the bearings.

Designs vary widely with the duty. One realization is shown in Fig. 4.9. The 4-pole p.m. rotor spins within an 8-slot stator winding in two sections, A

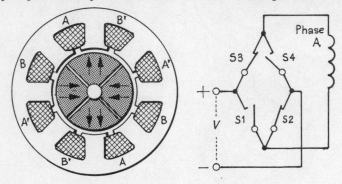

4.9 Four-pole brushless machine

and B. These correspond roughly to the 'phase' windings of a 2-ph synchronous machine. Commutation switching is by means of four semi-conductors per phase according to the circuit diagram, in the combinations listed (1 = on, 0 = off):

| Interval: | | (i) | (ii) | (iii) | (iv) | (v) | (vi) | (i) |
|-----------|-----|-----|------|-------|------|-----|------|-----|
| Switches: | S1 | 1 | 1 | 0 | 1 | 0 | 0 | 1 |
| | S2 | 0 | 0 | 0 | 0 | 1 | 0 | 0 |
| | S3 | 1 | 0 | 0 | 1 | 1 | 0 | 1 |
| | S4 | 0 | 1 | 0 | 0 | 0 | 0 | 0 |

## 4.9 ELECTROMOTIVE FORCE

We consider a conventional 2-pole cylindrical-rotor machine, Fig. 4.10, with d-axis stator poles and an armature with a commutator and brushes. We assume a uniform distribution of full-pitch armature coils, complete electrical and magnetic symmetry, and a no-load condition of zero armature current. The e.m.f. between the commutator brushes is derived from the Faraday law, eq. (1.11); but it is necessary to specify the gap-flux distribution. For simplicity it is first taken to be *sinusoidal*.

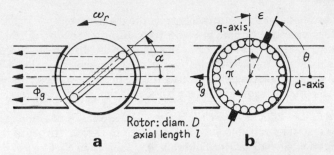

4.10 Turn and winding e.m.f.s

## Sinusoidal flux distribution

Let the gap flux per pole, $\Phi_g$, have a maximum density $B_m$ on the d-axis. The mean density over a pole-pitch is $B = B_m(2/\pi)$, whence
$\Phi_g = B\,(\frac{1}{2}\pi D)\,l = B_m Dl$. The flux linking the armature turn in Fig. 4.10(*a*)
is clearly $\Phi_\alpha = \Phi_g \sin \alpha$.

*Turn E.M.F.* Eq. (1.11) can now be applied to find the instantaneous e.m.f. $e_t$ in a single turn rotating at angular speed $\omega_r$ and situated momentarily at angle $\alpha$:

$$e_t = \omega_r \Phi_g \cos \alpha + (\mathrm{d}\Phi_g/\mathrm{d}t)\sin \alpha = e_{tr} + e_{tp}$$

The component $e_{tr}$ is due to rotation, the pulsational component $e_{tp}$ being present only if the flux has a time rate of change.

*Winding E.M.F.* Between brushes, Fig. 4.10(*b*), a fixed number of turns in series occupy always the same position with respect to the flux axis. The total $N_a$ turns are divided by the brushes into two parallel sets each of $\frac{1}{2}N_a$ turns. The e.m.f. between brushes is then

$$e = \omega_r \tfrac{1}{2} N_a \Phi_g \,\frac{1}{\pi} \int_{\theta - \pi}^{\theta} \cos \alpha \cdot \mathrm{d}\alpha + \tfrac{1}{2} N_a \frac{\mathrm{d}\Phi_g}{\mathrm{d}t} \int_{\theta - \pi}^{\theta} \sin \alpha \cdot \mathrm{d}\alpha$$

$$= \omega_r N_a \Phi_g (1/\pi)\sin \theta - N_a(\mathrm{d}\Phi_g/\mathrm{d}t)(1/\pi)\cos \theta = e_r + e_p \qquad (4.6)$$

The rotational and pulsational e.m.f. components are affected by the brush position and by whether or not $\Phi_g$ is time-varying.

(i) With $\theta = \frac{1}{2}\pi$ rad the brush- and q-axes coincide. Maximum summation of rotational turn-e.m.f.s is achieved, and the greatest rotational e.m.f. $e_r$ between brushes. But variation of $\Phi_g$ can induce no pulsational e.m.f. between brushes because the armature and field winding axes are in space quadrature, giving zero magnetic coupling between them. Thus $e_p = 0$.

(ii) With $\theta = 0$ or $\pi$ rad the brush- and d-axes coincide. Then $e_r = 0$, because in each armature circuit, one-half of the turns move in an N-pole and the other half in an S-pole field. Coincidence of the armature and field winding axes, on the other hand, results in maximum pulsational e.m.f.

between brushes in response to a change in the d-axis flux.

(iii) The axis of the brushes in Fig. 4.10($b$) is shown as *shifted* from the q-axis by an angle $\epsilon = (\frac{1}{2}\pi - \theta)$ rad. For a sinusoidally distributed flux, the result is a reduction of $e_r$ by the factor cos $\epsilon$. At the same time the field and armature windings are coupled by a flux proportional to sin $\epsilon$, so that a pulsational e.m.f. will appear between brushes if the pole-flux varies.

*Armature E.M.F. with Constant Field Flux*    The 'normal' condition in the steady no-load state is with $\theta = \frac{1}{2}\pi$ rad, i.e. with the brush axis set in the q-axis direction, and a constant field flux $\Phi_g$. Then from eq. (4.6)

$$e = E_r = \omega_r N_a \Phi_g (1/\pi) = 2nN_a\Phi_g = nZ_a\Phi_g \qquad (4.7)$$

where $n = \omega_r/2\pi$ [r/s] is the rotor speed, and $Z_a = 2N_a$ is the total number of active armature conductors. $E_r$ is the summation through one armature path of the individual turn-e.m.f.s, and thus depends on the *average* e.m.f. per turn and consequently on the average flux density. Thus only the *magnitude* of $\Phi_g$ is significant and not its distribution. In this respect, therefore, the assumption of a sinusoidal gap-flux distribution is irrelevant.

*Heteropolar Lap and Wave Windings*    Still considering only the rotational e.m.f. for constant field flux and zero armature current, eq. (4.7) applies to a 2-pole machine regardless of the type of full-pitch winding. In a multipolar machine with $2p$ poles and an armature winding having $2a$ parallel paths between brushes, a conductor in one revolution passes through a field of $2p$ poles, each with a flux $\Phi_g$; i.e. through a total flux $2p\Phi_g$. The number of turns in series in each path is $N_a/2a$. Applying eq. (4.7) gives the e.m.f. between successive positive and negative brushes as

$$E_r = (\omega_r/\pi)\,(N_a/2a)\,(2p\Phi_g) = 2(p/a)nN_a\Phi_g = (p/a)nZ_a\Phi_g \qquad (4.8)$$

The simplex lap winding has $a = p$, the simplex wave winding has $a = 1$. Their e.m.f.s are therefore

$$
\begin{aligned}
\text{Lap:} &\qquad E_r = 2nN_a\Phi_g = nZ_a\Phi_g \\
\text{Wave:} &\qquad E_r = 2pnN_a\Phi_g = pnZ_a\Phi_g
\end{aligned}
\qquad (4.9)
$$

For the same speed, flux per pole, armature turns and current per path, the conversion power $E_r I_a$ is the same whether the armature is lap- or wave-connected. The wave winding gives $p$ times the e.m.f. of the lap, but only $1/p$ of the current output. Choice depends on the rating and on the operating conditions. In large machines it is usually advantageous to use a lap winding in order to reduce the current per path and consequently the conductor cross-section; but with traction motors, brush inspection and maintenance are eased by adopting the multipolar wave winding with two brush-arms only. For small machines the wave winding is preferred (unless the operating voltage is low) because fewer armature turns are needed.

## 4.10 E.M.F. WAVEFORM

A more practical (though still idealized) gap-flux density distribution than
the sinusoidal is that in Fig. 4.11($a$) for a machine with pole excitation but
no armature current. From eq. (1.7), the rotational e.m.f. $e_{tr}$ in a full-pitch
armature turn is $2B_g lu$, and its waveform is the same as that of the flux
density, except for 'rectification' as the conductors pass the brush axis. The

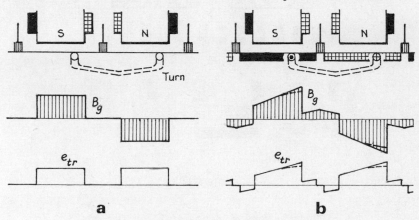

a                        b

4.11 Motional e.m.f. in armature turn

e.m.f. $E_r$ between brushes is the sum of the turn-e.m.f.s in series. When
armature current flows, as in ($b$), the armature m.m.f. distorts the flux
pattern and the turn-e.m.f. waveform. This should make no difference to $E_r$
as, ignoring saturation, the mean $B_g$ remains unchanged; but in a practical
machine the pole and teeth in the region of high density are subject to
greater saturation, and the flux density is reduced as indicated by the dotted
line. As a result there is a small reduction in $E_r$. But the turn-e.m.f. is still
peaked, increasing the p.d. between adjacent commutator sectors, with
consequent risk of sparkover.

*Compensation* Field distortion can be mitigated by pole-face compensating
(neutralizing) windings, Fig. 4.12($a$), carrying a current equal or proportional
to that in the armature conductors on the other side of the gap but oppositely
directed. If the two current sheets match, the armature reaction m.m.f. is
neutralized over the pole-arc. The effect is indicated in ($b$). The magnitude of
the torque is not affected, but it is transferred from the pole flanks to the
pole-face. The effect could alternatively be obtained by bevelling the pole-face
as in ($c$), but uniform flux density is then obtained only for one level of
armature current and one direction of rotation.

    Neutralization by compensating winding is not effective over the
interpolar space in which the commutation process takes place. Separate
commutating poles are provided in industrial machines to set up an appropriate
flux for forced commutation.

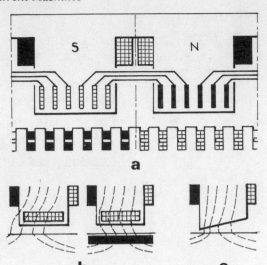

**4.12 Compensation**

### Ripple e.m.f.

Armature windings in slots cannot be considered as 'uniformly distributed' unless there are at least 20 slots per pole-pitch. Small and miniature machines can accommodate only a few slots, making the winding e.m.f. sensitive to rotor position. For this and other separate but interacting reasons, the armature e.m.f. between brushes contains a superimposed ripple component.

*Axis Swing*  As the number of commutator sectors is finite, the effective brush axis shifts to and fro about a mean position as the sectors move under the brushes in a conventionally commutated armature. Even the varying current density over the brush periphery may produce minor swings of the brush axis. In polarized machines, Fig. 4.8, there is clearly a substantial swing of the rotor m.m.f. axis; and for electronic commutation, Fig. 4.9, the swing may be typically $45°$. These conditions generate e.m.f. ripple, and also introduce cyclic fluctuations in torque.

*Slot Position*  Fig. 4.13($a$) shows three idealized gap-density distributions, all for the same total flux per pole and respectively sinusoidal, rectangular, and distorted rectangular. The slot conductors are assumed to be concentrated at the slot centres. Two successive positions of an armature with 5 slots per pole are drawn in (i) and (ii). The e.m.f. between brushes is the summation of the contributions of individual conductor groups. Consider the *sinusoidal* flux-density distribution: for the slot positions in (i) all five slots contribute e.m.f., whereas in (ii), for which the rotor has moved to the left by one-half slot-pitch, only four groups contribute. The brush e.m.f.s are therefore proportional to

(i) $\sin 18° + \sin 54° + \sin 90° + \sin 126° + \sin 162°\ \ \ \ = 3.24$
(ii) $\sin 36° + \sin 72° + \sin 108° + \sin 144°\ \ \ \ \ \ \ \ \ \ \ = 3.08$

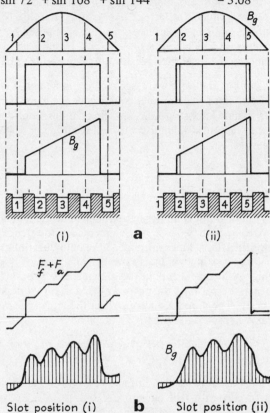

4.13 Ripple effects

The fluctuation is 5%. For the *rectangular* and *distorted* distributions the summations give 3.0 for (i) and 4.0 for (ii), a fluctuation of 14% of the mean. In a small machine it would be an advantage to aim at a sine-distributed flux density by chamfering the pole-shoes. Considerable reduction in slot-position ripple can be achieved by *skewing*, i.e. by angling the slots from the axial direction by about one slot-pitch.

*Tooth Ripple* In the foregoing the magnetic effect of slotting has been ignored. The armature m.m.f. $F_a$ is not the smooth triangle shown in Fig. 2.8(*b*) for uniform distribution of the rotor winding, but is 'stepped' as in Fig. 4.13(*b*) because of the segregation of the conductors in slots. The combination $F_f + F_a$ of field and armature m.m.f.s is correspondingly stepped. The gap density $B_g$ is determined at any point by the resultant m.m.f. there and the reluctance of the gap presented to it; and as the latter varies considerably between slot-openings and tooth-tips, the gap density shows strong ripples. Moreover, the flux per pole will depend on the number of teeth under the pole-arc. Compare the $B_g$ distributions (i) and (ii) in (*b*): the former has four 'humps', the latter three, causing the flux to vary cyclically with rotor position. Pole-flux pulsation is opposed by the closed-circuit field winding, in which counter e.m.f.s will be induced, but it cannot prevent the ripples in the local flux density which introduce pole-face losses.

*Rectified A.C. Supply*   The supply voltage to a d.c. machine if derived from an a.c. network contains a.c. components of frequency some multiple of the mains frequency. These impose current and flux ripples, considerably suppressed by the winding inductances but still resulting in some cyclic variation in the brush voltage.

## 4.11 MAGNETOMOTIVE FORCE

All current-carrying windings and permanent magnets in a machine contribute to its gap-flux pattern. The two essential elements are the field system and the armature. Industrial machines normally have commutating-pole windings, and some have compensating windings as well. The action of the m.m.f. sources in setting up the useful flux in the airgap region is basic to the energy-conversion process; the paths of the non-useful leakage flux depend on the structural geometry. The m.m.f. components in the airgap are now considered in terms of their magnitudes per pole.

*Field M.M.F.*   This is provided by permanent magnets, or by one or more exciting windings to give $F_f = \Sigma(N_f I_f)$.

*Armature M.M.F.*   For $2p$ poles and $2a$ parallel armature paths, a total of $N_a$ turns and an input current $I_a$, the current per path is $I_a/2a$ and the total m.m.f. $N_a I_a/2a$. The m.m.f. is distributed as in Fig. 2.8(*b*), its maximum value per pole being $F_a = N_a I_a/4ap$ on the brush axis. The distribution is triangular for a uniformly distributed winding, but stepping becomes prominent with reduction in the number of slots per pole.

*Compensating M.M.F.*   To neutralize $F_a$ over the pole-arc $b$ the m.m.f. $F_n = F_a(b/Y)$ is required.

*Commutating-pole M.M.F.*   Compole windings are required to oppose uncompensated m.m.f. in the commutation region and to set up a flux density $B_c$ to assist rapid reversal of the armature current. The m.m.f. is given by $F_c = F_a - F_n + kB_c$, where $k$ is a function of the compole magnetic circuit.
   The combined effect on the airgap of the several m.m.f. components is indicated in Fig. 4.14 for an industrial machine operating in the *motor* mode. For a *generator* the current directions and the m.m.f.s to which they contribute are reversed for the armature, compole and compensating windings. The combination of field and armature m.m.f.s (omitting the effects of slotting) is shown in (*a*). Addition of the compole m.m.f. gives (*b*),

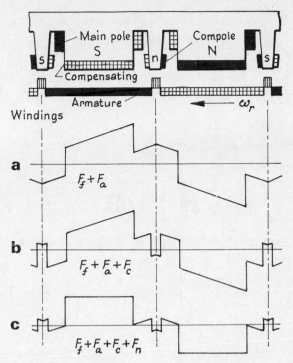

4.14 Gap m.m.f.s for motor mode

showing the local reversal in the commutating zone. Adding the compensating m.m.f. in (c) shows elimination of the armature-reaction distortion over the pole-arcs, and indicates the considerable reduction in compole m.m.f. required when a compensating winding is fitted.

### Brush shift

With the brush- and q-axes coincident, Fig. 4.15(a), the peak armature m.m.f. $F_a$ is a q-axis quantity. Shifting the brush axis by an electrical angle $\epsilon$, as in (b), shifts the axis of $F_a$, which can be regarded as arising from the combination (c) of current sheets QQ magnetizing on the q-axis and DD on the d-axis. The corresponding m.m.f. components are $F_{aq} = F_a(1 - 2\epsilon/\pi)$ and $F_{ad} = F_a(2\epsilon/\pi)$. Two significant effects result from brush-shift.

*Direct-axis Magnetization*  Fig. 4.15(c) shows that with a 'backward' shift (i.e. against the direction of rotation) $F_{ad}$ opposes $F_f$ and so reduces the airgap flux per pole. A forward shift has the opposite effect. These statements refer to the motor mode.

*Commutation*  In a motor without compoles, their effect can be simulated by a backward shift, which brings the commutating zone into the fringe field of

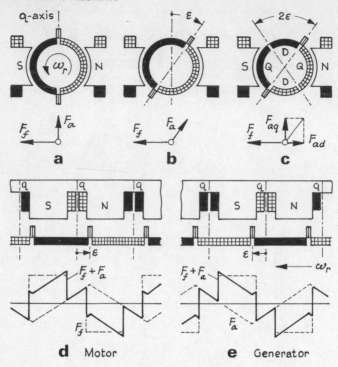

4.15 Brush-shift effects

the preceding main pole, as indicated in (*d*) of Fig. 4.15. In a generator the armature-current direction is reversed and a forward shift (*e*) is necessary. However, as the shift angle required varies with the load current, commutation assistance by brush-shift is suitable only for small unidirectional machines, and in normal industrial practice compoles are always provided instead.

In the discussion above, armature coils are assumed to be of full pitch. In chorded windings with a pitch less (rarely more) than a pole-pitch, conductors within the angle $2\epsilon$ have, in part, currents of opposing directions, reducing the demagnetizing effect.

### Armature excitation

Let the brush-shift angle be $\epsilon = \pi/2$ rad by aligning the brush-axis with the d-axis of the field system, and let the armature at rest be supplied with a current $I_d$. The armature m.m.f. $F_a = F_{ad}$ now acts directly on the poles, and it is possible to produce the working flux in the airgap without the aid of field windings, Fig. 4.16(*a*). If now the armature is driven, it rotates in its self-produced flux. Rotational e.m.f.s are generated in the armature turns, but no motional e.m.f. appears across the brushes as the coil e.m.f.s. are

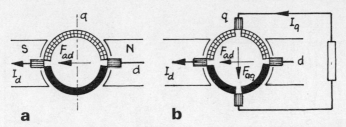

4.16 Armature excitation

self cancelling. An e.m.f. could, however, be picked off by a pair of q-axis brushes (*b*), and it could circulate a current $I_q$ in a load circuit, the armature carrying the superposed currents $I_d$ and $I_q$. This action is used in *cross-field* machines for power control and amplification.

### 4.12 LEAKAGE AND INDUCTANCE

The resultant airgap m.m.f., Fig. 4.14, enables the useful flux distribution to be assessed. There remains the non-useful flux that follows paths directly between main poles and around the armature slots and end-windings. Both main and leakage fluxes influence the performance of a machine.

*Leakage Flux* The flux $\Phi$ in a magnetic path across which an m.m.f. $F$ is expended is $\Phi = F\Lambda = F/S$, as in eq. (1.1). Because of the complex geometry of leakage-flux paths it is not easy to estimate the permeance $\Lambda$ or the reluctance $S$. Normally it is assumed that leakage flux paths extend between ferromagnetic surfaces, of infinite permeability and having some pre-determined distribution of magnetic potential. Attention is then concentrated on the 'air' (nonmagnetic) parts of each path. The flux pattern is first established by means of a flux plot, after which it is divided into a number of series or (more usually) parallel sections of area $a$ and length $l$. For $x$ sections in parallel the leakage permeance is

$$\Lambda = \mu_0 \Sigma(a_x/l_x) = \mu_0 \Sigma(\lambda_x) \tag{4.10}$$

and the problem reduces to the estimation of $\lambda = (a/l)$ for each section, $\lambda$ being termed the *permeance coefficient*. A geometrical approach to typical leakage estimation is given here, although digital computation can accurately be applied to quite complicated magnetic field problems without recourse to geometrical or experimental flux-plotting techniques.

### Main pole

The leakage flux is roughly mapped in Fig. 4.17. Four regions can be considered: (i) $\Phi_1$ and $\Phi_2$ respectively between the facing edges and outer sides of the pole-shoes, and correspondingly (ii) $\Phi_3$ and $\Phi_4$ for the pole shanks.

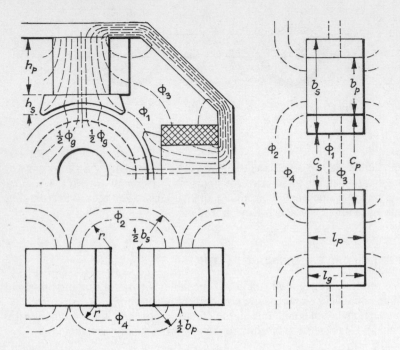

4.17 Pole leakage flux

Then the total leakage flux leaving the two facing sides and the two end surfaces of one pole structure is the sum of

(i) $\Phi_{sl} = 2\Phi_1 + 4\Phi_2$    and    (ii) $\Phi_{pl} = 2\Phi_3 + 4\Phi_4$

Between adjacent shoes the magnetic p.d. is the m.m.f. required to drive the working flux $\Phi_g$ from shoe to shoe through the gap-teeth-core-teeth-gap path. Between adjacent pole-shanks, however, the magnetic p.d. falls as the leakage path considered nears the yoke, becoming equal to the yoke m.m.f. at the root. It is adequate to assume that the magnetic p.d. between the shanks averages one-half of that between the shoes.

The total leakage flux per pole is assessed from estimated permeance coefficients. For $\Phi_1$ and $\Phi_3$ the coefficients are based on the relevant surface areas and on one-half of the mean distances $c_s$ and $c_p$, giving $h_s l_s / \frac{1}{2} c_s$ and $h_p l_p / \frac{1}{2} c_p$. For $\Phi_2$ and $\Phi_4$ the path is taken as a straight line connecting a pair of quarter-circles of radius $r$ ranging from zero to $\frac{1}{2} b_s$ or $\frac{1}{2} b_p$, the path lengths being expressed in terms of $r$ and their areas being obtained by integration. The leakage fluxes are then $\Phi_s = F_l(\mu_0 \lambda_s)$ for the shoes, and $\Phi_p = \frac{1}{2} F_l(\mu_0 \lambda_p)$ for the poles, where $F_l$ is the m.m.f. per pole for gap, teeth and armature core. Each permeance coefficient has the form

$$\lambda_l = 2lh/c + 2.9\, h \lg[1 + (\tfrac{1}{2}\pi b/c)] \tag{4.11}$$

with $l$, $h$ and $c$ being $l_s$, $h_s$ and $c_s$ for the shoes, and $l_p$, $h_p$ and $c_p$ for the pole shanks. Typically, $\Phi_{sl}$ is $20F_l/10^8$ and $\Phi_{pl}$ about four times as great in an industrial machine. Compoles shorten the leakage paths and increase the leakage flux, modifying its distribution.

*Inductance*  For a useful gap flux $\Phi_g$ produced by a field current $I_f$ in a pole winding of $N_f$ turns, the mean flux per pole approximates to $\Phi_p = \Phi_g + \Phi_{sl} + \tfrac{1}{2}\Phi_{pl}$, and the field-coil inductance is $L_f = N_f\Phi_p/I_f$. It differs from the 'ideal' value in eq. (2.11) because of slotting, leakage and saturation, and it varies considerably with the level of pole excitation.

### Armature

The armature has a fixed axis of magnetization, whether or not it rotates. Its m.m.f. develops a flux disposed symmetrically about the q-axis and does not affect the total gap flux (provided that the brushes are not shifted and that saturation is ignored). Fig. 4.18 is, for convenience, drawn for a 4-pole

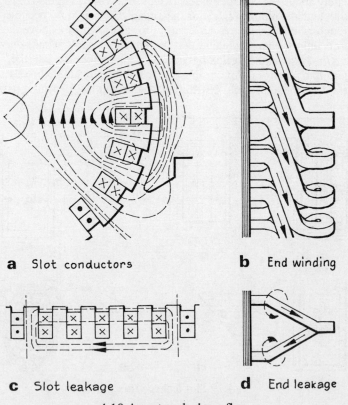

**a**  Slot conductors      **b**  End winding

**c**  Slot leakage      **d**  End leakage

4.18 Armature leakage flux

armature, so that the d-axes are mechanically at $90°$. It shows in $(a)$ an approximate leakage flux distribution for a slotted cylindrical armature due to its own current in the absence of compole and pole-face compensating windings. Although the flux lines are largely completed circumferentially through the polar arcs, the leakage flux also penetrates into the high-reluctance interpolar regions. Some of the flux encircles only the slots, contributing the slot-leakage component shown separately in $(c)$. Conditions in the overhang $(b)$ are complex, for the two-layer interleaving of the end-conductors introduces mutual coupling between neighbouring coil-ends. The length $l_o$ of the overhang in $(d)$ is a factor that affects the overhang leakage.

*Slot Leakage*  In an armature let all the conductors in one slot, Fig. 4.19($a$), have the same current direction, as will be the case in any slot beyond the commutating zone. The currents set up a slot flux which completes its closed path through the surrounding teeth and core. Such patterns have been analysed rigorously by Hague [9], but here we adopt the simplification $(b)$ in which the flux is directed in straight lines across the slot. The permeance coefficient $\lambda$ per unit axial length of slot is obtained by summing a number of components. Using the dimensions in $(c)$, we have $h_2/w_s$ for the region immediately above the conductors, $2h_3/(w_s + w_w)$ for the wedge, $h_4/w_s$ for the lip, and an empirical $l_g/y_s$ for the airgap. The height $h_1$ occupied by the conductors must be treated differently, for although flux above the conductors links them completely, that within height $x$ in $(d)$ links only the fraction $x/h_1$. Further, such flux has a lower density because only the fraction $x/h_1$ of the total slot current is available to provide the m.m.f. For a slot current of 1 A, uniformly distributed, the flux in the elemental path d$x$ is

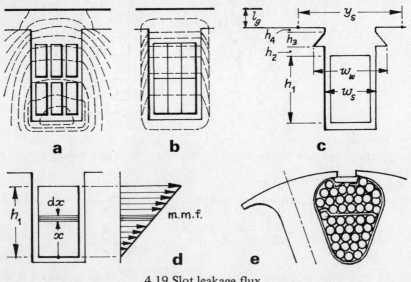

4.19 Slot leakage flux

$d\Phi = \mu_0(x/h_1 w_s) \cdot dx$ and it links only the fraction $x/h_1$ of the total conductor area. The permeance coefficient is therefore the integral of $(x^2/h_1^2 w_s) \cdot dx$, giving $h_1/3w_s$.

Summing the coefficients, the overall slot permeance per unit axial length is

$$\lambda_s = \frac{h_1}{3w_s} + \frac{h_2}{w_s} + \frac{2h_3}{w_s + w_w} + \frac{h_4}{w_s} + \frac{l_g}{y_s} \tag{4.12}$$

With a net slot length $I_s$ the total slot permeance is $\mu_0 I_s \lambda_s$. The slot leakage would be reduced if the slots were wide and shallow, but the teeth would be narrow (raising the tooth flux density and loss), the effective gap length would be increased and gap-flux ripple would be aggravated.

The same method is applicable to small cylindrical machines with slots shaped as in Fig. 4.19(e). Disc and moving-coil armatures are not slotted and have no associated steel core; their leakage flux is much smaller than for a slotted winding, but is more difficult to estimate.

*Overhang Leakage* Several workers have tackled this awkward problem. Carpenter [10] employs the concept of images: his expression, (i) below, is in terms of the length $l_o$ and the equivalent radius $r$ of a coil. A classic form (ii) is in terms of $l_o$ and the perimeter $q$ of the coil. The expressions are

$$\text{(i)} \ \lambda_0 = (1/2\pi) \ln(l_0/2r) \quad \text{(ii)} \ \lambda_0 = (1.15/2\pi \ [\ln(l_0/q) + 0.07]$$

*Slot and Overhang Leakage* It is usual to express the overhang permeance in per-unit length of slot. Thus eq. (4.12) with the addition of (i) or (ii) above is $\lambda_a = \lambda_s + (l_o/l_s)\lambda_o$. Then an armature coil of $z$ conductors in series, each carrying a current $I$ and having a slot length $l_s$, has a leakage flux

$$\Phi = 2\mu_0 I z \, l_s \lambda_a \tag{4.13}$$

*External Leakage* In Fig. 4.18(a) it is seen that leakage flux other than that associated with the slots and overhang crosses the airgap. In Sect. 2.4 this flux, in an idealized 2-pole armature without a compensating winding, was evaluated on the assumption that the armature winding could be taken as a uniform current sheet on the rotor surface, eliminating slots altogether and disregarding the overhang. The analysis led to eq. (2.12) for the armature inductance. The conditions are those in (i) of Fig. 4.20, in which (a) shows the current sheet, (b) the armature m.m.f. over the pole-face, and (c) the resulting external armature leakage flux, which crosses the gap radially, out of the armature in one half of the pole-arc and into it in the other.

When, as in (ii), there is a pole-face *compensating* winding, the external armature flux pattern is radically changed. Assuming that the armature and compensating current sheets within the pole-arc are identical, the flux of their combination must be circumferential, as could be inferred by comparing the current-sheet pair with that of the classic 'long solenoid'.

For a 2-pole machine with $N_a$ armature turns and an input current $I_a$, the m.m.f. of that region of the winding within the pole-arc is $F = \frac{1}{2} N_a I_a (b/Y)$.

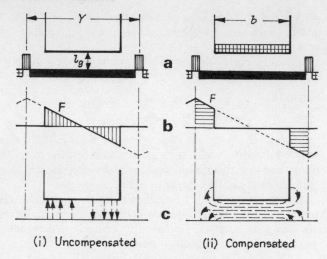

(i) Uncompensated        (ii) Compensated

4.20 External armature leakage flux

The circumferential path length is $b = Y(b/Y) = \frac{1}{2}\pi D(b/Y)$. The m.m.f. per unit path length is $H = F/b$ and the flux density is $B = \mu_0 H$ (assuming the flux to be confined to the airgap). The field energy density is $\frac{1}{2}B^2/\mu_0$ in a gap volume $(\pi D l l_g)(b/Y)$ concerned. Then the total field energy is $w_f = \frac{1}{2}L_a I_a^2$, whence

$$L_a = \frac{1}{2}N_a^2\, l l_g (b/Y)\mu_0/\pi D \tag{4.14}$$

This inductance, being due to the armature and compensating windings in combination, refers to the armature circuit and not to the armature alone. The total armature-circuit inductance requires the addition of the effects of armature and compensating winding slots, armature overhang, and compoles.

*Inductance Effects*  The overall armature-circuit inductance delays rapid changes of load current and affects the main-field saturation level. But even with a strictly constant load current there still remains the commutation process, in which coil currents are rapidly reversed. This effect, which is purely local, is considered separately and discussed in Sect. 5.3.

# 5 Commutation

## 5.1 COMMUTATION DEVICES

The essential switching and position-sensing functions of the *conventional* commutator have been described in Sect. 1.4. The process of commutation involves passing a current through the sliding contact between a brush and a circular assembly of copper bars (commutator sectors), each bar being connected to a pair of coil-ends through 'risers'. As the bars pass under the brush, the current in the coils to which they are attached is reversed. The brush/commutator assembly is an electro-mechanical converter which operates by mechanically making and breaking electrical contact between brush and sectors, a process that inherently occurs at any speed of rotation. But it limits the design of the machine in that its construction and dimensioning require the rotor diameter/length ratio to be high. As the commutator is invariably carried on the rotor, it must withstand centrifugal force and remain stable so as to retain a smooth contact surface; its temperature-rise must be limited to avoid softening and distortion; and it must be protected from excessive dirt and moisture. The brushes have to be maintained in holders, with a sliding fit to enable brush pressure to be applied for proper surface contact. The armature winding must be well subdivided to minimize the leakage inductance of the subdivisions and so limit the e.m.f. induced when the current is suddenly reversed. As the commutator sectors are separated by a very thin layer of insulation, the voltage between adjacent sectors must be low.

In *electronic* commutation, the making and breaking of contact between the d.c. supply and the armature coils is performed by solid-state switches. The extensive subdivision of the armature winding is not essential, and the use of fewer winding sections improves utilization of switches. The number of winding subdivisions may be as few as three. On purely economic grounds, the electronically commutated machine cannot compete with the conventional: it must have other compensating advantages, which are usually found in the duty that the machine is called upon to perform in speed-control and closed-loop systems.

In the following, the technology of conventional commutation is discussed in Sect. 5.2–5.8, and thereafter the problems that arise in 'assisted' and 'complete' electronic commutation are considered.

## 5.2 CURRENT REVERSAL

Consider first the idealized case of an armature winding, Fig. 1.11, with single-turn full-pitch coils and a brush width equal to that of a commutator sector. Fig. 5.1 shows the conditions in a particular coil (heavy line) as its

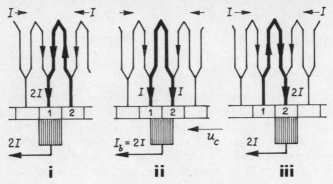

5.1 Idealized commutation process

sectors pass under a brush from which a constant current $I_b$ flows. Initially (i) the coil current is $I = \frac{1}{2}I_b$, and finally (iii) it is $-I$, the process of reversal taking a time $t_c$. If $I_b$ is distributed uniformly over the brush surface to give a constant current density, the currents in the connections to the coil undergoing commutation are proportional to the interface surface areas between sectors and brush, giving *straight-line commutation,* Fig. 5.2. Failure to

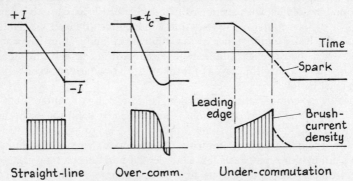

5.2 Coil-current change and brush-current density

achieve this condition results in *over-* or *under-commutation.* In the former, the change is too fast, causing high current density near the leading edge of the brush. In the latter, reversal is incomplete when the sector leaves the brush, and the effective commutation time is extended by a *spark*. Non-uniform current density results in higher $I^2 R$ loss, temperature-rise and brush wear. Sparking damages the commutator, leading to still greater brush wear.

## Resistance commutation

If the coil inductance were negligible, division of the brush current $I_b$ between adjacent sectors would be governed only by the combined resistance of the coil, the risers and the brush contact. The latter predominates: let it be $r_b$ for the whole area $a_b$ of the brush surface, and assume that it is *ohmic*. Then at a given instant the contact areas, Fig. 5.3(*a*) are $a_1$ and $a_2$ (where $a_1 + a_2 = a_b$). The corresponding contact resistances are $r_1 = r_b(a_b/a_1)$ and $r_2 = r_b(a_b/a_2)$. The two currents are inversely proportional to the resistances,

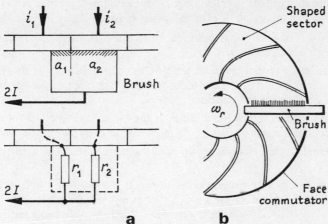

**a**          **b**

5.3 Resistance commutation

i.e. directly proportional to $a_1$ and $a_2$. As these change linearly with time, so also do the currents, giving straight-line commutation. This effect, termed *resistance commutation*, is relied upon for acceptable commutation in small machines without compoles. For constructional convenience some small motors have 'face-plate' commutators, Fig. 5.3(*b*), moulded or made from printed-circuit board. These can readily be shaped so that, with thin carbon-fibre brushes (Sect. 5.5), the rate of change of the contact areas $a_1$ and $a_2$ can be optimized to give a more nearly constant current density and improve the efficacy of the resistance commutation.

## Commutating-coil e.m.f.s

At operating speed, the time $t_c$ during which a commutating coil is short-circuited is typically 2 ms or less. The average rate of change of coil current is therefore of the order of 100 kA/s for a brush current of 100 A. As the armature coils have appreciable inductance, each commutating coil develops an inductive e.m.f. $e_{pc}$ that opposes the current change. It is normally necessary to introduce into the coil a motional e.m.f. $e_{rc}$ to neutralize $e_{pc}$. For this purpose commutating poles (compoles), Fig. 3.5, are fitted to provide a flux in the commutating zone, of the *same* polarity as that of the main pole ahead (in the direction of rotation) for a *generator,* and *opposite* for a *motor*. Complete neutralization at every instant is not possible, but provided that the difference does not exceed a volt or two it can be dealt with by the resistance-commutation process. However, the magnitude of $e_{pc}$ should not exceed $10 - 12$ V, imposing significant constraint in the design of the machine.

### Practical commutation

Commutation is, in fact, a much more complicated process than would appear from the foregoing, for the following reasons:

(i) The brushes normally span between 2 and 4 sector-widths, and not necessarily a whole number.

(ii) There may be several coil-sides per slot, and the coils may be short-pitched; commutation of top coil-sides will not take place at the same instant as for those at the bottom.

(iii) Besides the e.m.f. of self induction in the commutating coil, mutual e.m.f.s occur as a result of current change in other coil-sides in the same slot.

(iv) The rapid current changes set up eddy currents in the conductors and teeth.

(v) The brush-contact resistance is nonlinear. It is often taken to be inversely proportional to current, giving a roughly constant volt-drop between 1 and 2 V. It is subject to unpredictable variations due to surface irregularities, changes of temperature and humidity, ambient air pressure and aerodynamic lift.

### 5.3 INDUCTIVE E.M.F.S IN A COMMUTATING COIL

Grimshaw [13] shows that if the brushes are set in the q-axis (or 'neutral axis'), only changes in slot and overhang leakage fluxes need to be considered in calculating the e.m.f. in a coil undergoing commutation. The e.m.f. results from changes in the *self* flux and in the *mutual* flux set up by currents in adjacent coils that are being simultaneously commutated. Writing p for $d/dt$, then the inductive e.m.f. in coil 1 carrying $i_1$ and having a self inductance $L_{11}$ is $e_{pc} = L_{11} p\, i_1 + \Sigma(L_{1x} p\, i_x)$, where $L_{1x}$ is the mutual inductance between coil 1 and another coil $x$.

The *steady* flux pattern of an armature is illustrated in Fig. 4.18. Here we have to consider that part of the leakage flux associated with coils undergoing commutation: it is concerned with the self and mutual linkages in slot and overhang fields of conductors in which currents are being abruptly reversed. The approach is that in Sect. 4.11, developing permeance coefficients from which an inductance parameter can be derived.

## Self inductance

*Slot* The slot-flux pattern in Fig. 4.19(*a*) shows that, for the same current in each conductor, the bottom coil-sides are linked by the flux produced by the upper coil-sides as well as by their own self-flux. Adopting the simplified slot dimensions in Fig. 5.4, the permeance coefficients per unit length of slot can

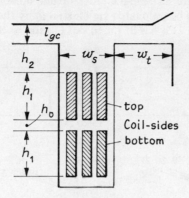

5.4 Slot dimensions

be written down by modifying eq. (4.12):

Top:  $\lambda_{st} = (h_1/3w_s) + (h_2/w_s) + (w_t/2l_{gc})$

Bottom:  $\lambda_{sb} = \lambda_{st} + (h_1/w_s) + (h_0/w_s)$  (5.1)

The term $w_t/2l_{gc}$ is an approximation to the effect of leakage flux from the teeth of width $w_t$ that twice crosses the compole airgap length $l_{gc}$ and passes through the compole shoe.

*Overhang* Expressed per unit length of *slot*, taken for simplicity to be the same as the gross core-length $l$, the overhang permeance coefficient can be taken as

Overhang:  $\lambda_o = (l_o/l)\,(1/2\pi)\,\ln(l_o/2r)$  (5.2)

as described in Sect. 4.11.

*Total* The overall permeance coefficient per unit length of slot is $\lambda_{11} = (\lambda_{st} + \lambda_{sb} + 2\lambda_o)$. For a coil of $z$ turns, the inductance due to self flux is

$$L_{11} = \mu_0 l\,\lambda_{11} z^2 = L_{st} + L_{sb} + L_o$$  (5.3)

To limit the self inductance (and therefore the inductive e.m.f.) it is customary to employ single-turn coils for machines large enough to accommodate the multi-sector commutator necessary.

## Mutual inductance

The mutual inductance between a given coil and others in the *same* pair of slots can be determined as for the self inductance, the values in some cases being identical.

*Slots*　For brevity, write 't.c. and b.c.' for 'top and bottom coil-side'. The mutual inductances to be accounted are illustrated by the example in Fig. 5.5. Consider coil 1: it has its t.c. in slot (1) and its b.c. in slot (5). In slot

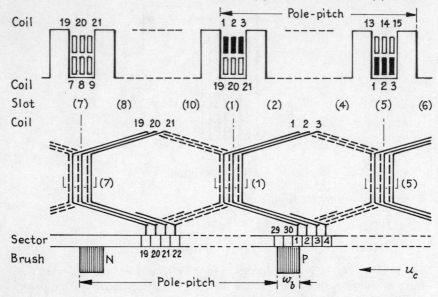

5.5 Two-pole short-pitch lap winding

(1) its t.c. is flanked by the t.c.s of coils 2 and 3, with which it has a mutual inductance $L_{tt}$, and further a mutual inductance $L_{tb}$ with the b.c.s of coils 19, 20 and 21 that lie below it. Similarly, in slot (5) the b.c. of coil 1 has mutual inductances $L_{bb}$ with its flanking b.c.s of coils 2 and 3, and $L_{bt}$ with the t.c.s of coils 13, 14 and 15 above it. In general, a coil has the several mutual inductance coefficients

$$L_{tt} = L_{st} \quad L_{bb} = L_{sb}$$
$$L_{tb} = L_{bt} = \mu_0 l \left[ (h_1/2w_s) + (h_2/w_s) + (w_t/2l_{gc}) \right] z^2$$

(5.4)

*Overhang*　Top- and bottom-layer coil-sides cross in the overhang at an angle near 90°, so that only coil-sides in the same layer contribute to the mutual inductance. Thus $\lambda_{tto} = \lambda_{bbo} = \lambda_o$, giving

$$L_{oo} = \mu_0 l \, \lambda_o z^2$$

(5.5)

## Inductive e.m.f.

Summing the self- and mutual-inductance effects for a given coil, its total inductive e.m.f. (or 'reactance voltage') is found.

Consider again the simple lap winding of Fig. 5.5. It has $C = 30$ coils and commutator sectors. There are 6 c.s. per slot and therefore $S = 10$ slots. The coils are short-pitched by 1 slot, so that coil 1 spans from slot (1) to slot (5) and its inductive e.m.f. is

$$e_{p_1} = (L_{sb} + L_{st} + L_o)p\,i_1 \qquad\qquad \text{self (slot and overhang)}$$

$$+ L_{tt}(p\,i_2 + p\,i_3) + L_{bb}(p\,i_2 + p\,i_3) \qquad \text{mutual t.c. and b.c. (slot)}$$

$$+ L_{tb}(p\,i_{19} + p\,i_{21} + p\,i_{20}) + L_{bt}(p\,i_{13} + p\,i_{14} + p\,i_{15})$$
$$\text{mutual t.c./b.c. and b.c./t.c. (slot)}$$

$$+ L_{oo}(p\,i_2 + p\,i_3) \qquad\qquad\qquad \text{mutual (overhang)}$$

All the changes $pi$ are the same, but do not occur at the same instant. Their timing depends on the arrangement of the winding.

## Current changes

The *idealized* commutation duration $t_c$ of a given single coil is $t_c = w_b/u_c$, or $t_c = (w_b - w_i)/u_c$ if the small intersector insulation thickness $w_i$ is allowed for. But in practice the *effective* commutation time $T_c$ during which inductive e.m.f.s occur in a coil as a result of current-changes in associated coils and of short-pitching, has a major significance.

*Lap Winding* In Fig. 5.5, when sector 2 just reaches brush P, the current $i_1$ in coil 1 begins to reverse along the current/time line 1t in Fig. 5.6. After a

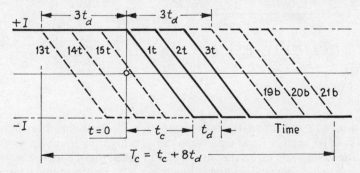

5.6 Current/time graph

delay $t_d = w_c/u_c$ for sector 2 to pass the leading edge of the brush, the current $i_2$ in coil 2 begins to change along the line 2t, and similarly for coil 3. The total time of commutation for the t.c.s of coils in slot (1) is thus $t_c + 2t_d$. In general, with $m$ coil-sides per slot (i.e., $\tfrac{1}{2}m$ per layer) the commutation time is $T_d = t_c + (\tfrac{1}{2}m - 1)t_d$.

Were the coils of full pitch, the b.c.s of the coils considered would commutate at brush N simultaneously. But the first b.c. in slot (1) is part of coil 19, which does not start to commutate until sector 20 reaches brush N, i.e. later by 3 $t_d$, as indicated by the dotted lines 19b, 20b and 21b in Fig. 5.6.

The b.c. of coil 1, in slot (5), has e.m.f.s induced by the current changes in coils 13, 14 and 15. Coil 13 is seen to have completed commutation one-half of a sector pitch before $t = O$, and the current-change lines for coils 13, 14 and 15 precede those for 1, 2 and 3 by a time corresponding to the short-pitching.

The total commutating time for the t.c.s of slot 1 is thus $T_c = 2 \times 3t_d + (3 - 1)t_d + t_c$. In general, for a winding short-pitched by $q$ slots and with $m$ coil-sides per slot ($\frac{1}{2}m$ per layer),

$$T_c = 2q(m/2)t_d + (\tfrac{1}{2}m - 1)t_d + t_c = [m(q + \tfrac{1}{2}) - 1]\,t_d + t_c \qquad (5.6)$$

*Wave Winding*  Where only two brushes are used with a $2p$-pole winding, commutation conditions are similar to those in a lap winding, except that $p$ coils in series will be commutated simultaneously. To shorten the axial length of the commutator, however, it is usual to provide $2p$ brushes. Fig. 5.7

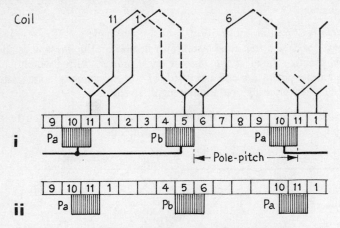

5.7 Two-pole wave winding

shows a 4-pole 11-sector armature winding, with the negative brushes omitted for clarity. In (i), sectors 5 and 6 are about to be bridged by brush Pb to initiate commutation in coil 6 (because 6 and 11 are in series between sectors 5 and 6, and coil 6 is short-circuited through sector 11 by the external inter-connection between the two positive brushes Pa and Pb). One-half sector-pitch later, (ii) shows sectors 11 and 1 (with coils 1 and 6 in series between them) about to be bridged by brush Pa. Coil 6 continues in short-circuit until sector 11 has moved away from brush Pa. But brush Pb has already started commutation of coil 6 by the time corresponding to one-half sector-pitch. The total commutation time is therefore $T_c = \frac{1}{2}(m - 1)t_d + t_c$. In general

$$T_c = [\tfrac{1}{2}m - (a/p)]\, t_d + t_c \tag{5.7}$$

The positive brushes are connected both externally and through coils in the armature winding. These paths have dissimilar impedances, so that there may occur cyclic variations in the current sharing, a phenomenon known as *selective commutation.*

### Total inductive e.m.f. and waveform

Combining the e.m.f. equation with the current-change diagram gives the magnitude and waveform of the self- and mutually-induced e.m.f. in a commutating coil. The method is illustrated in Example 5.1 below.

*Eddy Currents* The high rates of change of current and leakage flux may generate eddy currents in the conductors and in the surrounding armature core steel. Their effect is to reduce the inductance and to delay the sudden rise of e.m.f. The time-constants concerned are of the order of $0.1\, t_c$. The introduction into the slots of short-circuited damper windings to reduce inductive e.m.f.s has been suggested, but they occupy useful space. Taylor [14] proposes the use of *flux traps,* comprising copper foil about 0.02 mm thick placed adjacent to the conductors. The energy that causes sparking during commutation is that remaining after the rapid change of current. The flux trap 'freezes' the flux present as the commutator sector leaves the brush, so that the remaining energy is not available to sustain a spark. The trap occupies but little slot space, but it does give rise to additional loss.

*Banding Wire* Older machines may have the armature coils retained in their slots by steel banding wire, which significantly increases the leakage flux at the slot openings. If the banding becomes magnetically saturated, sudden flux changes occur as the coil currents pass through zero during commutation, and short peaks in the coil inductive e.m.f.s may occur. In modern machines, banding is made of fibreglass cord instead of steel wire.

EXAMPLE 5.1: Data for a 300 kW, 250 V, 6-pole 600 r/min shunt generator are as follows.

*Winding*: slots, $S = 81$; coils, $C = 162$ (single-turn); coil-sides per slot, $m = 4$; coil-span, 13 slots; lap connection, Fig. 5.8; full-load current, $300/250 = 1.20$ kA; field current, 12 A; armature current, 1212 A; current per path, $1212/6 = 202$ A.

*Armature*: diameter, $D = 0.64$ m; length, $l = 0.28$ m; overhang length per coil-side, $l_o = 0.43$ m; tooth width, $w_t = 15.3$ mm; compole gap length, $l_{gc} = 6.5$ mm; slot dimensions, Fig. 5.8.

*Commutator*: diameter, $D_c = 0.405$ m; peripheral speed, $u_c = 19.1$ m/s; sector width, $w_c = 7.85$ mm (including insulation); brush width, $w_b = 19.0$ mm. Obtain the waveform and mean value of the e.m.f. induced in a coil during commutation.

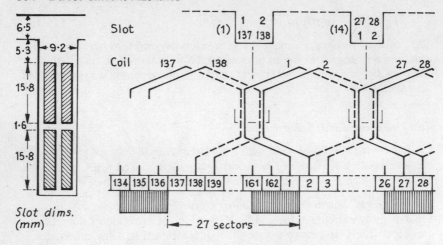

Slot dims.
(mm)

5.8 Slot and winding: Example 5.1

The commutating times are $t_c = w_b/u_c = 0.99$ ms, $t_d = w_c/u_c = 0.41$ ms, and $T_c = t_c + 3t_d = 2.22$ ms. The current-change diagram, Fig. 5.9(a), can now be

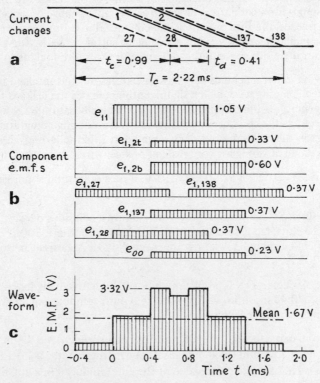

5.9 Inductive e.m.f.s in coil 1: Example 5.1

drawn. The rate of change of coil current is $pi = 2I/t_c = 408$ kA/s. The permeance coefficients are

Slot: linking t.c.s., eq. (5.1)

$$\lambda_{st} = (15.8/3 \times 9.2) + (5.3/9.2) + (15.3/2 \times 6.5) = 2.33$$

Slot: linking b.c.s., eq.(5.1)

$$\lambda_{sb} = 2.33 + (15.8/9.2) + (1.6/9.2) = 4.22$$

Overhang: eq. (5.2)

$$\lambda_o = (0.43/0.28) \, (1/2\pi) \, \ln(0.43/0.017) = 0.79$$

*Self inductance* With $\mu_0 l z^2 = \mu_0 \times 0.28 \times 1^2 = 0.35 \times 10^{-6}$,

$$L_{st} = 0.35 \times 2.33 = 0.82 \, \mu H \quad L_{sb} = 0.35 \times 4.22 = 1.48 \, \mu H$$

$$L_o = 0.35 \times 0.79 = 0.28 \, \mu H$$

whence from eq. (5.3)

$$L_{11} = L_{st} + L_{sb} + L_o = 2.58 \, \mu H$$

With $pi = 408$ kA/s, $e_{p11} = L_{11}pi = (2.58 \times 10^{-6})(408 \times 10^3) = 1.05$ V

*Mutual inductance* $L_{tt} = L_{st} = 0.82 \, \mu H, L_{bb} = L_{sb} = 1.48 \, \mu H$, and from eq.(5.4)

$$L_{tb} = L_{bt} = 0.35 \times 10^{-6} [(15.8/2 \times 9.2) + (5.3/9.2) + (15.3/2 \times 6.5)]$$
$$= 0.91 \, \mu H$$

The mutual e.m.f.s can now be calculated. They arise from the t.c. and b.c. couplings of coil 1 with coil 2 ($e_{1,2t}$ and $e_{1,2b}$); between coil 1 and coils 27, 28, 137 and 138 respectively ($e_{1,27}, e_{1,28} \ldots$); and the coupling in the overhang:

$$e_{1,2t} = L_{st}pi = 0.82 \times 10^{-6} \times 408 \times 10^3 = 0.33 \text{ V}$$

$$e_{1,2b} = L_{sb}pi = 0.60 \text{ V}$$

$$e_{1,27} = e_{1,28} = e_{1,137} = e_{1,138} = L_{tb}pi = 0.37 \text{ V}$$

$$e_{oo} = 2L_o pi = 2 \times 0.28 \times 10^{-6} \times 408 \times 10^3 = 0.23 \text{ V}$$

Plotting the e.m.f. components as related to the corresponding current changes, Fig. 5.9(a), gives (b) and the summation (c) for the waveform of the total inductive e.m.f. in coil 1 during the commutation period.

*Average Value* The mean inductive e.m.f. in coil 1 during commutation is readily obtained from diagram (c) to be 1.67 V.

**Approximate mean e.m.f.**

The permeance of a slot is a function of its shape but not of its actual dimensions. For a depth/width ratio of 3.5, the total slot leakage flux $\phi_s$ per ampere of slot current is about 4 $\mu$Wb/m. Thus with $m$ coil-sides per slot and a current $I$ per conductor the slot flux is $\phi_s = 4mzI10^{-6}$ Wb/m. The overhang leakage flux approximates to $\phi_o = 0.8$ $\mu$Wb/m per ampere in the group of coils forming one layer. Thus the overhang leakage flux can be taken as $\phi_o = 0.4mzI10^{-6}$ Wb/m. The total for a coil-side is $\phi_l = \phi_s + \phi_o$ $= mzI (4l + 0.4l_o) 10^{-6}$ Wb for the appropriate lengths of slot and overhang, and twice this for a complete coil. In commutation, $\phi_l$ reverses in a time $T_c$, and the average inductive e.m.f. is

$$e_{pc} = 4mzI (4l + 0.4l_o)/10^6 T_c \tag{5.8}$$

EXAMPLE 5.2: Use the empirical expression in eq. (5.8) to estimate the mean e.m.f. in a commutating coil of the machine in Example 5.1.

$$e_{pc} = 4 \times 4 \times 1 \times 202(4 \times 0.28 + 0.4 \times 0.43)/10^6 \times 2.2 \times 10^{-3}$$
$$= 1.90 \text{ V}$$

This empirical value, for a slot of depth/width ratio of 4.3, is in reasonable agreement with 1.67 V as calculated in Example 5.1.


## 5.4 MOTIONAL E.M.F. IN A COMMUTATING COIL

Compoles are fitted in the space between the main poles to generate in the commutating coils a motional e.m.f. that, ideally, cancels the inductive e.m.f. ('reactance voltage') at every instant during the commutation process.

Three component fluxes exist in the interpolar space, due respectively to (i) the armature m.m.f., (ii) fringing flux from the main poles, and (iii) the compole m.m.f. In combination these provide the flux in which the motion of the armature develops the neutralizing e.m.f. in the commutating coils. The resulting flux distribution is complicated: Erdelyi [15] has attempted a flux-plotting and computer-analysis method of dealing with the problem. An approximate estimate can, however, be made by treating the three components separately.

*Armature M.M.F.* In Sect. 4.10 the armature m.m.f. per pole is shown to be $F_a = N_a I_a/4ap$, and Fig. 4.14 shows that $F_a$ must be opposed by the compole m.m.f., whether or not the main poles carry compensating windings.

*Fringing Flux* Although with ideal symmetry the main-pole flux density is zero on the q-axis, the commutating zone extends over slots on either side of the axis, and these move in the fringing field. In small machines this may be exploited by brush shift away from the q-axis in the appropriate direction.

*Compole M.M.F.* The peak flux density $B_{cm}$ in the gap under the middle of the compole must be such as to generate a peak e.m.f. $e_{rc}$ equal (or nearly so) to the peak of the inductive e.m.f. $e_{pc}$. Thus $B_{cm} = e_{pc}/l_c u$ for a compole of axial length $l_c$. The compole flux is $\Phi_c = B_{cm}l_c w_z/k_c$, where $w_z$ is the width of the commutating zone at the armature surface, and $k_c$ is the form-factor of the flux distribution.

### Estimation of compole m.m.f.

The magnetic circuit, Fig. 5.10, in which the main and compole fluxes are indicated separately, shows that in the yoke and armature core the two are additive or subtractive. The total m.m.f. per compole is obtained by summing the component m.m.f.s for each part of the circuit, using the methods described in Sect. 3.7. As the main flux is greater than the compole flux under normal operating conditions, some component m.m.f.s may be *negative*. Further, the number of teeth under a compole shoe will usually vary as the armature rotates, causing the permeance of the compole airgap to fluctuate. For this reason the compole gap length $l_{gc}$ may be made 1.3 to 1.5 that of the main gap length $l_g$.

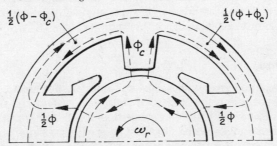

5.10 Main-pole and compole flux paths

Both $e_{pc}$ and $e_{rc}$ are proportional to the peripheral speed, the former because the speed determines the commutation time. As $e_{pc}$ is also proportional to the conductor current, the compoles are excited in series with the armature. The number of turns may well be small, and to secure an integral number, the axial length of the compole (or its shoe) is chosen in the design stage. Further adjustments may have to be made after a test on load. If compoles are fitted they *must* be excited: otherwise the low-reluctance path across $l_{gc}$ would encourage excessive and deleterious flux in the commutating zone produced by the armature m.m.f.

*Approximate Calculation* If the ferromagnetic parts of the magnetic circuit are not highly saturated, the m.m.f. required per compole is $F_{cp} = KB_{cm}l_{gc}/\mu_0$, where $K$ is an empirical factor typically between 1.1 and 1.3.

*Brush Shift* With a compole machine the brushes must be located very close to the q-axis. Brush shift causes the compole flux to aid or oppose the main flux, giving the effect of series compounding. In a motor, a forward shift increases the main flux and reduces the speed, while a backward shift reduces the main flux and could lead to instability. In a generator, forward shift reduces and backward shift increases the generated armature e.m.f.

## Compole design

*Leakage*   Considerable compole flux leaks to the adjacent main pole of opposite polarity, and may have a magnitude comparable to, or even greater than, the useful flux. To minimize leakage, the axial length of the compole may be 20 to 30% shorter than the armature core length.

*Saturation*   To avoid saturation, which would impair the proportionality between useful compole flux and armature current, the outer width of the pole should be greater than the inner to accommodate the leakage flux. A tapered compole is readily achieved if the stator magnetic circuit is completely laminated.

*Compole Shoe*   The width and profile of the shoe (or pole tip) must yield the required distribution of the flux density over the commutating zone. The width should preferably embrace an integral number of armature slot-pitches in order to minimize gap-permeance fluctuations.

## Special windings

Where the armature has a number $m$ of coil-sides per slot greater than 2, the commutation conditions for the several coils are not identical. The result may be a blackening of every $\frac{1}{2}m$th commutator sector. In a *split-pitch winding* with $m = 4$, the two t.c.s in one slot form coils with b.c.s in separate slots, to give respective coil spans differing by $+\frac{1}{2}$ and $-\frac{1}{2}$ of a slot pitch, making the commutating conditions the same for each coil.

Brush-current sharing is improved, and with it the commutation, by the provision of *equalizer connections* (Sect. 4.4). The same effect is given by a *frog-leg winding,* comprising a lap and a wave winding in the same slots and connected to the same commutator sectors. The wave winding acts as an equalizer to the lap winding.

Occasionally *duplex windings* are employed. They embody two separate windings, connected to alternate sectors and parallelled by the brushes. The mean voltage between sectors is then one-half of the mean voltage per coil.

## 5.5 BRUSHGEAR

Choice of brush material, design of brush holders and proper maintenance are essential for sparkless commutation.

## Materials

Since the beginning of this century, carbon-based brushes for industrial machines have replaced the earlier forms using copper wire or gauze. The mechanical and physical properties of carbon are relatively insensitive to temperature-rise. Carbon does not melt nor weld, has a low coefficient of thermal expansion, a high heat capacity, a resistance to thermal shock and an adequate thermal conductivity. As its density is relatively low, it can be made to follow minor irregularities of the commutator surface.

'Soft' brushes are formed from graphite, 'hard' brushes from coke and charcoal. The material is finely ground, mixed with a tar or resin binder and pressed into blocks. After heat treatment at 200–1000°C, the blocks are ready for use as *hard-carbon* or *natural-graphite* brushes, unless impregnation is required as for traction or airborne machines. A further treatment at 2000–3000°C purifies the material and converts all amorphous carbon into graphite, giving the *electrographitic* range of brushes. For high-current densities and low volt-drop, the original mix includes metal powder to provide *metal-graphite* brushes.

Brush grades are so numerous that a choice for a machine may be delayed until it is on test, or even after a period of service. The table gives typical properties, namely the contact drop $v$ per brush, the working current density $J$, and the brush pressure $p$, the appropriate peripheral speed $u_c$ of the commutator, and the coefficient $\mu_b$ of brush friction:

| Type | $v$ V | $J$ mA/mm$^2$ | $p$ mN/mm$^2$ | $u_c$ m/s | $\mu_b$ |
|------|-------|---------------|---------------|-----------|---------|
| Natural graphite | 0.7 – 1.2 | 100 | 14 | 50 – 60 | 0.1 – 0.2 |
| Hard carbon | 0.7 – 1.8 | 65 – 85 | 14 – 20 | 20 – 30 | 0.15 – 0.25 |
| Electrographitic | 0.7 – 1.8 | 85 – 110 | 18 – 21 | 30 – 60 | 0.1 – 0.2 |
| Metal-graphitic | 0.4 – 0.7 | 100 – 200 | 18 – 21 | 20 – 30 | 0.1 – 0.2 |

*Natural Graphite* Has good lubricating properties for high speeds and low noise level. Is fragile, and impurities may cause commutator wear. With non-conducting resin binder, the higher volt drop helps commutation in small machines without compoles.

*Hard Carbon* Robust and cheap. Wears sector insulation at the same rate as the sector copper. Suits fractional-kilowatt machines and larger low-speed machines with easy commutation conditions.

*Electrographitic* Wide application to industrial machines. Low rate of wear. Conditions of vibration (as in traction) can be countered by brush pressures up to 70 mN/mm$^2$.

*Metal-graphite*   Low contact drop and high current density suitable for low-voltage high-current machines (automobile-starter and battery-vehicle motors, and some tachometer generators).

*Carbon Fibre*   Constructed of fine carbon fibres with the flexibility of a paint brush. Contact drop $2-3$ V, and higher friction coefficient. Brush positioning tends to be imprecise. Suitable for small machines without compoles. May be placed at the trailing edge of a conventional brush, or bonded thereto, to alleviate sparking.

**Brush dimensions**

The required contact area is found from the current per brush and the recommended current density. The peripheral width (*thickness*) is settled by the number of commutator sectors to be covered. To give flexibility in following commutator surface irregularities, the axial length (*width*) required may be provided by a set of mechanically separate brushes of width not more than three times their thickness.

To ensure the stability of a brush in its holder, the radial *height* of the brush must be several times its thickness, with provision for wear which may range from 1 mm to 5 mm per 1000 h depending on brush material and duty. Standard dimensions for brushes are laid down in BS 96, but alternatives are available.

**Brush holders**

A holder is essentially a metal box, insulated from the frame and rigidly mounted close to the commutator surface. The brush must be able to slide in the box and to be pressed against the commutator surface by a stainless-steel or phosphor-bronze spring. In brush manufacture, a flexible copper 'pigtail' is incorporated to provide a terminal connection. The spring should, ideally, provide a pressure independent of brush wear. The simple *helix,* Fig. 5.11(*a*), may lose half its effective pressure and is confined to low-power machines. The spiral *torsion* spring is much less sensitive to brush wear.

Split brushes in up to four parts with a common compression spring, or tandem brushes with individual springs, may sometimes be used to improve the flexibility of brush contact in large machines.

*Stability*   In *non-reversing* machines the brush box is given either a *trailing* or a *reaction* rake, Fig. 5.11(*b*). The spring, friction and radial forces in the trailing box combine to maintain contact between the brush and the left-hand face of the box. But the good stability is obtained at the price of frictional resistance to the sliding of the brush to accommodate commutator surface irregularities, and the trailing angle should be restricted to 15° or less. In the reaction box the brush slides more easily: the right-hand side of the brush is in contact with the box, but there is an opposing friction force. To maintain the side pressure the reaction angle should not be less than 30–40°.

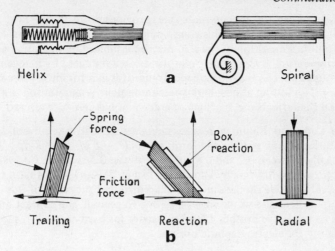

Helix     **a**     Spiral

Spring force    Box reaction    Friction force

Trailing     Reaction     Radial

**b**

5.11 Brush details

The reaction type is preferred if the brushes are narrow.

For machines that must run in either direction, stability conditions are unaffected if the brush box is radial. The tendency of the brush to lie 'diagonally' is countered by the use of brushes of greater height set in close-fitting boxes. However, holders angled at 15–20°, giving 'trailing' in one direction and 'reaction' in the other, have been found satisfactory.

*Brush Staggering* To avoid grooving of the commutator surface, brush sets are set in slightly different axial positions. Each part of the surface should sweep past the same number of positive and negative brushes because these give different individual rates of wear.

### Contact phenomena

The process of current transfer between brush and commutator surfaces is a complicated one. Some technological effects (volt-drop, wear, friction and temperature-rise) are readily measurable, and are found to be functions of brush material, commutator peripheral speed, brush pressure, magnitude and direction of current, and ambient conditions.

*Surface Properties* Even with very smooth surfaces, actual contact occurs only at relatively few discrete points, the remaining areas being separated by gaps of the order of 0.5 $\mu$m. Passage of current through the contact points of appreciable resistance sets up a p.d. which, between the non-contact areas, develops between them an electric field of very high intensity, e.g. 200 kV/cm, enough to initiate field emission, augmented by wear particles. These phenomena, together with mechanical friction, result in the formation of a skin or 'patina' on the commutator surface. The skin, composed mainly of particles of carbon and cuprous oxide, provides protection against commutator surface wear and is considered to be desirable for good commutation. Formation of the skin may be inhibited by operation at too low a temperature, by ambient atmospheres contaminated by halogen and sulphurous gases, and by atmospheric dryness (as in a totally enclosed machine). Too low an atmospheric pressure impairs commutation – a condition that affects d.c. machine operation in high-altitude aircraft.

*Contact Volt-drop* Empirical relations between volt-drop $v$, current density $J$ and brush pressure $p_b$ take the forms $v = k_1 J^x$ and $v/J = r = k_2/p_b^y$, where $k_1$ is usually between 0.2 and 0.6, and $y$ between 0.5 and 1.0. Curves of volt drop at pressures differing by about 10% in Fig. 5.12($a$) show that if two brushes, side-by-side on the same arm, had different pressures of this order, their current densities would differ by 30%. In general, the contact drops at positive and negative brushes differ, markedly for hard-carbon brushes, but only slightly for electrographitic brushes.

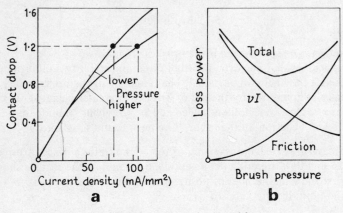

5.12 Brush contact drop and loss

*Friction* The friction coefficient varies with the brush current and with speed, but is not sensitive to current direction.

*Copper Picking* Particles of copper are sometimes found adhering to a brush face. The reason is obscure.

*Copper Dragging* Fine filaments of copper sometimes grow from the trailing edge of a brush, tending to bridge adjacent sectors. The cause appears to be mechanical, and it may be obviated by bevelling the sharp edges of the sectors.

*Loss* The contact loss $vI$ falls with increasing brush pressure, but the friction loss rises. Their sum has a minimum, Fig. 5.12($b$), which is also the condition of minimum wear. An empirical expression for the friction loss is $p_b \mu_f u_c$ mW/mm² of brush contact area at a brush pressure $p_b$, giving typically between 50 and 90 mW/mm².

## 5.6 COMMUTATOR FLASHOVER

Minor sparking is common: an arc between successive brush-arms or from a brush to the frame is less frequent but far more damaging. Such a flashover is most likely under transient load conditions, but it may occur on steady load, or even on no load.

### Steady state

The mica-based insulation between commutator sectors is thin (e.g. 0.5–1.5 mm) and is recessed by a similar amount. It is normally quite sufficient to withstand the voltage between sectors, but if the recesses become filled with copper and carbon dust, fragile conducting paths are formed, which carry current and collapse, initiating an arc. If the arc is carried round the commutator, the result is a terminal-to-terminal flashover.

### Transient state

Large and sudden changes of load current cannot be matched by the compole flux, which is delayed by induced eddy currents or limited by saturation. As a result, the armature is momentarily under- or over-commutated, and heavy sparking may lead to a flashover. Such conditions can also result from an interruption and re-establishment of the supply voltage to a motor.

### Prevention

The basic factor is the p.d. between adjacent commutator sectors. For industrial machines this should not exceed a mean of 20 V and a peak of 25–40 V. Additionally, the peripheral voltage gradient around the commutator should be limited to about 30 V/cm to avoid ionization of the skin of air at the commutator surface.

High-speed circuit-breakers can be used to limit flashover damage. Where heavy transient loads are part of the duty of a machine, flash barriers are fitted. These are arc-resisting sheets set between and near to the brush arms, with their inner edges very close to the surface of the commutator.

## 5.7 DESIGN LIMITATIONS

The conversion power of a heteropolar cylindrical-rotor machine is shown in Sect. 2.3 to be

$$P = \pi^2 D^2 lBAn \tag{5.9}$$

in terms of the main dimensions $D$ and $l$, the mean gap flux density $B$, the armature electric loading $A$, and the speed $n$. These represent the mechanical, magnetic and thermal parameters. Commutation imposes constraints on the choice of the variables in respect of

(i) a limit on $E_{pc}$, the e.m.f. induced in a commutating coil;

(ii) the permissible mean voltage $E_c$ between adjacent sectors; and

(iii) the commutator construction, its peripheral speed, and the number and width of sectors that it contains.

Limitations (i) and (ii) necessitate the adoption of single-turn coils in all but the smaller range of industrial machines.

In assessing the maximum power $P$ for a conventional industrial machine, we adopt the following typical limiting parameters:

| | | | |
|---|---|---|---|
| Field: mean gap flux density | $B$ | = 0.80 T |
| Armature: electric loading | $A$ | = 60 kA/m |
| peripheral speed | $u$ | = 40 m/s |
| Commutator: e.m.f. in commutating coil | $E_{pc}$ | = 10 V |
| mean p.d. between sectors | $E_c$ | = 20 V |
| peripheral speed | $u_c$ | = 40 m/s |
| minimum sector width | $w$ | = 4 mm |

*Speed*  From eq. (5.9), $D^2 l = P/\pi^2 BAn \cong 2(P/n)10^{-6}$ W per r/s. For economic physical dimensions, a high speed $n$ is desirable.

*Diameter*  $E_c = 2Blu = 2\pi BDln$, so that eq. (5.9) can be written $P = \frac{1}{2}\pi DAE_c \cong 19 \times 10^6 D$ watts. The sector p.d. criterion indicates that a machine with a diameter of 1.0 m may have a power approaching 2 MW.

*Core Length*  As $E_c = 2 Blu$, then $l = E_c/2Bu$. With both $B$ and $u$ taking upper limits, this gives $l = 0.31$ m irrespective of the diameter. But a smaller diameter will permit a greater core length, and will decrease the inertia, an advantage if the duty demands rapid speed change or reversal.

*Inductive E.M.F.*  The rate of change of current during commutation is proportional to the peripheral speed. The slot permeance is proportional to core length and the overhang permeance to the pole-pitch. A multipolar design reduces the latter permeance and is also advantageous in reducing the flux per pole and the physical size of the field system; but the basic frequency $f = pn$ of flux reversals in the armature teeth and core is raised, and with it the core loss. A compromise is to choose $f$ between 25 and 40 Hz. As a result, $2p$ may be between 10 and 20 for high-power, large-diameter, low-speed designs.

*Voltage*  For an armature voltage $V_a$, the number of commutator sectors between successive brushes must not be less than $V_a/E_c$. The commutator periphery is therefore $2p(V_a/E_c)w$ and the peripheral speed is $u_c = 2p(V_a/E_c)wn$. The armature voltage becomes $V_a = E_c u_c/2pnw$. Taking $pn = f = 33$ Hz,

$$V_a = 20 \times 40/2 \times 33 \times 0.004 = 3000 \text{ V}$$

This is the maximum voltage for which a d.c. machine can be built. Traction motors for this working voltage were successfully operated more than 50 years ago, but large modern industrial machines work at no more than 1000 V.


## 5.8 COMMUTATOR TECHNOLOGY

The many sectors of a conventional industrial commutator demand a diameter $D_c$ approaching 0.8 of the rotor diameter $D$. The sector material is typically a silver-bearing copper with a softening temperature of 350°C: it has good thermal and electrical conductivity, and encourages skin formation. Hard micanite sheets separate the sectors. Their thickness is 0.5—1.0 mm, and their outer edges are undercut by the same amount to allow for sector wear.

The sector/insulation assembly may be *arch-bound* or *wedge-bound*, Fig. 5.13, the former being the more common. Each sector is firmly held by

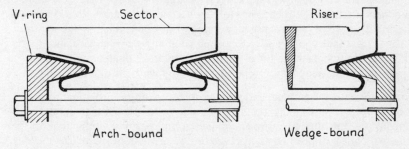

Arch-bound        Wedge-bound

5.13 Commutator clamping

its neighbours and by V-shaped clamping rings secured by bolts. A small clearance is left as indicated, so that tightening the bolts forces the sectors radially inward. The assembly is carried on a sleeve or spider mounted on the shaft, or on an extension of the armature spider. The wedge-bound method is similar, except that no clearance is provided at the V-rings. A recent development is to dovetail the sectors into a steel hub, the copper being bonded to, and insulated from, the steel by fused glass. Long commutators for high-speed or high-current machines may have steel rings shrunk on to the commutator over a band of suitable insulating material.

*Risers* These comprise copper strips, soldered and rivetted to the sectors or integral therewith, the outer ends being soldered to the armature coil-ends. In small machines the coils may be soldered directly to the sectors. It should be feasible, after disconnecting the risers, to withdraw the commutator assembly intact from the armature.

*Temperature-rise* Excessive temperature-rise will distort the commutator and cause the joints to deteriorate. The risers have a fanning effect, but it may require to be augmented by the provision of a fan located at the further end of the armature.

*Seasoning* This is an important constructional operation. The commutator assembly is heated to 150–200°C, run at normal speed, and allowed to cool. The bolts are then tightened, and the process repeated until stability has been achieved. Finally the surface is machined, using a diamond-tipped tool.

## Maintenance

Regular inspection and maintenance are essential to ensure that no blackening, burning or other irregularity has developed on the commutator surface, that the brushes move freely in their holders, and that no accumulation of dirt or carbon dust has occurred. The cause of any deficiency in these respects must be sought, lest increasing damage occur.

*Machine Faults*  These can often be detected by their effect on the commutation. Mechanical faults include vibration, bearing wear (indicated by change in the compole gap) and clogged ventilation ducts. Electrical faults are generally attributable to bad riser joints or to open- or short-circuiting of an armature coil. Bad joints cause blackening or burning, starting one sector ahead of the faulty joint and perhaps repeated at pole-pitch intervals due to the resulting unbalance in the armature currents. An open circuit in an armature coil develops severe sparking that aggravates as the load is raised.

*Brushgear Faults*  Sparking, and undue brush or commutator wear, result from incorrect positioning, pressure or grade of the brushes, giving rise to unequal current sharing between parallelled brush-arms or individual brushes in a set. Zones of differing appearance on the brush contact surface usually indicate a change of current density during the commutation cycle, indicating an imperfect distribution of the compole flux. Brush noise is caused by vibration, the sources of which may be brush 'chatter' in their boxes and protruding sector insulation. Armature and commutator unbalance may also cause noise, and the various sources may often be diagnosed by observation of the noise pitch.

*Bedding*  New brushes have to be 'bedded' on the commutator surface. A simple method is to pass first a medium-grade and then a fine-grade cloth between the brush and the commutator at rest, taking care to pull the cloth in the direction of rotation. A quicker alternative for a large machine is to run it on no load and apply to the commutator surface a 'bedding stone', which releases abrasive particles.

## Testing

*Brush Setting*  This must be accurate, particularly for compole machines. In the 'kick neutral' test on a stationary machine, all the service brushes are removed; then two special brushes, each ground to a chisel edge along the axial centre, are placed in brush boxes of opposite polarity and connected to a low-reading centre-zero voltmeter. When an established field current is suddenly quenched, a momentary deflection of the voltmeter occurs unless the test brushes are set in the q-axis. The test is repeated for a few slightly different positions of the armature to establish an optimum position. An alternative method is to find by means of probes (e.g. ordinary lead pencils) the two adjacent sectors between which no momentary voltage is observable when a field current is interrupted. This determines the d-axis, and the brush setting can be measured therefrom.

*Black Band*  This is an informative test for the efficacy of the compoles. It requires the compole current to be independently variable. For a range of fixed armature currents $I_a$, the compole current $I_{cp}$ is reduced until sparking occurs, then raised until sparking reappears. From a plot of the results, Fig. 5.14, the range of sparkless ('black') commutation is obtained. If the envelope is asymmetrical at low load, the brushes are incorrectly located;

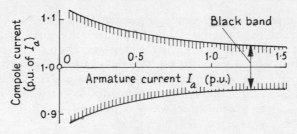

5.14 Black-band test

if it tilts with increase of $I_a$, the machine is over- or under-commutated; and if it is very narrow at high load levels, the compoles are saturated. Magnetic shims can be used between a solid compole body and the yoke to which it is bolted, to vary the commutating flux. Because the compole m.m.f. has to oppose the armature m.m.f. as well as to develop the commutating flux, the per-unit change in gap length required is several times as great as the per-unit change in $I_{cp}$ as inferred from the black-band test. Where shims are not applicable (as in a laminated-stator machine), a reduction of $I_{cp}$ is possible by means of a diverter resistor; but unless it has a matching inductance as well, it may be ineffective under conditions of transience.

*Contact Drop*  Uniformity of brush current density may be checked with the machine running by measuring and plotting the volt drop between the commutator and the brush at a few points along its width, as in Fig. 5.15. Correct commutation should give a horizontal straight line, but under- or over-commutation will yield rising or falling curves. Though less informative than the black-band test, the contact-drop test is easy to perform and is a useful check.

*Armature Faults*  A suspected fault in the armature winding can be detected and located by passing a current through the stationary winding and measuring the p.d. between sectors around the commutator. The p.d.s are identical if the winding is sound. A very low p.d. indicates a short-circuited coil; an open-circuited coil will result in a zero reading between all pairs of

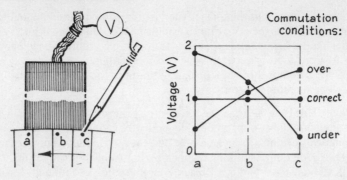

5.15 Contact-drop test

sectors in the path affected except that pair terminating the faulty coil. If a high-voltage insulation-tester indicates the presence of an earth fault, the sector-p.d. procedure can be extended to measuring the p.d. of each sector to earth, which will reverse as the earthed coil is passed. A sector in a parallelled part of the winding may also show a zero, and the test must be repeated with a changed position of the armature. The earthed coil will always show a zero, but the others will not.

## 5.9 ASSISTED COMMUTATION

Attempts have been made to avoid some of the limitations discussed in Sect. 5.7 by 'assisting' the conventional commutator with the help of solid-state diodes and thyristors.

### Diode-assisted

The commutator is constructed in two parts, each with alternate active and idle sectors, Fig. 5.16. Current is fed into the armature winding at a brush arm by a pair of brushes, each in series with a diode and connected together at the input terminal. In position ($i$), all current passes through brush A. As the commutator moves to position ($ii$), coil X is short-circuited through both brush/diode units. In the closed circuit so formed, movement of coil X through the compole flux generates an e.m.f. $e_r$ and a circulating current $i_c$ which causes the current in B to rise, and that in A to fall towards zero. The current in A cannot reverse, because such a direction is blocked by the diode. Thus the whole current $2I$ has been transferred to B. A like process occurs at the brush position where the current leaves the armature, except that the diodes have the reverse blocking direction.

The essential condition is that the commutating e.m.f. $e_r$ must be greater than the induced e.m.f. $e_p$ in order that the current in A shall reach zero before its sector parts contact with the brush. The diodes inhibit over-commutation: the careful balance of $e_p$ by $e_r$, essential to conventional

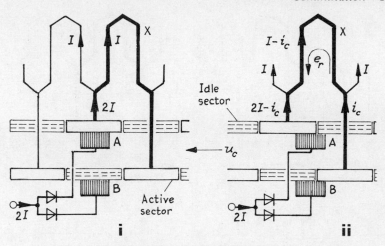

5.16 Diode-assisted commutation

commutation, is unnecessary. The short-core and single-turn coil restrictions
are lifted, permitting fewer sectors (e.g. 4 per pole-pitch for a 500 V
machine), a mean p.d. between active sectors of 60 V, a commutating e.m.f.
of 100–200 V, and a rigid commutator construction.

However, the system has some drawbacks. The commutating e.m.f. in
motors is zero at starting and too small at low speeds. While resistance
commutation is some help, it may be necessary to introduce an auxiliary
commutating e.m.f. from some external source. Further, a machine cannot
pass from a motor to a generator mode unless each 'motor' diode is
parallelled by a 'generator' (reversed) diode, and selective switching
provided. A major practical difficulty is shown by the transfer of brush B
from an idle sector in (*i*) to an active sector in (*ii*), at which uneven brush
pressure, aggravated by the commutating e.m.f., may cause destructive
sparking. Some amelioration is gained by substituting carbon-fibre brushes.

### Thyristor-assisted

The entering-edge sparking can be avoided by replacing diodes by thyristors,
the triggering of which is delayed until a brush is in *full* contact with an active
sector. By itself this prevents flow of the circulating current $i_c$, so each brush
is split into two, Fig. 5.17. Strictly, each part-brush should have its own
thyristor, but it is found in practice that current transfer between parts can
be entrusted to the resistance-commutation process.

*Commutation* The process is set out in Fig. 5.17 for a machine in which each
sector width is equal to the span of a double brush, with straight-line
commutation assumed. The part brushes are designated A1 and A2, B1 and
B2. Four successive positions (*abcd*) of a coil with respect to the brushes are
shown:

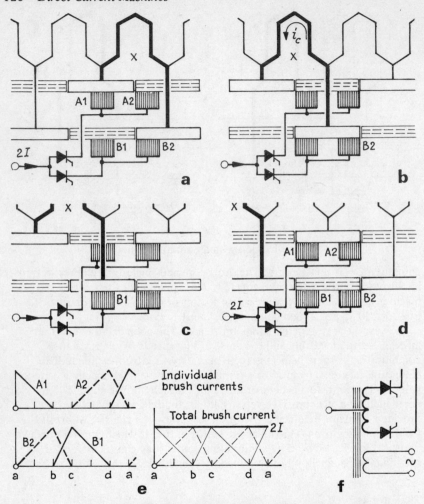

5.17 Thyristor-assisted commutation

(*a*) B2 has been previously blocked to prevent sparking as the commutator moves to present to it an active sector. A2 and B1 are on idle sectors. The whole current $2I$ flows through A1.

(*a-b*) During progress to position (*b*), B2 is triggered to conduct, coil X is short-circuited through A1 and B2, and a circulating current $i_c$ generated by the compole flux reduces $I_{a1}$ to zero and raises $I_{b2}$ to $2I$. Some current may flow through B1 as the active sector begins to cover it, but the parallel brush B2 makes sparking unlikely.

(*b*) All current flows through B2.

(*b-c*) A2 is blocked, and current transfers from B2 to B1 by resistance commutation.

(*c*) All current flows through B1.

(*c-d*) The compole reduces $I_{b_1}$ to zero and raises $I_{a_2}$ to $2I$.

(*d*) All current flows through A2. A1 is triggered.

(*d-a*) Current is transferred from A2 to A1, which was triggered at position (*d*).

The individual brush currents, and their sum (which is a constant $2I$) are shown in Fig. 5.17(*e*).

The commutating e.m.f. must exceed the inductive e.m.f. in a coil, but with assisted commutation its precise magnitude and waveform are less critical. The compole m.m.f. component producing the commutating flux need not vary with the armature current: indeed, a constant flux is advantageous for low speeds and loads, and for transients. However, the commutating e.m.f. vanishes at starting, calling for some external aid. The method in Fig. 5.17(*f*) is to inject an e.m.f. by transformer action from an available a.c. supply at mains frequency, provided that the mean injected voltage is sufficient to effect commutation within one half-period and that three half-periods are less than the time required for commutation.

Thyristor-assisted commutation for industrial machines is rare, although experimental equipments have been described, e.g. by Bates and Stanway [16]. Not unexpectedly, characteristic surface markings have been observed on commutators, due possibly to erosion of contact areas by transient currents at the end of a commutating period. Differences of coil inductance might be the cause. The use of carbon-fibre brushes is a palliative, but more effective is the fitting of a suppressor circuit to each thyristor.

## 5.10 BRUSHLESS COMMUTATION

Thyristors can be arranged to eliminate the conventional commutator entirely, allowing the machine to be 'inverted' by placing the field windings on the rotor and by controlled switching of the currents in a stationary armature winding, using sensors responding to rotor position to maintain a unidirectional torque. In the interests of economy, the number of switches is substantially fewer than the number of sectors that would be chosen for conventional commutation.

Consider the arrangement in Fig. 5.18. Let thyristors 1 and 7 be carrying $2I$ and all others be blocked. As the compoles approach the locations of thyristors 2 and 8, the latter are triggered, so that coils B and G are short-circuited. The circulating currents generated by the compoles then reduce $i_1$ and $i_7$ in thyristors 1 and 7 to zero, when they turn off, and raise $i_2$ and $i_8$ to $2I$.

Brushless electronic commutation has not been exploited for industrial machines, but it can be advantageous for *small* motors. It avoids sparking and consequent radio interference, eliminates brush wear and commutator maintenance, and is free from the difficulties that can arise in a conventional commutator from dirt, moisture and low atmospheric pressure. It allows more

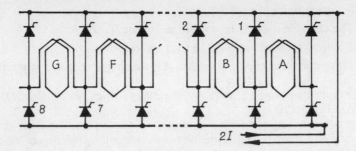

5.18 Brushless commutation

freedom in the choice of design parameters and maximum speeds, especially in closed-loop control systems. The field can be substantially simplified by omitting compoles and adopting a permanent-magnet main-pole construction.

### Position sensors

These determine the appropriate switching instants. The most common are magnetic, optical or Hall-effect devices.

*Magnetic* Pick-up coils on the stator are energized by a flux produced by main or auxiliary rotor poles in accordance with the rotor position. The method is capable of operating the switching without the use of amplifiers, but the flux distribution and the pick-up coil inductance limit the build-up rate of the position signals.

*Optical* A lamp with a shutter mounted on the rotor shaft is used to illuminate photo-cell sensors. The device has small size and low mass, and the response is fast.

*Hall-effect* Hall cells respond to the rotor poles, driving a direct-voltage output corresponding to the magnetic polarity. The signal build-up time is retarded by leakage flux, and an auxiliary supply is required for the cell current.

Magnetic sensing is employed for the upper ratings. Optical devices are advantageous for aerospace motors. Hall generators, which are small and relatively cheap, are commonly chosen for the low-power range.

### Switching

The requirements include adequate current-handling capability and the need, in some cases, for action in both motor and generator modes. Transistors are used to switch very small machines, typical arrangements being shown in Fig. 5.19: (*a*) refers to unidirectional, (*b*) and (*c*) to bi-directional operation. A network for a small machine with four stator pole windings ('phases') and a d.c. supply, Fig. 5.20(*a*), has two Hall sensors H carrying a constant current

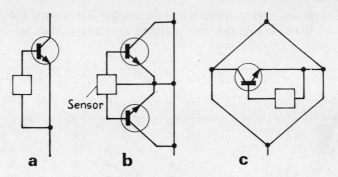

5.19 Transistor switching

and responsive to the rotor pole fluxes. The stator current pattern remains constant (but shifts with rotor position), giving relatively smooth running. But power loss occurs because the transistors Tr do not operate in a purely on-off mode, and the stator windings are under-utilized. A better winding utilization is afforded by the network (*b*), in which a stepped voltage is

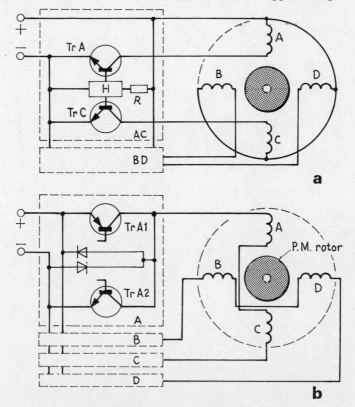

5.20 Transistor-commutated four-phase motors

applied to the windings to reduce harmonic content and improve the efficiency. The table gives the 'on' conditions for the transistors for intervals of one-eighth of a revolution.

| Rotor position | Ta1 | Ta2 | Tb1 | Tb2 | Tc1 | Tc2 | Td1 | Td2 |
|---|---|---|---|---|---|---|---|---|
| 1 |   | ● | ● |   | ● |   |   | ● |
| 2 |   | ● | ● | ● | ● |   | ● | ● |
| 3 |   | ● |   | ● | ● |   | ● |   |
| 4 | ● | ● |   | ● | ● | ● | ● |   |
| 5 | ● |   |   | ● |   | ● | ● |   |
| 6 | ● |   | ● | ● |   | ● | ● | ● |
| 7 | ● |   | ● |   |   | ● |   | ● |
| 8 | ● | ● | ● |   | ● | ● |   | ● |

# 6 Steady-State Performance

## 6.1 APPROACHES

A conventional heteropolar d.c. machine can operate in both generator and motor modes, and in most cases can change from one mode to the other automatically. It is, however, convenient to consider the modes separately, because for a generator the main interest is the variation of the terminal voltage with load and speed, while for a motor the characteristic features are the relation of torque to speed and the adaptability to various forms of mechanical load.

In general, the stator may carry two d-axis windings (shunt-field and series-field) and two q-axis windings (compensating and compole). The rotor has a single winding fed through brushes to give its m.m.f. a q-axis orientation. In *cross-field* machines the rotor has a second pair of brushes, set in the d-axis. It is the choice of interconnection between these several windings that gives the d.c. machine its wide range of performance characteristics.

A partially physical approach to steady-state performance is preferred to the purely analytical because it enables magnetic saturation and other significant practical effects to be taken into due account. The basis is the *magnetization characteristic* relating the useful flux to the field m.m.f. that produces it, Fig. 3.14. If the m.m.f. is produced by a single field winding carrying a current $I_f$, and if the machine is driven at a constant speed, the q-axis armature rotational e.m.f. $E_r$ is directly related to the d-axis gap flux, so that the magnetization characteristic can be drawn in terms of $E_r$ and $I_f$. In the absence of saturation, the $E_r/I_f$ relation is the straight 'air line' corresponding to the airgap reluctance. At higher levels of exciting current, magnetic-circuit saturation absorbs an increasing proportion of the total field m.m.f., and in consequence the mean gap flux density in industrial machines is limited to about 0.8 T and the density under the pole shoes to 1.1 T. In the armature teeth the flux density may then approach 2 T.

On load, *armature reaction* (in the absence of a compensating winding) gives the effect illustrated in Fig. 4.11, distorting the gap flux distribution. With unsaturated poles there is no change in the *total* gap flux. But this condition is rare, and under normal conditions of saturation there is, at full-load armature current, a reduction typically of 4–8% in the total gap flux. The effect is nonlinear, and difficult to include in calculations of performance other than by empirical methods based on tests.

## 6.2 PERFORMANCE

The relevant performance equations for a $2p$-pole cylindrical heteropolar machine, electrically and magnetically symmetrical and with $N_a$ armature turns connected in $2a$ parallel paths by the application of q-axis brushes, are

Armature terminal voltage   $V_a = E_r \pm I_a r_a$

Rotational e.m.f.   $E_r = [(N_a/\pi)(p/a)]\, \omega_r \Phi = K \omega_r \Phi$

Electrical torque   $M_e = K I_a \Phi$   (6.1)

Speed   $\omega_r = E_r/K\Phi$

Converted power   $P = E_r I_a = M_e \omega_r$

The $+$ and $-$ signs in the first equation refer respectively to motor and generator modes.

*Armature Reaction*  Consider a generator driven at a constant speed $\omega_r$ and excited to produce a constant armature voltage $V_a$. On no load (zero armature current) the field current is $I_{f0}$ as indicated in Fig. 6.1. Curve A is a plot of the field current *calculated* from the open-circuit magnetization

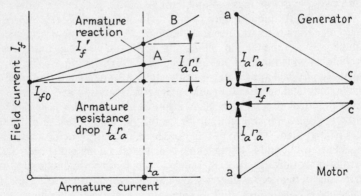

6.1 Armature volt-drop and reaction effects

characteristic for a range of armature currents $I_a$, using the relation $E_r = V_a + I_a r_a$; curve B is the *observed* field current for the same range. The ordinates between A and B thus give, for any $I_a$, the armature-reaction effect expressed as equivalent to a reduction $I_f'$ in the field current. Both the resistance-drop and armature-reaction effects could be taken simultaneously into account by assuming the actual armature resistance $r_a$ to be increased to $r_a'$. but $r_a'$ will be a function of both $I_a$ and $I_f$.

*Armature Triangle*  Resistance and reaction can alternatively be combined in an 'armature triangle', shown for generator and motor modes in Fig. 6.1. The vertical side ab represents the volt drop (or rise) $I_a r_a$; while cb gives the reaction demagnetizing effect in terms of $I_f'$, which is again dependent on the magnetic-circuit conditions as determined by $I_f$ and $I_a$. Provided that appropriate values can be chosen, the armature-triangle and equivalent-resistance methods should give the same results.

*Load Characteristic* Evaluation can be graphical using the armature triangle, or numerical using the adjusted armature resistance. The accuracy is limited, but normally adequate.

## 6.3 GENERATORS

Although direct-current supplies are commonly derived from a.c. supplies by rectification, d.c. generators are built with outputs of a few watts up to several megawatts for applications such as the following.

(1) Electrochemical plant for electrodeposition and metal refining.
(2) Diesel-electric locomotives, large rolling mills and bucket elevators where individual d.c. motors require to give special output characteristics.
(3) Battery charging for standby or emergency supply.
(4) Ship and train auxiliaries, isolated experimental stations and aircraft.
(5) Testing plants.
(6) Variable-speed wind- or water-turbine generation.
(7) Synchronous-machine excitation.
(8) Automatic control systems.

In most of these fields of application, the d.c. generator may suffer substantial competition from solid-state rectifier alternatives.

## 6.4 GENERATOR: SEPARATE EXCITATION

Permanent-magnetic and current-excited forms are shown in Fig. 2.11. With the latter, a wide range of control features is obtainable. The field winding normally has many small-section turns per pole and a current that does not exceed a small fraction of the full-load armature current. Economical variation of the field current is obtained by use of a series rheostat or, if variation down to zero is desired, by a voltage divider or controlled rectifier. The machine equations for the steady state are

$$V_f = I_f r_f \quad V_a = K\omega_r \Phi - I_a r_a \tag{6.2}$$

where $\Phi$ is a nonlinear function of the field current. Reversal of the field current or of the direction of rotation will reverse the polarity. With zero field current there is a small generated voltage resulting from residual magnetism.

*Load Characteristics* Given the open-circuit or magnetization curve for constant speed, Fig. 6.2(a), the load characteristic can be found. A vertical line is drawn from the base line corresponding to a specified field current $I_f$ to cut the magnetization curve, and the triangle for a given

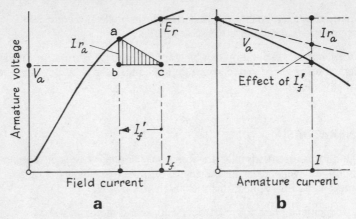

6.2 Separately excited generator: characteristics

armature current $I_a$ is located, thus determining the terminal voltage $V_a$. The process repeated for a range of load currents yields the load characteristic ($b$), which shows an armature terminal voltage that falls (typically 5–10%) between no-load and full-load operation. It can be inferred that on short circuit the armature current is likely to be high.

*Amplification*   In that a small variation of the power input to the field winding can result in a large variation of the armature power output, the separately excited generator can be regarded as a rotating amplifier. The field voltage is $V_f = I_f r_f$. The rotational e.m.f. of the armature is $E_r = K\omega_r\Phi$ from eq. (6.1), and if $K\Phi$ is expressed as $GI_f$, where $G$ is a constant if saturation is negligible (but not otherwise), then the voltage and power gains are

$$m_v = E_r/V_f = G\omega_r/r_f \quad m_p = m_v(I_a/I_f) = G\omega_r I_a/I_f r_f$$

For high gains the field resistance should be low; but this will increase the field time-constant $L_f/r_f$ and make the response sluggish, a drawback that can be overcome by more elaborate arrangements, such as the amplidyne.

## 6.5 GENERATOR: SHUNT EXCITATION

The machine is self-excited by connecting the field circuit across the armature terminals, Fig. 6.3($a$). The equations for a machine with an armature voltage and current $V_a$ and $-I_a$, a field voltage and current $V_a$ and $I_f$, and an input current $-I$ (i.e. an *output* current $+I$) are

$$V = V_a = V_f = I_f r_f \quad -I = -I_a + I_f$$

$$V_a = E_r - I_a r_a = K\omega_r\Phi - (I + I_f)r_a$$

(6.3)

There are three characteristic operating features: (i) the self-excitation

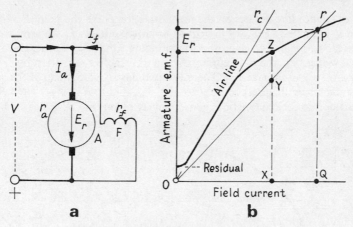

6.3 Shunt generator: self-excitation

process, (ii) the dependence of stability on saturation, and (iii) the voltage drop on load due to armature resistance and reaction, its effect on the field current and the self-limited short-circuit current.

*Self-excitation Build-up* Consider an unloaded machine driven at a constant speed, with the field circuit open. There is no field current and consequently no rotational armature e.m.f. If the field circuit is now closed on to the armature terminals, no exciting current will flow in it and the armature voltage remains zero. Self-excitation depends on the presence of *residual* magnetism in the magnetic circuit, and when the machine is run up for the first time, it must be separately excited to establish residual flux of the appropriate polarity. The constant-speed open-circuit magnetization characteristic, Fig. 6.3($b$), relates $E_r$ to $I_f$. The line OP is the voltage/current relation for a resistance $r = r_a + r_f$. For a field current OX the rotational e.m.f. is XZ, which exceeds the voltage XY absorbed by the field-circuit resistance $r$. The remaining voltage YZ is available to increase the field current at a rate determined by the field-current inductance. Over the whole of the magnetization curve between O and P, the generated e.m.f. exceeds $I_f r$ so that build-up to $I_f = $ OQ is possible. The condition then stabilizes at the no-load operating point P.

If the field-circuit resistance line has a slope greater than that of the air line, build-up is inhibited. The air line for a given speed has, in terms of resistance, a slope corresponding to a *critical* resistance $r_c$. For $r > r_c$ build-up is not possible, while for $r < r_c$ the excitation can increase at a rate depending on the voltage differences YZ and the inductance. As $r_c$ is directly proportional to speed, a machine run up from rest with the field circuit closed will not begin to excite until a speed is reached for which the appropriate $r_c$ is equal to the field-circuit resistance. It is, of course, essential that the field current generated by rotation in the remanent flux does not oppose that flux.

*Load Characteristic* The graphical derivation is given in Fig. 6.4. The field-circuit resistance line defines on the magnetization curve the point P and the no-load armature e.m.f. $E_{r0}$. For a given armature current $I_a$ the armature triangle can be located to give the corresponding e.m.f. $E_r$, the armature terminal voltage $V_a$, the field current $I_f$ and its effective reduction by $I_{f'}$ as the result of armature reaction. There are two possible locations of the triangle, corresponding respectively to $E_{r1}$, $V_{a1}$ and to $E_{r2}$, $V_{a2}$. The voltage/armature-current characteristic is consequently re-entrant, as indicated in (*b*).

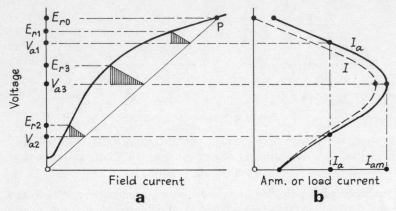

6.4 Shunt generator: load characteristics

The maximum armature current $I_{am}$ occurs for the largest armature triangle, for the condition $E_{r3}$, $V_{a3}$. The short-circuit current is not zero because of the residual flux. The full-line characteristic in (*b*) is for the armature current $I_a$: the dotted characteristic is for the output (load) current $I$, obtained by subtracting from $I_a$ the appropriate value of the field current $I_f$.

A numerical derivation for the load characteristic is based on the effective armature resistance $r_{a'}$ to incorporate the effect of the armature reaction, and the field m.m.f. per pole $F_f = N_f I_f$. The terminal voltage is $V = V_f = V_a$, with $V_f = I_f r_f = (F_f/N_f) r_f$ and $V_a = E_r - (I + I_f)r_{a'}$, whence

Field current     $I_f = (E_r - Ir_{a'})/(r_f + r_{a'})$

Output current    $I = \left[ E_r \dfrac{N_f}{r_f + r_{a'}} - F_f \right] \dfrac{r_f + r_{a'}}{N_f r_{a'}}$     (6.4)

Terminal voltage     $V = (E_r - Ir_{a'}) \, r_f/(r_f + r_{a'})$

Assuming that $r_{a'} \ll r_f$ (which may not be valid if armature reaction is prominent), eqs.(6.4) reduce to

Output current     $I = [E_r(N_f/r_f) - F_f] \, r_f/N_f r_{a'}$

Terminal voltage     $V = E_r - Ir_{a'}$     (6.5)

$E_r$ corresponding to a selected $F_f$ is obtained from the magnetization curve, whence the output quantities $V$ and $I$ are calculated.

EXAMPLE 6.1: A 17.5 kW 110 V shunt generator has a field-circuit resistance of 100 $\Omega$ and 1000 field turns per pole. The effective armature resistance is $r_a' = 0.10\ \Omega$. The magnetization characteristic at rated speed is given in Fig. 6.5($a$). Determine the critical field-circuit resistance $r_c$ and the load characteristic $V/I$.

*Critical Resistance* The tangent to the linear part of the magnetization curve (i.e., the air line) gives $r_c = 190\ \Omega$.

*Load Characteristic* Inserting values in eqs. (6.5), there result $I = (10E_r - F_f)$ and $V = E_r - 0.1I$. The following table is then drawn up for a series of values of $F_f$ and $E_r$.

| $F_f$ | (A-t): | 1330 | 1100 | 900 | 700 | 600 | 500 | 300 | 100 |
|-------|--------|------|------|-----|-----|-----|-----|-----|-----|
| $E_r$ | (V): | 133 | 126 | 119 | 108 | 100 | 88 | 57 | 20 |
| $I$ | (A): | 0 | 160 | 290 | 380 | 400 | 380 | 270 | 100 |
| $V$ | (V): | 133 | 110 | 90 | 70 | 60 | 50 | 30 | 10 |

The output voltage/current characteristic is plotted in Fig. 6.5($b$). The full-load point is $V = 110$ V, $I = 160$ A. Beyond the maximum-current point the operating conditions are unstable.

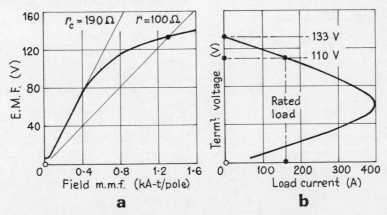

6.5 Shunt generator: load characteristics: Example 6.1

Reducing the field current of a shunt generator by means of a rheostat in the field circuit enables a family of load characteristics, Fig. 6.6, to be obtained, but reduction by more than about 20% is impracticable. Nevertheless, the simplicity of the machine makes it useful for applications that require a limited voltage range, economically controllable. For closely constant voltages a simple form of feedback control can be applied.

*Tuned Generator* A shunt machine cannot generate a stable output voltage when an attempt is made to operate it on the linear part of the magnetization curve. But it can do so if its field-circuit resistance $r_f$ is slightly *greater* than the critical $r_c$ and a control-field winding K is provided on the stator to add a

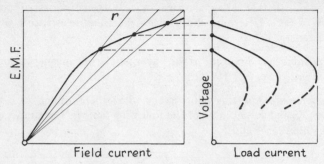

6.6 Shunt generator: field control

constant (but adjustable) m.m.f. $F_k$, Fig. 6.7. The operating point P is now located by the intersection of the shifted line of resultant m.m.f. and the magnetization characteristic. Let the control field K have $N_k$ turns/pole, a resistance $r_k$ and a current $I_k = V_k/r_k$ to give an m.m.f. $F_k = N_k I_k$. From Fig. 6.7, the point P is given by $E = I_f r_c$ and also by $E = (I_f - k I_k)r_f$, where $k = N_k/N_f$. The two expressions lead to $I_k = I_f(r_f - r_c)/kr_f$. Then the ratio of

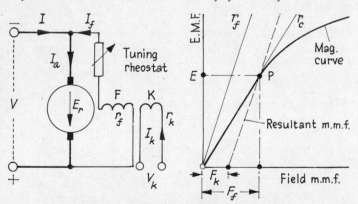

6.7 Tuned generator

the armature and control-field voltages, i.e. the voltage gain, is

$$\frac{E}{V_k} = \frac{I_f r_c}{I_k r_k} = \frac{k r_f r_c}{r_k(r_f - r_c)}$$

For a high gain, the main field-circuit resistance $r_f$ must be close to the critical resistance $r_c$. Reversal of the control current will reverse the armature voltage; thus a large change in output voltage is achieved with a small change in control voltage. The arrangement gives a simple and effective high-power amplifier (e.g. single-stage Rototrol or Magnavolt), albeit with a rather slow response.

## 6.6 GENERATOR: SERIES EXCITATION

The field and armature windings are connected in series, Fig. 6.8. The field winding has comparatively few turns, and its resistance $r_s$ is comparable with

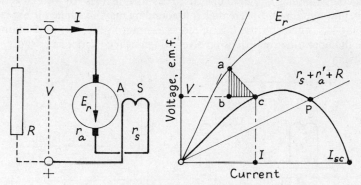

6.8 Series generator: load characteristics

that of the armature, $r_a$. The circuit is not completed until the generator terminals are closed through a load resistance $R$, and self-excitation is possible only if the slope of the line representing $(r_s + r_a + R)$ is less than the slope of the air line of the magnetization characteristic for the given speed of drive.

*Load Characteristic* This can be obtained graphically from the e.m.f./field-current magnetization curve by locating the armature triangle for a given current $I$. The vertical side $ab$ is relatively large as it includes both $r_s$ and $r_a$. Alternatively, the output voltage $V = E_r - I(r_s + r_a')$ can be employed, substituting for $r_a$ the equivalent resistance $r_a'$ that includes an approximation to the armature-reaction effect. Drawing the total resistance line for $(r_s + r_a' + R)$ locates the operating point P for a load of resistance $R$. For low loads the terminal voltage is roughly proportional to the current, a 'booster' property useful in motor testing to provide IR drop compensation; but for a purely resistive load the conditions are unstable for the region to the left of the maximum-voltage point. A matter of importance when a series motor is to be braked regeneratively is that, for the generator mode, either its field or its armature connections must be reversed.

## 6.7 GENERATOR: COMPOUND EXCITATION

The load characteristic of a shunt generator can be modified by adding on the d-axis a winding in series with the armature, providing at rated current an m.m.f. $F_s$ not usually exceeding about 10% of the shunt-field m.m.f, $F_f$ at normal voltage. The shunt field winding can be connected, as in Fig. 6.9, either (*a*) across the output terminals ('long-shunt') or (*b*) across the armature ('short-shunt'), but there is little practical difference. The m.m.f. $F_s$ may assist $F_f$ ('cumulative compound') or oppose it ('differential compound').

Cumulative compounding, the more common arrangement, can compensate for internal volt drop in armature resistance and reaction to give an approximately constant terminal voltage over the normal range of load ('level compounding'), or even one that rises with increase of load current ('over-compounding'). Differential compounding may be applied if a steeply dropping load characteristic is required as a form of generator protection where the operating conditions involve sudden heavy overloads or short circuits.

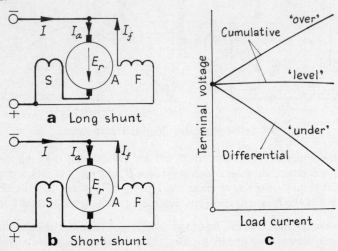

6.9 Compound generator: load characteristics

*Load Characteristic*  Self-excitation occurs as in a shunt machine. The load characteristic can be found graphically by use of the armature triangles, Fig. 6.10($a$). The vertical side $ab$ is $I_a(r_s + r_a)$ and the horizontal side $bc$ is $I_f' \pm I_a(N_s/N_f) = I_f' \pm xI_a$, where $x$ is the ratio $N_s/N_f$ of the series-field and shunt-field turns per pole. Locating the triangle, as in ($b$) for the cumulative case, between the magnetization curve and the shunt-field resistance line enables the terminal voltage $V$ to be obtained for a range of load currents and a constant shunt field current $I_f$.

For the *long-shunt* connection, the load characteristic is found analytically by replacing $r_a$ by $r_a'$ to include approximately the armature-reaction effect. The total field excitation is $F_f \pm F_s = N_f(I_f \pm xI_s)$, with the + sign for cumulative and the − sign for differential compounding. Writing $(r_s + r_a') = r'$, then

$$V = E_r - I_a r' = E_r - Ir' - I_f r' = I_f r_f$$

whence $I_f = (E_r - Ir')/(r' + r_f)$ and $V = (E_r - Ir')r_f/(r' + r_f)$.

For simplicity, let $r' \ll r_f$: then

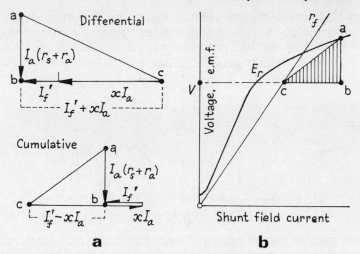

6.10 Compound generator: load characteristics

Output current $\qquad I = \dfrac{E_r[N_f(1 \pm x)/r_f] - F_f}{N_f[(1 \pm x)\,(r'/r_f) \mp x]}$ (6.6)

Terminal voltage $\qquad V = E_r - Ir'$

with the upper signs for cumulative and the lower for differential compounding. $E_r$ is found from the magnetization curve for a given $I_f$ (and therefore of $F_f = N_f I_f$). The corresponding terminal voltage $V$ and load current $I$ can then be calculated.

A similar analysis for the *short-shunt* connection, with again the assumption that $r' \ll r_f$, leads to

$$I = \frac{E_r[N_f/r_f] - F_f}{N_f[1 \mp x]} \qquad V = E_r - Ir'$$ (6.7)

In passing from a generator to a motor mode, the series field winding of a machine has its m.m.f. reversed. A cumulatively compounded generator becomes a differentially compounded motor, and vice versa.

## 6.8 GENERATOR: THREE-FIELD EXCITATION

A generator that supplies electric-welding equipments, or feeds motors likely to be stalled (as when they drive bucket excavators), has to operate under conditions of frequent short circuit. Its load characteristic should sustain the terminal voltage up to a specified normal output current but reduce it to zero as the current rises towards, say, twice normal. Although this requirement can be approximately met by strong differential compounding, it is more accurately achieved by a combination of shunt, series and separately excited field windings, Fig. 6.11. The separate winding P provides a constant m.m.f. and some degree of control. The shunt winding F augments the m.m.f. but

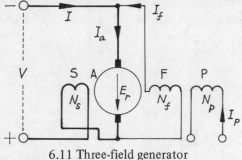

6.11 Three-field generator

has little effect on load on account of magnetic-circuit saturation; it helps to maintain the voltage at normal loads; and it is ineffective on short circuit. The series winding S is differentially connected and is designed so that the resultant field m.m.f. on short circuit is sufficient only to deal with the armature IR volt drop.

*Load Characteristic*   For the short-shunt connection in Fig. 6.11, the separately excited winding P develops an m.m.f. $F_p = N_p I_p$. The shunt and series field windings F and S have respectively $N_f$ and $N_s$ turns per pole with $N_s = x N_f$, resistances $r_f$ and $r_s$, and currents $I_f$ and $I$. The effective armature resistance is $r_a'$ and its current is $I_a = I + I_f$ when the output current is $I$. The total m.m.f. per pole is

$$F = F_p + N_f I_f - N_s I$$

For a resultant excitation that gives an open-circuit armature e.m.f. $E_r$ at constant speed, the voltage across the shunt-field winding is
$E_r - I_a r_a' = E_r - I r_a' - I_f r_a' = I_f r_f$, whence $I_f = (E_r - I r_a')/r_f$ on the assumption that $r_a' \ll r_f$. Then

$$F = F_p + N_f (I_f - xI) = F_p + N_f [(E_r - I r_a')/r_f - xI]$$

from which an expression for $I$ can be found. The terminal voltage is
$V = I_f r_f - I r_s = (E_r - I r_a') - I r_s = E_r - I(r_a' + r_s)$. The results are

Output current   $$I = \frac{E_r(N_f/r_f) + (F_p - F)}{N_f[(r_a'/r_f) + x]}$$

(6.8)

Terminal voltage   $$V = \frac{E_r N_f [x - (r_s/r_f)] - [F_p - F] (r_a' + r_s)}{N_f [(r_a'/r_f) + x]}$$

The load current and the shape of the load characteristic can be varied over a wide range by adjusting the separate excitation and the proportions of the turn ratios. The resistance of the shunt-field winding must, however, exceed the critical resistance, as otherwise the machine can self-excite and the separate excitation loses load control.

The transient response of the three-field generator is slow. In the case of certain loads, such as motors driving excavators, a slow response may be advantageous because a momentary current peak may give an impulse large enough to shift an obstruction.

EXAMPLE 6.2: A three-field generator has, when driven at 750 r/min, an open-circuit magnetization characteristic given by

| M.M.F./pole (kA-t): | 1.0 | 2.0 | 2.5 | 3.0 | 3.5 |
|---|---|---|---|---|---|
| E.M.F. (V): | 100 | 185 | 220 | 247 | 268 |

The field turns per pole are $N_p = 900$, $N_f = 2000$, $N_s = 2$. The resistances are $r_f = 220 \ \Omega$ and $r_s = 0.013 \ \Omega$. The effective armature resistance is $r_a = 0.10 \ \Omega$. The machine is driven at 750 r/min and the separate excitation is set to 1.0 A (giving $F_p = 900$ A-t/pole). Plot the load characteristic and indicate the m.m.f. contribution of each field winding.

The critical shunt-field resistance is 200 $\Omega$, so that in operation the separate excitation will not lose control of the load. The ratio $N_s/N_f = x = 0.001$. From eq. (6.8)

$$I = [900 - F + E_r(2000/220)]/2000 [(0.1/220) + 0.001]$$

$$= 3.1 E_r + 0.34(900 - F)$$

$$V = \frac{E_r 2000[0.001 - 0.013/220] - [900 - F]0.113}{2000[(0.1/220) + 0.001]}$$

$$= 0.65 E_r - 0.039 (900 - F)$$

Working with $F$ from the magnetization curve, points on the load characteristic and the m.m.f. contributions are calculated below.

| $F$ | (A-t): | 0 | 1000 | 2000 | 2500 | 3000 | 3500 |
|---|---|---|---|---|---|---|---|
| $E_r$ | (V): | 0 | 100 | 185 | 220 | 247 | 268 |
| $3.1 E_r$ | (A): | 0 | 310 | 577 | 682 | 766 | 831 |
| $0.34(900 - F)$ | (A): | 306 | −34 | −374 | −544 | −714 | −884 |
| Current $I$ | (A): | 306 | 276 | 203 | 138 | 52 | −53 |
| | | | | | | | |
| $0.65 E_r$ | (V): | 0 | 65 | 120 | 143 | 161 | 174 |
| $-0.039(900 - F)$ | (V): | −35 | 4 | 43 | 62 | 82 | 101 |
| Voltage $V$ | (V): | −35 | 69 | 163 | 205 | 243 | 275 |
| | | | | | | | |
| $I_f$ | (A): | −0.14 | 0.33 | 0.76 | 0.95 | 1.12 | 1.25 |
| $F_f = N_f I_f$ | (A-t): | −280 | 660 | 1520 | 1900 | 2240 | 2500 |
| $F_s = N_s I$ | (A-t): | −612 | −552 | −406 | −276 | −104 | 106 |

The results are plotted in Fig. 6.12.

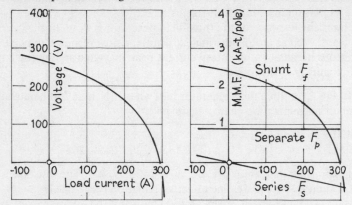

6.12 Three-field generator: characteristics: Example 6.2

## 6.9 GENERATORS: PARALLEL OPERATION

It is unusual in practice to run d.c. generators in parallel with each other, but common to parallel a generator with a battery. In railway traction a number of motors may be arranged in parallel to regenerate to the supply system. Problems of load-sharing and of stability require to be considered.

### Load sharing

The division of load between two generators, or between a generator and a battery, can be inferred from their load characteristics. Fig. 6.13 shows

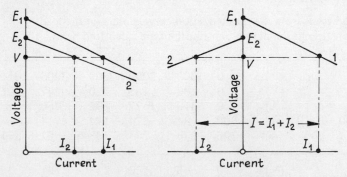

6.13 Generators: parallel operation

alternative combinations of the load characteristics of generator 1 and generator (or battery) 2 having respective no-load voltages $E_1$ and $E_2$ and linear fall of terminal voltage with increase of load current. Such characteristics can be expressed in the form $V = E - Ir$, where $r$ is an equivalent resistance. In parallel there is a common terminal voltage

$V = E_1 - I_1 r_1 = E_2 - I_2 r_2$, determining the individual contributions $I_1$ and $I_2$ to the total load current $I = I_1 + I_2$. It is readily shown that

$$V = IR \quad \text{and} \quad I = [E_1 r_2 + E_2 r_1]/[r_1 r_2 + (r_1 + r_2)R]$$

with $I_1 = (E_1 - V)/r_1$ and $I_2 = (E_2 - V)/r_2$. The steady-state operating voltage is $V$. If $V$ exceeds $E_2$, then machine 1 supplies the whole load and in addition will supply current to 2.

Let the load resistance be reduced. A rise in load current of $\Delta i = \Delta i_1 + \Delta i_2$ lowers the common terminal voltage by $\Delta v$. Then $\Delta i_1 = \Delta v/m_1$ and $\Delta i_2 = \Delta v/m_2$, where $m_1$ and $m_2$ are the (negative) slopes of the load characteristics $\Delta v/\Delta i$. The ratio of the two contributions is $\Delta i_1/\Delta i_2 = m_2/m_1$, so that a machine with a steeply falling characteristic will take a small share only of any additional load.

In the governing of prime movers it is usual for the speed to fall with increase of load, an effect that can be taken into account by adjusting the equivalent resistance $r$ or the slope $m$. Further, the e.m.f. of a machine is adjustable by varying the field excitation, making steady-state load sharing readily controllable.

## Stability

To ensure stable running in parallel, changes in load current or in field current or in prime-mover speed must produce conditions that act to counteract the disturbance. The features of importance are the speed governing, the load characteristics and the method of connection.

## Parallel connection

*Shunt or Separate Excitation* Fig. 6.13 applies to this case. A drop in speed due, say, to a faltering prime mover, will cause reduction in the load taken by the affected machine and allow the speed to rise and restore normal operation.

*Series Excitation* Fig. 6.8 shows that the series generator has a load characteristic with a rising slope. When two similar generators are connected in parallel at constant speed, a momentary increase in the current of one causes its e.m.f. to rise and therefore its current to rise still further, making stable operation impossible. Stability can, however, be achieved by either of the connections in Fig. 6.14(*a*) or (*b*). In (*a*) an equalizing connector of very low resistance parallels the field windings separately, to divide any change of current equally between them. The cross-connection (*b*) gives a more positive action in that an increase of current in one machine causes a rise of field current in the other, so that neither machine can 'snatch' the load. Practical application of series generators in parallel is substantially confined to the use of series traction motors as generators for braking. Here there is some hazard introduced by the possibility of differential wheel-slip causing a significant divergence between e.m.f.s.

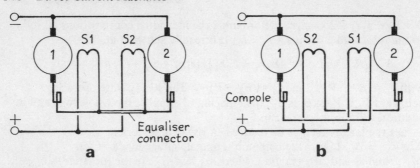

6.14 Series generators: parallel connection

*Compound Excitation*  In cumulative compounding, giving a level or potentially rising load characteristic, conditions are similar to those for series excitation. An equalizer connector, Fig. 6.14(*a*), must be employed.

## 6.10 MOTORS

The most common industrial workhorse is the a.c. cage induction motor. The compensating advantages of the more complex and more costly d.c. motor are its capability of developing a wide variety of torque/speed characteristics and its economical speed control. Electric traction, and many industrial drives, demand speed variation in both directions from zero speed to a maximum; in other cases, precise variation over a narrow range, or a speed held constant to considerable accuracy, may be required to secure optimum productivity from the driven plant. Automatic control systems make extensive use of the adaptability of the d.c. motor.

In the following, the basic steady-state running characteristics of the conventional forms of motor are discussed. The complications introduced for starting, braking, reversal and speed control are considered analytically in Chapter 7 and practically in Chapter 8.

## 6.11 MOTOR: SEPARATE AND SHUNT EXCITATION

In both of these, Fig. 6.15, the field current is independent of any action

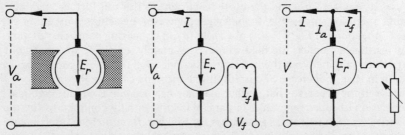

6.15 Permanent-magnet, separately excited and shunt motors

taking place in the armature. For the shunt motor, the same supply (nominally at constant voltage) is used for both armature and field circuits. Alternatively, the working flux in small and miniature machines is provided by a permanent-magnet system. With current excitation, the field current can be adjusted by a rheostat in the field circuit for speed control, but with p.m. excitation it is necessary to employ a series rheostat in the armature circuit.

Eqs. (6.1), with $I_a$ positive for the motor mode, give the relations:

Speed/current $\qquad\qquad \omega_r = (V_a - I_a r_a)/K\Phi$

Torque/current $\qquad\qquad M_e = K\Phi I_a$

$\qquad\qquad\qquad\qquad\qquad\qquad\qquad\qquad\qquad\qquad$ (6.9)

Torque/speed $\qquad\qquad M_e = K\Phi(V_a - K\Phi\omega_r)/r_a$

Converted-power/speed $\quad P = M_e\omega_r = K\Phi\omega_r(V_a - K\Phi\omega_r)/r_a$

The field current is $I_f = V_f/r_f$, and in a plain shunt motor $V_f = V_a = V$, the common terminal voltage.

If saturation could be neglected it would be possible to write $K\Phi = GI_f$ and $E_r = GI_f\omega_r = G(V_f/r_f)\omega_r$. The relations in eq. (6.9) then linearize to

Speed/current $\qquad \omega_r = (V_a - I_a r_a)r_f/GV_f$

$\qquad\qquad\qquad\qquad\qquad\qquad\qquad\qquad\qquad\qquad$ (6.10)

Torque/speed $\qquad M_e = GI_f I_a = (GV_f/r_f r_a)[V_a - G(V_f/r_f)\omega_r]$

EXAMPLE 6.3: Neglecting saturation and armature reaction, draw the steady-state performance characteristics for a 10 kW separately excited motor with an armature voltage $V_a = 100$ V, for field voltages $V_f$ of 100, 75 and 50 V. The motional inductance is $G = 0.90$ and the resistances are $r_f = 100$ Ω, $r_a = 0.10$ Ω.

Using eqs. (6.10) with $V_a = 100$ V, there results

$$\omega_r = 11.1(1000 - I_a)/V_f \quad M_e = 0.009 \, V_f I_a \quad P = 0.1 I_a(1000 - I_a)$$

The speed/current, torque/current, torque/speed and power/speed relations are plotted in Fig. 6.16. It is seen that speed falls by only a small fraction between no-load and full-load conditions, justifying the term 'constant-speed' characteristic. For a given field current, the torque is directly proportional to the armature current. The converted-power curves have a maximum value $V_a^2/4r_a = 25$ kW at one-half of no-load speed. This has little practical significance except where momentary overloads have to be sustained, as the armature $I^2 R$ loss is more than 6 times that for rated load.

EXAMPLE 6.4: For the machine of Example 6.3, draw the steady-state performance characteristics for a constant field voltage $V_f = 100$ V and armature voltages $V_a$ of 100, 75 and 50 V.

With numerical values in eqs. (6.10)

$$\omega_r = 0.11(10V_a - I_a) \quad M_e = 0.9 I_a \quad P = 0.1 I_a(10V_a - I_a)$$

The curves are given in Fig. 6.17. Armature-voltage control can give speeds down to zero. The torque depends only upon the armature current.

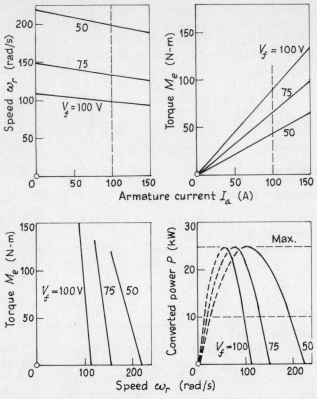

6.16 Separately excited motor: field control: Example 6.3

*Practical Characteristics* The idealized characteristics in Examples 6.3 and 6.4 are modified in practice by saturation and armature reaction effects. The armature triangle can be used with the magnetization curve, obtained at a constant test speed $\omega_m$, to predict the speed/current relation, Fig. 6.18. With an armature voltage $V_a$ and a field current $I_f$, the triangle for a chosen armature current $I_a$ is located with its apex $c$ at the point $(V_a, I_f)$, determining the value of $E_r$ and the effective field current $I_f - I_f'$. The speed is then given by $\omega_r = \omega_m(E_r/E_m)$. From constructions for a range of $I_a$ the speed/current characteristic is obtained.

The reduction of the gap flux by armature reaction reduces the torque. It also increases the speed, so counteracting the effect of the armature IR volt drop. To avoid a rise of speed with current (which might lead to instability), a shunt motor may be fitted with a weak cumulative series field winding, termed a *stabilizing winding*, to assist the shunt field and to counteract field weakening at higher loads.

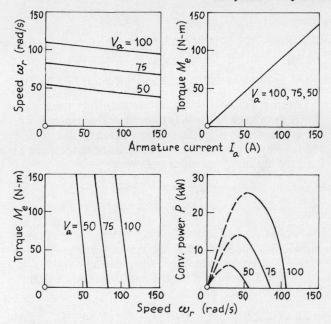

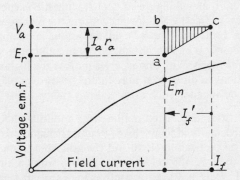

6.17 Separately excited motor: armature control: Example 6.4

6.18 Shunt motor: determination of speed/current characteristic

## 6.12 MOTOR: SERIES EXCITATION

The series motor has a wide range of application in ratings from a few watts to several hundred kilowatts, and for small 'universal' machines that can be both a.c. and d.c. fed.

Taking the terminal voltage as $V = V_s + V_a$, the current as $I = I_s = I_a$ and the total resistance as $r = r_s + r_a$ for the series-connected field winding S and armature A, eqs. (6.1) give

Speed/current                 $\omega_r = (V - Ir)/K\Phi$

Torque/current               $M_e = K\Phi I$

Torque/speed                  $M_e = K\Phi(V - K\Phi\omega_r)/r$     (6.11)

Converted-power/speed   $P = M_e\omega_r = K\Phi\omega_r(V - K\Phi\omega_r)/r$

The flux is a nonlinear function of the current, its effect being much more prominent in the series than in the shunt machine. Nevertheless, the essential features of steady-state operation can be appreciated if saturation is neglected and the flux assumed to be proportional to the current. Then $E_r = G\omega_r I$ and the linearized equations reduce to

Speed/current                 $\omega_r = (V - Ir)/GI$

Torque/current               $M_e = GI^2$               (6.12)

EXAMPLE 6.5: Neglecting saturation and armature reaction, draw the steady-state performance characteristics for a 100 V 10 kW series motor operated on terminal voltages $V$ of 100, 75 and 50 V. The resistances are $r_a = 0.15\ \Omega$, $r_s = 0.05\ \Omega$, and the motional inductance is $G = 0.01$. Eqs. (6.12) give the relations

$$\omega_r = 100(V - 0.2I)/I \quad M_e = 0.01I^2 \quad P = (V - 0.2I)I$$

which are plotted in Fig. 6.19. The torque is large at starting and at low speeds, and drops steeply with increasing speed, a characteristic particularly suited to traction and cranes. In fractional-kilowatt motors the high low-load speeds (e.g. 10 000 r/min, 1000 rad/s) may be advantageous in comparison with the maximum of 3000 or 3600 r/min obtainable from an induction motor on normal supply frequencies. With larger series motors the high no-load speed precludes applications to drives that may fail or be thrown off. A weak shunt field may have to be added in such cases to limit the no-load speed.

*Practical Characteristics*   Saturation nonlinearity modifies the idealized characteristics, particularly at heavy load. Practical characteristics can be assessed from the magnetization curve and the armature triangle as for the shunt motor, Fig. 6.18. As the field current is now variable, the apex $c$, located at point $(V, I)$, is moved to the left as load current is reduced, and the vertical side $ba$ is now $I(r_s + r_a)$. The speed for a given current is $\omega_r = \omega_m(E_r/E_m)$. At currents of full-load level and upward, the speed will be up to 20% greater than that given by the idealized condition. The torque is less than the idealized value and tends, for currents exceeding about one-half of full-load value, to become asymptotic to a straight line through the origin.

The magnetization curve, necessary for most practical treatments, is sometimes not available, but it can readily be obtained from the speed/current curve. Let the magnetization curve be required for a speed $\omega_m$: find a current $I_m$ on the speed/current characteristic at which the speed is $\omega_m$, then the e.m.f. $E_m = V - I_m r$ gives a point on the magnetization curve. Another current $I_n$ at a speed $\omega_n$ and an e.m.f. $E_n = V - I_n r$ is found; the corresponding point on the magnetization curve is $E_m = E_n(\omega_m/\omega_n)$.

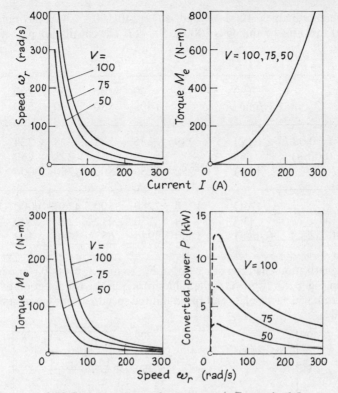

6.19 Series motor: voltage control: Example 6.5

EXAMPLE 6.6: A 240 V series motor has a field winding S with 25 turns/
pole, resistances $r_a$ = 0.05 Ω and $r_s$ = 0.03 Ω, and a speed/current
characteristic given by

| Current | (A): | 50 | 100 | 150 | 200 | 300 |
|---------|------|------|-----|-----|-----|-----|
| Speed | (r/min): | 1600 | 830 | 600 | 500 | 405 |

The armature supplied at 240 V takes 10 A (for core and mechanical losses)
when the field S is separately excited to give a speed of 1600 r/min. (*a*)
Derive the magnetization characteristic for a speed of 600 r/min. (*b*) Estimate
the number of turns per pole of an additional shunt winding F to limit the
no-load speed to 1600 r/min, and the speed/current relation that results. The
circuit resistance of the winding F is 100 Ω.

(*b*) *Magnetization Characteristic* The rotational e.m.f. $E_r$ = 240 − 0.08$I$ is
found for each current $I$ and speed $n$. Then the e.m.f. for 600 r/min is
calculated from $E_m = E_r (600/n)$.

(*b*) *Speed Limitation* As a plain series motor running at 1600 r/min the
machine requires a field m.m.f./pole of $F_s = N_s I$ = 50 × 25 = 1250 A-t, but
on no load $F_s$ = 10 × 25 = 250 A-t. The shunt field m.m.f. $F_f$ must supply

the difference, namely 1000 A-t, and as $I_f$ = 240/100 = 2.4 A, the number of shunt turns/pole required is 1000/2.4 = 417. The calculation proceeds as follows:

| Current $I$ | (A): | 10 | 50 | 100 | 150 | 200 | 300 |
|---|---|---|---|---|---|---|---|
| Speed $n$ | (r/min): | | 1600 | 830 | 600 | 500 | 405 |
| $E_r = 240 - 0.08\,I$ | (V): | 239 | 236 | 232 | 228 | 224 | 216 |
| $E_m = E_r(600/n)$ | (V): | | 89 | 168 | 228 | 269 | 320 |
| $F_s = N_s I$ | (A-t): | 250 | 1250 | 2500 | 3750 | 5000 | 7500 |
| $F_s + F_f$ | (A-t): | 1250 | 2250 | 3500 | 4750 | 6000 | 8500 |
| $E_m$ | (V): | 89 | 150 | 212 | 260 | 290 | 330 |
| $n = 600(E_r/E_m)$ | (r/min): | 1600 | 944 | 657 | 526 | 463 | 393 |

The magnetization characteristic relating $E_m$ to m.m.f. for 600 r/min is shown in Fig. 6.20. It provides the values of $E_m$ in the third section of the Table, from which the plain and speed-limited speed/current curves are obtained.

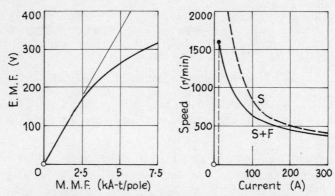

6.20 Series motor: characteristics: Example 6.6

## 6.13 MOTOR: COMPOUND EXCITATION

The machine in Example 6.6 is compounded in that it has both series and shunt field windings, but it retains the substantially 'inverse-speed' characteristic of a series motor. The normal compound motor has shunt characteristics modified by a comparatively weak cumulative or differential series winding, Fig. 6.21.

*Differential Compounding* With increase of load, the series winding reduces the gap flux and causes the speed to rise above that of a plain shunt machine, an effect that can be used to counter the speed drop caused by armature resistance. An approximately constant speed (*level compounding*) can be obtained over the range of normal load, but on overload the enhanced effect of the series winding will cause the speed to rise, and instability may occur. Further, at starting the slower build-up of the shunt field may allow the series winding to rotate the armature in the wrong direction.

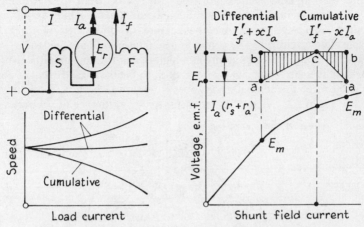

6.21 Compound motor: determination of characteristics

*Cumulative Compounding* The augmented fall of speed with load makes the machine useful for goods lifts, hoists, certain machine-tools, and the drive of impulsive and fluctuating loads that can be equalized by the kinetic energy stored in a flywheel mounted on the shaft.

*Practical Characteristics* The magnetization curve at speed $\omega_m$ and the armature triangle are employed in the manner indicated in Fig. 6.21 to find the e.m.f. $E_r$. The speed for the chosen current is then $\omega_r = \omega_m(E_r/E_m)$.

## 6.14 MOTORS: PARALLEL AND SERIES OPERATION

*Parallel* Load sharing with shunt motors connected electrically in parallel can be treated as for parallelled shunt generators, Sect. 6.9, and there will be comparable restrictions as to satisfactory division of the load. Series motors, with their steeply falling speed/current characteristics, share the load satisfactorily, even if their characteristics are slightly different. Series motors operating on a common load are widely used in railway traction service and can accommodate the small and inevitable differences in wheel diameter which cause their speeds to differ.

*Series* Motors in series share the load in proportion to their rotational e.m.f.s. Virtually the only common application is to series traction motors.

## 6.15 CROSS-FIELD MACHINES

The full potentiality of the d.c. machine can be realized by the logical extension of the d-axis field and q-axis armature to a form in which there are windings on both axes of the stator, and brush-pairs on the rotor forming both q- and d-axis m.m.f. directions. Such a *cross-field* machine uses armature excitation (Sect. 4.10) on both axes as features in the essential working principle. The machine can be driven as a generator with special load characteristics, or more commonly as a high-power rotating amplifier.

To limit the gap reluctance on both axes, to facilitate the fitting of compoles and to free the commutating zones from unwanted flux, the poles are split as in Fig. 6.22, giving a 2-pole machine the appearance of a 4-pole

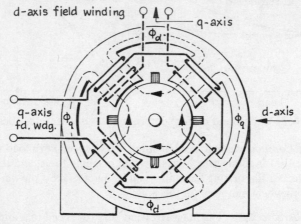

6.22 Cross-field machine

structure. Field windings provide for d- and q-axis fluxes. Airgaps are as small as is mechanically practicable to attain adequate flux from the armature m.m.f.s, and both stator and rotor are laminated in order to avoid eddy-current damping of the windings and secure more rapid response.

Superposition of the two axis currents in the single armature winding limits the rotor rating to about 0.9 of that for the same winding in a conventional machine, and the spaces to accommodate the compoles reduce the effective polar area to about 0.6–0.7 of that for a normal machine in the same frame. The output rating is therefore 0.5–0.6 of the normal capability of a given frame, so that cross-field machines are not usually built for outputs exceeding 15–20 kW.

Saturation and commutation, to which a cross-field machine is notably sensitive, affect the performance, but to include them specifically makes an analysis exceedingly complicated. As saturation levels are usually kept low, it is sufficient in a basic treatment to assume that d- and q-axis currents, m.m.f.s and fluxes can be considered independently and their results superposed.

## Metadyne transformer

The simplest cross-field machine, Fig. 6.23, has no field windings, the stator providing only a low-reluctance path for the fluxes generated by the rotor currents. The machine is driven at a constant speed $\omega_r$. Axis quantities are

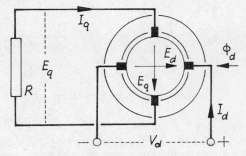

6.23 Metadyne transformer

distinguished by the subscripts d and q. Assuming steady-state conditions and at first ignoring the volt drops in armature-winding resistance, an applied voltage $V_d$ sets up a d-axis current $I_d$, the m.m.f. of which develops a d-axis flux $\Phi_d$. Rotation generates an e.m.f. $E_q$ across the q-axis brushes of value $KI_d$. If the q-axis brushes are connected to a load of resistance $R$, the output current $I_q = E_q/R$ develops an m.m.f. and a flux $\Phi_q$ in which the armature rotates to generate $E_d = KI_q$ across the d-axis brushes, opposing $V_d$ and equal to it. Thus $V_d = E_d = KI_q$ and $V_q = E_q = KI_d$. The output current is therefore $I_q = V_d/K$. This means that with a constant d-axis applied voltage $V_d$, the q-axis output current $I_q$ is a constant and independent of $R$; the machine acts as a constant-voltage/constant-current converter with $V_d I_d = V_d(V_q/K) = V_q I_q$, i.e. output power equal to input power. The input is derived from an external source, the driving motor providing no power other than mechanical loss.

Introducing the armature resistance $r_a$, then

$$V_d = G\omega_r I_q + I_d r_a \qquad V_q = G\omega_r I_d - I_q r_a$$

$$M_e = GI_d I_q - GI_q I_d = 0$$

If the load resistance is large, $V_q$ may demand a d-axis flux such as to saturate the magnetic circuit and impair the constant-current property. Nevertheless there is a useful working range, and in the past the metadyne transformer has been used for starting series railway motors to eliminate the loss in starting rheostats.

## Metadyne generator

If, as in Fig. 6.24(*a*), a d-axis field winding F is added and the q-axis brushes are short-circuited, an output is obtainable from the d-axis armature brushes. The m.m.f. $F_f$ produces a flux $\Phi_d$ that generates a rotational e.m.f. $E_q$ and,

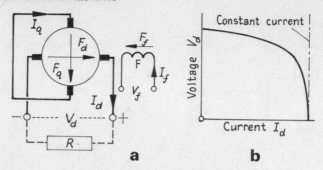

6.24 Metadyne generator

because of the short-circuit, a current $I_q = E_q/r_a$, the m.m.f. of which produces a q-axis flux $\Phi_q$. In turn, $\Phi_q$ generates an e.m.f. $E_d$ across the d-axis brushes and a current $I_d$ in a load resistance $R$ connected across them. The m.m.f. of $I_d$ opposes $F_f$. Increasing $R$ reduces $I_d$, allowing $\Phi_d$ to rise and restore the current $I_d$. For a given field m.m.f. $F_f$ therefore, conditions tend to stabilize at a constant current. But for higher load resistance, maintenance of $I_d$ demands a comparable increase in $V_d$ and the consequent flux drives the magnetic circuit into saturation. The practical characteristic, Fig. 6.24($b$), approaches a limiting voltage at zero current.

Several additional d- and q-axis windings are commonly fitted for improving the characteristics. Those indicated in Fig. 6.25($a$) are:

*d-axis*: F, main field; N, neutralizing; Sd, series.
*q-axis*: M, ampliator; Sh, shunt; Sq, series.

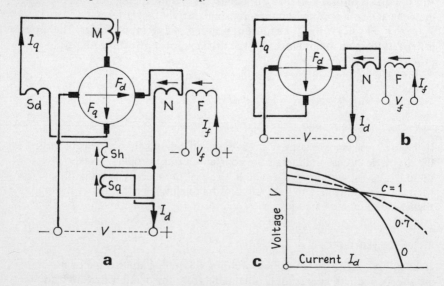

6.25 Metadyne generator

The winding N neutralizes, partly or wholly, the demagnetizing effect of the d-axis armature m.m.f., and has a major influence on the characteristics of the generator.

*Main Field Winding* F supplies the control field.

*Neutralizing Winding* Carrying the d-axis output current, N develops an m.m.f. in opposition to that of the armature winding D. The d-axis flux is therefore more dependent on the current in F, giving it a greater measure of control. The effect on the load characteristic is indicated in (*c*) in terms of the degree of neutralization $c = N_n/N_a$, the ratio of the effective turns of the windings N and D. Complete neutralization ($c = 1$) gives the machine a characteristic like that of a separately excited generator: neutralization has changed the behaviour from constant current to constant voltage, and in this form a very small current in winding F can control a large output, resulting in a high power amplification. Such a fully neutralized (or fully compensated) generator is called an *amplidyne*. With no neutralization ($c = 0$) the characteristic is that in (*c*). Partial neutralization, e.g. $c = 0.7$, gives an output voltage/current relation intermediate between the extremes.

*Ampliator Winding* The q-axis flux is developed by the current $I_q$. If produced solely by the armature winding Q, the flux is limited by the permissible armature $I^2R$ loss. The same flux can be attained for a smaller current by adding the q-axis series-connected stator winding M having a cumulative sense. If M, the ampliator winding, has the same number of effective turns as Q, then M and Q in combination will have roughly twice the resistance of Q and four times its inductance. This is of no great significance in the steady state, but as the time-constant (the inductance/resistance ratio) is doubled, the transient response is slowed.

*Shunt Winding* The q-axis winding Sh enables the characteristic to be modified. The resistance of the winding must be greater than the critical resistance associated with the q-axis to prevent self-excitation in the absence of control current in F. The winding Sh has no effect when the output terminals are short-circuited, and only minor effect on no load when the q-axis magnetic circuit is saturated. At intermediate loads it will raise the output voltage, as in the three-winding generator.

*Series Windings* Sq provides compounding. Sd in series with the q-axis brushes may be fitted to compensate for the falling brush-contact resistance with increasing current.

*Compole Windings* Commutation conditions at the d-axis brushes are similar to those in an ordinary d.c. machine so that conventional compole windings in series are required. At the q-axis brushes the conditions are easier, particularly if the current is minimized by fitting windings M and Sq. Except in large machines it may be possible to dispense with q-axis compoles.

## Analysis

The basic characteristics of the metadyne generator can be derived by an analysis of the effects of windings, F, N, D and Q only, with the assumption of a linearized (unsaturated) magnetic circuit. The system considered is that in Fig. 6.25($b$). The applied voltage equation is written for each of the three circuits for the steady state, with $V_f$ applied to the main field circuit, zero to the q-axis circuit (because it is short-circuited) and the output terminal voltage $V$ to the d-axis circuit. As $I_d$ is an output current, its input equivalent is $-I_d$. The m.m.f. of the neutralizing winding N opposes that of the armature winding D. The resistances of the D and Q armature windings are each $r_a$. Motional inductances such as $G_{xy}$ refer to the e.m.f. induced in winding X due to rotation in the flux produced by winding Y. The voltage equations are then

Field circuit $\qquad\qquad V_f = r_f I_f$

Closed q-axis circuit $\quad 0 = r_a I_q + G_{qf}\omega_r I_f + G_{qd}\omega_r I_d + G_{qn}\omega_r(-I_d)$

Output d-axis circuit $\quad V = (r_a + r_n)I_d - G_{dq}\omega_r I_q$

In the q-axis equation, the neutralization coefficient $c = N_n/N_a = G_{qn}/G_{qd}$, so that the last two terms combine to $G_{qd}\omega_r I_d(1 - c)$, and thence

$$I_q = -\omega_r[G_{qf}I_f + G_{qd}I_d(1 - c)]/r_a$$

Substituting this in the output voltage equation gives

$$V = (r_a + r_n)I_d + G_{dq}\omega_r^2[G_{qf}I_f + G_{qd}I_d(1 - c)]/r_a$$

In terms of output current

$$V = \frac{G_{dq}G_{qf}\omega_r^2}{r_a}I_f - \left[\frac{G_{dq}^2\omega_r^2(1-c)}{r_a} + (r_a + r_n)\right]I_d$$

$$= KI_f - [K(G_{dq}/G_{qf})(1 - c) + (r_a + r_n)]I_d \qquad\qquad (6.13)$$

where $K = G_{dq}G_{qf}\omega_r^2/r_a$. With complete neutralization ($c = 1$)

$$V = KI_f - (r_a + r_n)I_d \qquad\qquad (6.14)$$

and the machine behaves like a separately excited generator, eq. (6.2), with a roughly constant *voltage*. As $\omega_r^2$ is a factor in $K$, the polarity is independent of the direction of rotation.

With winding N omitted, there is no neutralization ($c = 0$) and

$$V = KI_f - [K(G_{dq}/G_{qf}) + r_a]I_d$$

Let the load be a resistance $R$: then $V = I_d R$ from which

$$I_d = KI_f/[K(G_{dq}/G_{qf}) + R + r_a] \cong I_f(G_{qf}/G_{dq}) \qquad\qquad (6.15)$$

The output *current* is now roughly constant and independent of load resistance and speed for a given field current $I_f$. The Rosenberg generator (Sect. 12.5) was an early application of the un-neutralized metadyne generator, and was widely used as a variable-speed, axle-driven, battery-charging machine for train lighting with a polarity independent of the direction of the train.

With partial neutralization (e.g., $c = 0.7$) a characteristic intermediate between constant voltage and constant current results. Typical voltage/current characteristics are shown in Fig. 6.25(c).

## Amplidyne

This term is applied to a metadyne generator with complete or near-complete neutralization ($c \to 1$), a condition for which a very small field current can control a large output. From eq. (6.14), neglecting the internal IR volt drop, $V = KI_f = KV_f/r_f$. The voltage gain is

$$V/V_f = K/r_f = G_{dq}G_{qf}\omega_r^2/r_a r_f \tag{6.16}$$

If the speed is high, as it can be in small machines, the power gain $VI_d/(V_f^2/r_f)$ can approach $10^5$. The chief application of the amplidyne is in control systems, for which its steady-state operation is secondary to its transient response.

EXAMPLE 6.7: An amplidyne, with 95% compensation ($c = 0.95$), has a q-axis armature e.m.f. constant $G_{qf}\omega_r = 100$ V per A in the control field winding F, and a constant $G_{dq}\omega_r = 20$ V per A in the armature. The winding resistances are $r_f = 20\ \Omega$, $r_a = 0.5\ \Omega$, $r_n = 0.1\ \Omega$. For a control-field current of 0.1 A, calculate (a) the open-circuit output voltage and the short-circuit output current, (b) the output current and voltage, and the power gain, for a load resistance $R = 50\ \Omega$.
The constant $K$ in eq. (6.13) is $(G_{dq}G_{qf}\omega_r^2)/r_a = 4000$.
(a) Open-circuit voltage ($I_d = 0$): $V = KI_f = 4000 \times 0.1 = 400$ V
Short-circuit current ($V = 0$):

$$I_d = 400/[(800 \times 0.05) + 0.60] = 400/40.6 = 9.85 \text{ A}$$

(b) Output current: $I_d = 400/(40.6 + 50) = 4.42$ A
Output voltage: $V = I_d R = 221$ V. Output power, $P_o = 221 \times 4.42 = 977$ W; control power, $P_i = 20 \times 0.1^2 = 0.2$ W. Power gain

$$P_o/P_i = 977/0.2 = 4880.$$

The voltage/current characteristic is linear between (400 V, O A) and (O V, 9.85 A).

## Application

Several forms of cross-field machine were developed and widely applied in the years between 1930 and 1960 to provide high-power high-gain amplification in control systems, notably in the field of naval gunnery and military assault vehicles as well as in industrial drives. Since then the commercial scope of these electromagnetically interesting machines has declined as a result of the development of static electronic devices that can fulfil the same need.

# 7 Transient Performance

## 7.1 APPROACHES

Flux-current interaction has formed the basis for the evaluation of steady-state performance. For transient operation a more analytic approach is needed. The electromagnetic part of the machine system is represented by electric circuits with self, mutual and rotational inductance as well as resistance; and the mechanical part by friction, damping, elasticity and inertia as well as load torque. Together the two parts form an equivalent electromechanical system susceptible to the analysis of dynamic performance.

Such a model is an idealization. It assumes electric-circuit linearity with complete or partial neglect of the effects of slotting and skew, of core and stray loss (except empirically), of eddy-current and commutation phenomena, and of practical mechanical features such as constructional asymmetry, vibration, noise and critical speed. In particular, magnetic-circuit saturation cannot readily be taken into account in a linear theory: in series machines, for example, the magnetic nonlinearity is far too great, and in shunt generators the self-excitation phenomenon is a function of both saturation and remanence, with which a linear theory cannot cope. Brush contact-drop may be a nonlinearity important in low-voltage and cross-field machines, and the distribution of the current density over the contact surface of brushes may render indeterminate the precise orientation of a supposedly q-axis armature m.m.f. Further, several of the parameters, both electrical and mechanical, are subject to variation with temperature.

In the *numerical* evaluation of machine performance it is possible fairly easily to deal with several nonlinearities by a digital-computer simulation, using stepwise solution with numerical integration. But *analytical* solutions must be scrutinized for practical validity by a physical appreciation of the ways in which the linear model differs from reality. In general, the differential equations involve voltage, speed and current that may be functions of time; and products of such terms are nonlinear in the mathematical sense, necessitating step-by-step solution. Simplifying and approximating techniques may sometimes be employed to make solutions more tractable. For example, the time-constants of electromagnetic transients are often much shorter than those of the mechanical system and may be evaluated on the assumption that the speed is momentarily constant; and finite-difference equations can be set up for perturbations of a sustained cyclic character. Effects known from experience to be of minor importance may be omitted to clarify the physical implications of a more tractable analysis.

Transients occur in starting, braking, speed control, load change, switching and faults. *Industrial machines* are considered here as an introduction to motor control technology in Chapters 8 and 9. *Control machines,* analytically identical but having different responses of interest, are considered in more detail in Chapter 11.

## 7.2 MODEL

Elementary conventional d.c. machines have two basic electric circuits, F for the working flux magnetizing on the d-axis, and A for the armature magnetizing on the q-axis. The instantaneous power $p_f = v_f i_f$ into the field circuit supplies the conduction loss $i_f^2 r_f$ and the rate of change of field energy $dw_f/dt$. The armature input power $p_a = v_a i_a$ supplies the corresponding components $i_a^2 r_a$ and $dw_a/dt$, together with the conversion power $e_r i_a = M_e \omega_r$. Provided that the flux/current relations can be assumed to be linear, magnetic linkage can be expressed in terms of constant inductances $L_f$ and $L_a$, yielding the system equations

Field: $\qquad\qquad v_f = (r_f + L_f p) i_f$

Armature: $\qquad\quad v_a = (r_a + L_a p) i_a + G \omega_r i_f \qquad\qquad (7.1)$

Torque: $\qquad\quad M_e = G i_f i_a = -(M_l + M_a + M_f + M_d)$

The torque terms, itemized in Sect. 1.5, are inputs. If a machine has a permanent-magnet field, $v_f$ is omitted and the motional e.m.f. term in $v_a$ reduced to $k_e \omega_r$.

### Series windings

Most industrial machines have *compoles,* the effect of which is a small change (usually an increase) in the effective armature-circuit inductance $L_a$. Where *compensating* windings are fitted, $L_a$ is reduced. *Series field* windings in the armature circuit, used alone or to modify performance characteristics (as in compounding a generator or providing a motor with a greater starting torque), modify the system equations to

Field $\qquad\qquad v_f = (r_f + L_f p) i_f + L_{fs} p\, i_a$

Armature: $\qquad\quad v_a = (r_a + L_a p) i_a + L_{fs} p i_f + G \omega_r (i_f + x i_a) \qquad (7.2)$

Torque: $\qquad\quad M_e = G(i_f + x i_a) i_a$

where $L_{fs}$ is the mutual inductance between the main and series field windings F and S, parameters $r_a$ and $L_a$ refer to the complete armature circuit, and $x$ is the turns-ratio $N_s/N_f$ of S to F. The rotational e.m.f. and the torque include components contributed by the fluxes of F and S which, on the linear assumption, can be superposed.

**Transfer function**

Motors and generators may serve as power elements in control systems and for analysis may be represented by transfer functions (t.f.s) relating an input control quantity to an output quantity, e.g. armature voltage to speed in a motor, or armature voltage to field voltage in a generator.

*Motor* As an example, consider a motor with a constant field current $I_f$, a constant-torque load $M_l$ and a total coupled inertia $J$. Writing $GI_f = K_f$ and $(r_a + L_a p) = z_a$, and equating the electromagnetic torque $M_e$ to the output mechanical torque demand, then eq. (7.1) gives $M_e = M_l + Jp\omega_r$, whence $I_a = (M_l + Jp\omega_r)/K_f$, and as the armature voltage is $z_a i_a + K_f \omega_r$, then

$$V_a = (z_a/K_f) [M_l + J \{ (K_f^2/Jz_a) + p \} \omega_r]$$

relating speed to armature voltage, with interrelated time-constants arising from the armature inductance and the inertia. In practical cases the relation is complicated by the fact that $M_l$ is normally itself a function of speed.

Such cases can readily be solved numerically by digital computation. But their analytical form obscures the basic response, so that it is more informative to employ simplifications. If, for example, the armature inductance $L_a$ can be neglected ($z_a = r_a$) and if the system is purely inertial ($M_l = 0$), then the voltage/speed relation has the single time-constant $\tau_m = Jr_a/K_f^2$. The Laplace transform of the t.f. becomes

$$\omega_r(s)/V_a(s) = 1/[K_f \tau_m (s + 1/\tau_m)]$$

For a suddenly applied armature voltage, with the machine initially at rest, the speed rises exponentially from zero to $\omega_1 = V_a/K_f$.

*Generator* The t.f. relating the armature rotational e.m.f. $e_r$ developed at constant speed by a separately excited generator with a voltage $v_f$ applied to the field circuit of resistance $r_f$ and inductance $L_f$ is

$$E_r(s)/V_f(s) = K_f/(r_f + L_f s) = K_f/L_f(s + 1/\tau_f)$$

The field current of an a.c. generator may be derived from *cascaded* d.c. generators on the same shaft. The field voltage $v_{f_1}$ of the pilot exciter 1 is controlled; its armature supplies the field of the main exciter 2, the armature of which feeds the a.c. generator field. The *open-circuit voltage* t.f., neglecting armature inductance and assuming a constant speed, is

$$E_{r_2}(s)/V_{f_1}(s) = K_{f_1} K_{f_2}/[L_{f_1} L_{f_2} (s + 1/\tau_{f_1}) (s + 1/\tau_{f_2})]$$

with two field time-lags. The *closed-circuit current* response of the a.c. generator field, $I_{a_2}(s)/V_{f_1}(s)$, will include a further time-lag due to the large inductance of the a.c. generator field winding.

## 7.3 MOTOR: STARTING

When a motor is started from rest and run up to a final speed $\omega_1$, its electrical supply must furnish the energy $w_a = \frac{1}{2}J\omega_1^2$ stored in the system inertia $J$. The $I^2 R$ energy loss in the armature-circuit resistance, whether or not it includes an external starting rheostat, is equal to the kinetic energy stored in the acceleration process. The energy input to the armature circuit will therefore be $2w_a$ plus that expended in supplying load torque and friction. Consider for simplicity a machine with a constant field excitation $I_f$ following the application to the armature of a constant voltage $V_a$, the load being purely inertial and the only loss being that in the armature-circuit resistance $r_a$. At any speed $\omega_r$ the rotational e.m.f. is $e_r = GI_f\omega_r = K_f\omega_r$. At the final steady-state speed $\omega_1$ the armature current is zero, so that $K_f\omega_1 = V_a$, whence $V_a = e_r(\omega_1/\omega_r)$. During run-up, $i_a r_a = V_a - e_r$ and the $I^2 R$ loss is $r_a i_a^2 = (e_r/\omega_r)i_a(\omega_1 - \omega_r)$. But $(e_r/\omega_r)i_a = M_e = J(d\omega_r/dt)$, so that the total conduction energy loss is

$$\int r_a i_a^2 \cdot dt = J \int_0^{\omega_1} (\omega_1 - \omega_r)\cdot d\omega_r = \frac{1}{2}J\omega_1^2$$

which is the stored kinetic energy. A comparable result applies to a series motor. The conduction loss can be reduced by current limitation where the voltage can be applied in steps (as in the series-parallel control of a pair of traction motors) or as a smooth rise (as in Ward-Leonard control). But the acceleration time will be prolonged.

### Separately excited motor

With a constant armature voltage $V_a$ applied to start the motor from rest, the equation of motion is obtained from $e_r = K_f\omega_r$ and

$$i_a = (V_a - e_r)/(r_a + L_a p) \qquad M_e = K_f i_a = M_l + Jp\omega_r$$

with $M_l$ expressed as an appropriate function of speed. If the armature inductance is small enough to be neglected, and if the load is purely inertial, the equation of motion is

$$p\omega_r + (K_f^2/Jr_a)\,\omega_r = (K_f/Jr_a)V_a$$

which has a standard form and solves to

$$\omega_r = (V_a/K_f)\,[1 - \exp(-K_f^2/Jr_a)\,t]$$

If in addition to inertia there is a *constant* load torque $M_l$, the solution is the same except that the factor $V_a/K_f$ is replaced by $[(V_a/K_f) - M_l \cdot r_a/K_f^2]$. In either case the run-up speed/time relation has an exponential form.

EXAMPLE 7.1: A separately excited motor with a field current $I_f = 1.5$ A develops an open-circuit armature voltage of 250 V when driven at 200 rad/s.

The armature resistance is $r_a = 1.0\ \Omega$ and its inductance is negligible. The motor drives a load of constant torque $M_l = 37.5$ N·m and the coupled inertia is $J = 2.50$ kg·m$^2$. Calculate the speed and current resulting from the application to the armature at rest (i) of a constant voltage $V_a = 240$ V, (ii) of a voltage $\frac{1}{2} V_a = 120$ V, increased suddenly to $V_a = 240$ V after 3.0 sec. Obtain the power/time relations for each case.

(i) The required parameters are

$$GI_f = K_f = 250/200 = 1.25 \text{ V per rad/s}$$

$$(V_a/K_f) - M_l \cdot r_a/K_f^2 = 192 - 24 = 168 \text{ rad/s}$$

$$K_f^2/Jr_a = 1.25^2/2.50 \times 1.0 = 0.625 \text{ s}$$

then $\omega_r = 168[1 - \exp(-0.625\,t)]$ and $i_a = 240 - 210[1 - \exp(-0.625\,t)]$.

These are plotted in Fig. 7.1(i) together with the power/time curves. The input is $p_i = V_a i_a$ and the power to drive the load is $p_o = M_l \omega_r$. The difference $p_i - p_o$ at any instant is the loss $i_a^2 r_a$ and the rate of energy supply to the rotating parts.

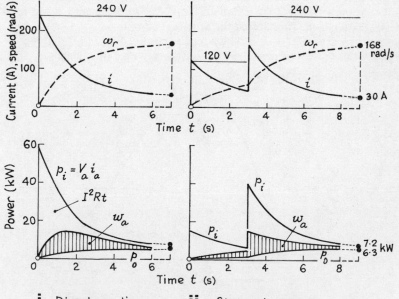

i  Direct-on-line    ii  Stepped-voltage

7.1 Separately excited motor: starting: Example 7.1

(ii) $K_f = 1.25$ as before. For the 120 V condition the limiting speed is 72 rad/s, whence $\omega_r = 72[1 - \exp(-0.625\,t)]$ and $i_a = 120 - 90[1 - \exp(-0.625\,t)]$. For $t = 3.0$ s, $\omega_r = 61$ rad/s and $e_r = 76$ V. When the armature voltage is changed to 240 V the step is equivalent to $V_a - e_r = 164$ V. Thus

$$\omega_r = 61 + 107[1 - \exp(-0.625\,t)] \text{ and } i_a = 30 + 134\exp(-0.625\,t)$$

Plots of these and the corresponding powers in Fig. 7.1(ii) show that although the run-up time is increased, a considerable reduction in peak input power and of $I^2R$ loss is achieved. In both (i) and (ii) the area of the acceleration-power/time curve is the stored kinetic energy $w_a = \frac{1}{2}J\omega_1^2 = 35.3$ kJ.

### Shunt motor

It is usual for the field and armature circuits to be switched simultaneously, and the field time-constant may be significant. In the following Example the armature inductance is abnormally large in order to emphasize the initial conditions.

EXAMPLE 7.2: The parameters of a shunt motor are as follows. Field: $r_f = 100\ \Omega$, $L_f = 1.0$ H. Armature: $r_a = 25\ \Omega$, $L_a = 0.50$ H, $G = 1.0$ V/A per rad/s. Load: $(M_f + M_d) + M_l = (\omega_r + 50)$ mN-m, $J = 10^{-4}$ kg-m$^2$. With the armature at rest, a step voltage $V = 50$ V is applied to field and armature in parallel. Derive the torque, speed and currents (i) initially, before the rotor starts to move, (ii) during acceleration, and (iii) in the final steady state.

(i) *Initial* The field time-constant is $\tau_f = L_f/r_f = 0.01$ s, and the field current is the exponential $i_f = 0.5[1 - \exp(-100\ t)]$. The rotor remains at rest until $(Gi_f)i_a = 50$ mN-m. Thus at first the armature current is $i_a = 2.0[1 - \exp(-50\ t)]$. It is shown by the plots in Fig. 7.2($a$) that the torque reaches 0.050 N-m at $t = 3.6$ ms. Thereafter rotation begins.

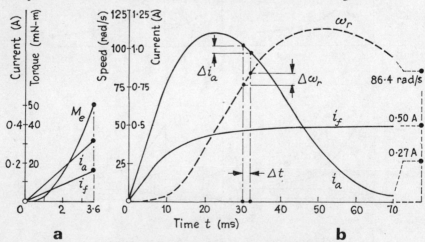

7.2 Shunt motor: starting: Example 7.2

(ii) *Acceleration* It is now necessary to solve in steps the two armature equations

$$V = (Gi_f)\omega_r + i_a r_a + L_a(\Delta i_a/\Delta t) = \omega_r i_f + 25i_a + 0.5(\Delta i_a/\Delta t)$$

$$M_e = (Gi_f)i_a = (50 + \omega_r)10^{-3} + (\Delta\omega_r/\Delta t)10^{-4}$$

together with $i_f = 0.5[1 - \exp(-100\ t)]$, for finite differences $\Delta\omega_r$, $\Delta i_a$ and $\Delta t$ of speed, armature current and time. The graphs in Fig. 7.2($b$) show the results for the first 70 ms. For example, at $t = 30$ ms the values are $i_f = 0.475$ A, $i_a = 1.04$ A, $\omega_r = 76.0$ rad/s, $M_e = 0.494$ N-m, and the load torque $(50 + 76)10^{-3} = 0.126$ N-m. Then $J(\Delta\omega_r/\Delta t) = 0.494 - 0.126 = 0.368$, and the acceleration is $\Delta\omega_r/\Delta t = 0.368 \times 10^4 = 3680$ rad/s$^2$. Applying the voltage equation

$$50 = 76 \times 0.475 + 25 \times 1.04 + 0.5(\Delta i_a/\Delta t)$$

gives $\Delta i_a/\Delta t = -24.2$ A/s. Using these values as the initial conditions for the next time interval, taken as $\Delta t = 2$ ms, the new values for $t = 32$ ms become

$i_f = 0.480$ A. $i_a = 1.04 - 24.2 \times 0.002 = 0.99$ A; $M_e = 0.480 \times 0.99$ $= 0.475$ N-m; $\omega_r = 76.0 + 3680 \times 0.002 = 83.4$ rad/s; $M_l + M_f + M_d$ $= (50 + 83.4)10^{-3} = 0.133$ N-m; $\Delta\omega_r/\Delta t = (0.475 - 0.133)10^4 = 3420$ rad/s$^2$; $V = 50 = 83.4 \times 0.48 + 25 \times 0.99 + 0.5(\Delta i_a/\Delta t)$, whence $\Delta i_a/\Delta t = -29.6$ A/s, from which the next step can be calculated.

(iii) *Steady State*  The field current is $I_f = V/r_f = 0.50$ A. For the armature, $(GI_f)\omega_r = V - r_a I_a$ and $(GI_f)I_a = M_l + M_f + M_d$, from which the steady-state condition is $I_a = 0.27$ A and $\omega_r = 86.4$ rad/s.

During run-up, Fig. 7.2, each current begins to rise at a rate $di/dt = V/L$. As the rotor accelerates and the field current approaches its steady value, the rotational e.m.f. exceeds the terminal voltage, causing the armature current to reduce and the speed to oscillate. After $t = 30$ ms the field current is almost constant and the armature current is then a measure of the electromagnetic torque. The step-by-step calculation, necessary because both speed and current are functions of time, can readily be programmed for digital computation.

## Series motor

Were flux proportional to current, we could write for a machine with resistance $r = r_a + r_f$ and inductance $L = L_a + L_f$ the equations

$$V = (r + Lp)i + G\omega_r i \qquad M_e = Gi^2$$

but in fact magnetic nonlinearity makes necessary a step-by-step calculation with due regard to the actual flux/current relation, using a digital method or the following graphical construction which can be applied to systems in which the mechanical time-constant is substantially greater than the electrical one.

*Graphical Method*  In Fig. 7.3($a$), the torque/speed *steady-state* performance characteristic $M_e$ of a series motor is plotted and reduced by the known frictional and load torques to leave the accelerating torque $M_a$. If $M_a$ is assumed to be constant over a small speed change from $\omega_p$ to $\omega_q$, then the time to accelerate between these speeds is

$$t = (J/M_a) \int_{\omega_p}^{\omega_q} d\omega_r = (\omega_q - \omega_p)/\alpha$$

where $\alpha = M_a/J$ is the acceleration. The construction shown gives $\Delta\omega_r/\Delta t = M_a/J$ for the slope of the speed/time curve $(c)$ over speed steps $\omega_p$ to $\omega_q$ using the slope of PQ in $(b)$. Pole P is placed so that $OP = J$ to the same scale that $OQ = M_a$. From the resulting speed/time curve it is possible to construct the current/time curve, using the motor steady-state characteristics.

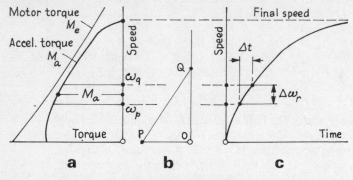

7.3 Graphical integration

## Rheostat starting

Ignoring inductance, a motor switched direct-on-line has an initial prospective current peak $V/r$. Only small high-resistance motors may be started in this way: a starter rheostat is normally connected in series with the armature to limit the initial current and to avoid undue mechanical shock to the shaft and load. The rheostat is cut out in steps as the machine runs up, the process being generally automatic. The usual method of rheostat calculation (Sect. 8.2) assumes that currents are determined only by resistance. For a shunt motor the armature voltage equation is then $V_a = K_f\omega_r + (R + r_a)i_a$, where $R$ is the resistance of the rheostat steps still in circuit. The armature current is assumed to vary between upper and lower current limits $I_1$ and $I_2$, giving the speed/current relation $(a)$ in Fig. 7.4. The presence of armature inductance and system inertia modifies this idealized behaviour to that in $(b)$. A current jump when a rheostat section is cut out can no longer be instantaneous, and inertia results in overshoot before the motor stabilizes at a point P where the electrical and load torques balance.

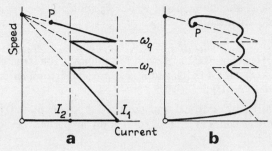

7.4 Rheostat starting

## 7.4 MOTOR: LOADING

When changes of supply voltage or field current or mechanical load are impressed on a motor in operation, the machine adjusts to the new condition through an electromechanical transient. Cases of interest include impulsive changes of load torque, load equalization, and cyclic fluctuations such as those encountered in compressors.

### Step change of load

If a constant load torque $M_l$ is suddenly applied to a shunt motor running at speed $\omega_0$ on no load, the machine has no torque to meet the demand, which must be furnished by a speed drop to release some of the stored kinetic energy. A change $\omega$ at a rate $d\omega/dt = M_l/J$ introduces a change $K_f\omega$ in the rotational e.m.f. and (neglecting armature inductance) $K_f\omega/r_a$ in the armature current, resulting in a torque $K_f^2\omega/r_a$. When this balances $M_l$ the speed stabilizes. The speed-change is exponential, given by $\omega = \omega_d[1 - \exp(-t/\tau_m)]$ where $\tau_m = Jr_a/K_f^2$ is the mechanical time-constant and the final steady-state speed is $\omega_1 = \omega_0 - \omega_d$.

Appreciable armature inductance will introduce an electrical time-constant $\tau_e = L_a/r_a$. The armature current is now $K_f\omega/(r_a + L_a p)$: it cannot rise in proportion to the speed-drop, and the deficiency means that the speed falls to a lower value followed by overshoot and oscillation. The torque equation is $K_f^2\omega/(r_a + L_a p) = M_l - Jp\omega$, whence the transform of the speed for a step function $M_l(s) = M_l/s$ becomes

$$\omega(s) = \frac{M_l}{J} \cdot \frac{s + r_a/L_a}{s[s^2 + s(r_a/L_a) + (K_f^2/JL_a)]} \tag{7.3}$$

yielding $\omega(t)$ as a damped oscillation.

EXAMPLE 7.3: A 750 kW 600 V 8-pole shunt motor has a no-load speed $\omega_0 = 23.0$ rad/s (220 r/min). The coupled inertia is $J = 1440$ kg-m$^2$; the armature parameters are $r_a = 18$ m$\Omega$, $L_a = 0.72$ mH. Neglecting mechanical loss and armature reaction, determine the speed, (i) without and (ii) with the effect of armature inductance, following the sudden application of a constant load torque $M_l = 60$ kN-m with the motor initially on no load.

Taking the no-load armature current to be negligible, then $K_f = V_a/\omega_0$ = 600/23 = 26.0 V per rad/s.

(i) $L_a = 0$. We have $\omega_d = M_l r_a/K_f^2 = 1.60$ rad/s and $\tau_m = J r_a/K_f^2 = 0.038$ s. Then $\omega = 1.60[1 - \exp(-0.026\ t)]$ with $t$ in millisec. The steady state speed on load is $\omega_0 - \omega_d = 21.4$ rad/s, and it is approached exponentially as indicated by the curve (i) in Fig. 7.5.

(ii) $L_a = 0.72$ mH. The electrical time-constant is $\tau_e = L_a/r_a = 0.04$ s. Inserting numerical values in eq. (7.3) and applying partial-fraction expansion gives the speed drop

$$\omega(t) = 1.6[1 - \exp(-0.026\ t) \cos(0.022\ t) + 0.96 \exp(-0.026\ t)$$
$$+ 0.96 \exp(-0.026\ t) \sin(0.022\ t)]$$

with $t$ in millisec. This is the decaying oscillation (ii) in Fig. 7.5. The steady-state speed is again 21.4 rad/s (240 r/min), but it is reached through a minimum of 20.9 rad/s.

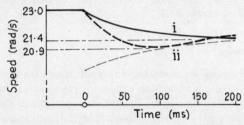

7.5 Shunt motor: step change of load: Example 7.3

*Series Motor* Even if saturation is neglected and the motor terminal resistance $r$ and inductance $L$ are taken as constants, the equations $V = (r + Lp)i + Gi\omega_r$ and $M_e = Gi^2$ in combination lead to nonlinear differentials. Approximations for a limited range can be obtained by linearizing the torque/speed relation, Fig. 7.6($a$), with $M_e = ki = a - b\omega_r$. In general, however, a valid analytical solution is not possible.

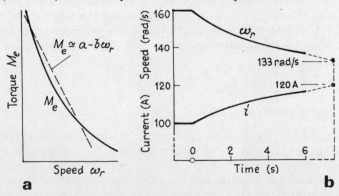

7.6 Series motor: step change of load: Example 7.4

EXAMPLE 7.4: A 200 V series motor runs at $\omega_1$ = 160 rad/s when delivering an output $P_1$ = 20.0 kW. The system inertia is $J$ = 8.0 kg-m$^2$. Assuming the machine to be *ideal*, find the speed and current when the load torque is suddenly increased by 44%.

*Initial.* $I_1$ = 100 A, $\omega_1$ = 160 rad/s, $G = V/I_1\,\omega_1$ = 0.0125, $M_1$ = 200 × 100/160 = 125 N-m.

*Final* $M_2$ = 125 × 1.44 = 180 N-m, $I_2$ = 100 $\sqrt{1.44}$ = 120 A, $\omega_2$ = 200/(0.0125 × 120) = 133 rad/s, $P_2$ = 24 kW.

*Transient* From $d\omega_r/dt = (M_e - M_l)/J$ we obtain

$$d\omega_r/dt = (400 \times 10^3/\omega_r^2) - 22.5$$

This is more readily integrated by inversion to $dt/d\omega_r$, or a solution by finite steps can be employed. The result is shown in Fig. 7.6(*b*). The rates of change may be slow enough for the inductance to have only a minor effect, but in a *practical* machine considerable modification would be imposed by losses, saturation and armature reaction.

## Cyclic change of load

The load torque on a ship's propeller shaft fluctuates rhythmically because of the action of the screw blades; in rolling mills, drive shafts are subject to load fluctuation and strain; in mine winders the cage/rope system may vibrate longitudinally and impose torque variations on the winder motor. These are typical of small changes about a steady mean condition.

Consider a pure inertial system with a steady condition represented by $V_f, I_f, V_a, I_a$ etc. If small fluctuations about the mean occur, we can assume that eq. (7.1) can be rewritten with $v_f$ replaced by $V_f + \Delta v_f$, $i_a$ by $I_a + \Delta_{i_a}$, etc. So long as the product of changes (e.g. $\Delta i_f \times \Delta i_a$) can be ignored, we can assume that the mean steady quantities have the changes

$$\Delta v_f = (r_f + L_f \mathrm{p})\Delta i_f$$

$$\Delta v_a = (r_a + L_a \mathrm{p})\Delta i_a + G\omega_r \Delta i_f + GI_f \Delta\omega_r \qquad (7.4)$$

$$\Delta M_e = GI_f \Delta i_a + GI_a \Delta i_f = J\mathrm{p}\Delta\omega_r$$

superimposed on them. Eq. (7.4) can be used to investigate the effect of changes in the variables, singly or in combination. If the changes are cyclic and sinusoidal and of frequency $\omega$, it is only necessary to replace p by j$\omega$ and to treat the finite differences as *phasors*. Gibbs [18] gives a comprehensive discussion of the method.

## 7.5 MOTOR: BRAKING

In some duties that require rapid retardation it is necessary to brake a motor during the operating period. A speed reduction from $\omega_1$ to $\omega_2$ of a system of inertia $J$ requires a braking power $p_b$ for a braking time $t_b$ such that

$$\int_0^{t_b} p_b \cdot dt = \tfrac{1}{2} J (\omega_1^2 - \omega_2^2) \tag{7.5}$$

In practice $t_b$ is shortened by core and mechanical losses and by the presence of any load torque. Electrical braking methods are:

*Dynamic*  The armature is disconnected from the supply and then closed across a resistor $R_e$, Fig. 7.7($a$). The machine acts as a generator, driven by the stored kinetic energy and dissipating power in the armature-circuit resistance.

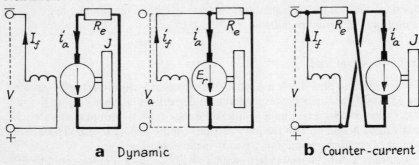

**a**  Dynamic                                    **b**  Counter-current

7.7 Shunt motor: braking connections

*Counter-current*  The armature terminal connections are reversed so that the supply voltage, now augmented by the rotational e.m.f., imposes a large current and a strong braking torque, Fig. 7.7($b$).

*Regenerative*  If the rotational e.m.f. is greater than the applied voltage, the machine generates and produces a braking torque by current reversal, maintained down to the speed at which the e.m.f. and voltage balance. The method requires a supply capable of accepting the generated power without undue rise of terminal voltage.

### Dynamic braking

Consider a separately excited or shunt machine with a constant field current $I_f$, an e.m.f. constant $(GI_f) = K_f$ and an armature of resistance $r_a$. For dynamic braking the armature is closed on to a resistor $R_e$ to give a total armature-circuit resistance $r_a + R_e = R$, making it legitimate to neglect the armature inductance. Then the rotational e.m.f. at a speed $\omega_r$ is $K_f\omega_r$ and the armature current is $K_f\omega_r/R$. The braking power is $p_b = (K_f\omega_r)^2/R$, and from eq. (7.5) we obtain, for a speed reduction from $\omega_1$ to $\omega_2$

$$t_b = (JR/K_f^2) \cdot \ln(\omega_1/\omega_2) \qquad \omega_2 = \omega_1 \exp[-(K_f^2/JR)\, t_b] \tag{7.6}$$

for a case in which only the inertia is significant. Core and mechanical losses and load torque may sometimes be represented approximately by the loss in a fictitious resistance $R_m$ in parallel with $R_e$, so that $R = r_a + R_e R_m/(R_e + R_m)$. By timing a given speed drop with a series of values of $R_e$, both $R_m$ and $J$ can be evaluated. For example, a 3.75 kW machine with $J = 0.21$ kg-m$^2$ and $r_a = 0.5\ \Omega$ gave the following braking times for a speed drop from $\omega_1$ to $\omega_2 = \omega_1/10$, for which $t_b = 2.3(J/K_f^2)R$:

| Resistance $R_e$ | ($\Omega$): | $\infty$ | 6.2 | 2.8 | 1.9 | 1.1 |
|---|---|---|---|---|---|---|
| Braking time $t_b$ | (s): | 2.0 | 1.2 | 0.8 | 0.6 | 0.4 |

From the results it can be inferred that $R_m$ is about 4.2 $\Omega$.

For fast braking, $R_e$ must be small to secure a high dissipation rate. The braking torque falls with speed and vanishes at standstill, so that a separate friction brake may be required to stop the motor and to hold the rotor at rest.

## Counter-current braking ('plugging')

In a machine with constant excitation the braking torque is $K_f i_a$. As a current-limiting resistance $R_e$ in the armature circuit is essential, it is valid to neglect the armature inductance. For a pure inertial load, $K_f i_a$ can be equated to $Jp\omega_r$. When plugging is initiated, reversal of the armature voltage is represented by the application of a step voltage $-V_a$ to the armature at operating speed, with the result that the current jumps to $-(V_a + e_r)/R$. Braking is maintained through standstill, and unless the armature is then open-circuited the motor will reverse.

EXAMPLE 7.5: A 200 V shunt motor with a constant field current of 2.5 A runs on no load at $\omega_1 = 100$ rad/s. The rotational inductance is 0.80 H, the armature resistance is 0.60 $\Omega$ and the system inertia is 4.0 kg-m$^2$. Estimate the braking times and currents for the motor to be retarded (i) to 40 rad/s and (ii) to standstill when braked on an external resistance $R_e = 4.4\ \Omega$ (a) dynamically, (b) by counter-current.
The data give $G = 0.80$, $(GI_f) = K_f = 2.0$, $R = r_a + R_e = 5.0\ \Omega$.

(a) *Dynamic* Apply eq. (7.6) with $JR/K_f^2 = 5.0$. The initial braking current is $e_r/R \cong V_a/R = 200/5.0 = 40$ A.

(i) $\omega_2 = 40$ rad/s. The braking time is $t_b = 5 \ln(100/40) = 4.6$ s. The speed, $\omega_r = 100 \exp(-0.2\, t)$, is plotted in Fig. 7.8(a) together with the generated current $i_a$ and the $I^2 R$ loss. The energy loss in resistance is the area of the loss/time curve, i.e. 16.8 kJ, equivalent to the reduction in stored kinetic energy.

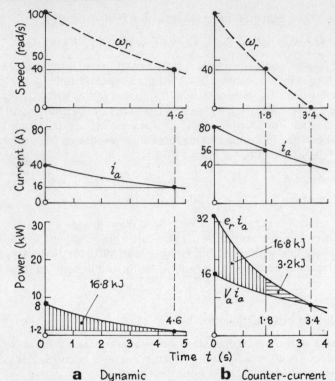

**a** Dynamic    **b** Counter-current

7.8 Shunt motor: dynamic and counter-current braking characteristics:
Example 7.5

(ii) $\omega_2 = 0$. The logarithmic expression gives $t_b = \infty$. Friction will therefore
determine the braking time. An estimate can be based on the assumption that
the motor stops when the torque at a low speed $\omega'$ is equal to the static
friction torque. Supposing this to be 2.0 N-m, then $\omega' = 2.5$ rad/s and
$t_b = 5 \ln(100/2.5) = 18$ s. To halve the braking time it would be necessary to
halve $R$ to 2.5 $\Omega$ and therefore reduce $R_e$ to 1.9 $\Omega$. The initial braking
current would then be 80 A.

(*b*) *Counter-current* Normal operating conditions are $M_e = K_f i_a = Jp\omega_r$,
and $V_a = K_f\omega_r + Ri_a$, whence $V_a = [K_f + (JR/K_f)p] \, \omega_r$. On no load with
$i_a = 0$, then $V_a = e_r = K_f\omega_1$. When the armature is disconnected the
rotational e.m.f. remains momentarily unchanged so that on re-applying $V_a$
in reverse, the initial current is $i_a = -(V_a + e_r)/R = -400/5 = -80$ A,
developing a braking torque $M_e = 2.0 \times 80 = 160$ N-m acting against the
direction of rotation, producing a retardation rate
$d\omega_r/dt = M_e/J = 40$ rad/s$^2$. The transform of the subsequent behaviour is
therefore

$$- V_a/s = K_f\omega_r(s) + (JR/K_f) \, [s\omega_r(s) - \omega_1]$$

Inserting numerical values and simplifying gives

$$\omega_r(t) = [200 \exp(-0.2\,t) - 100] \text{ rad/s}$$

Braking starts at $t = 0$ with an $I^2R$ loss of $80^2 \times 5.0 \times 10^{-3} = 32$ kW. The armature supply $V_a i_a = 16$ kW provides one half, the other half being derived from the rate of reduction of kinetic energy. Speed, current and $I^2R$ loss curves to a time base are given in Fig. 7.8($b$). Compared with those in ($a$), it can be seen that the more rapid braking is paid for by a considerable rise in energy loss.

(i) $\omega_2 = 40$ rad/s. The braking time is $t_b = 1.8$ s. As with dynamic braking, the kinetic energy converted is 16.8 kJ, but there is also an input of 24.4 kJ from the armature supply.

(ii) $\omega_2 = 0$. For braking to standstill, $t_b = 3.4$ s. The whole kinetic energy of 20 kJ is converted, and the input from the supply is 40 kJ. Thus the total energy dissipated in the braking operation is *three* times the stored kinetic energy.

### Regenerative braking

Regeneration is obtainable with shunt and separately excited motors, and with compound motors if the series compounding is relatively small. Series motors require a reversal of either the field or the armature connections, and are more controllable if there is an additional stabilizing field. The method is possible only if the rotational e.m.f. exceeds the armature applied voltage, a condition that can be produced by strengthening the field or by varying the armature voltage. The operating conditions are based on eq. (7.1).

EXAMPLE 7.6: The 200 V shunt motor of Example 7.5 runs on no load at $\omega_1 = 100$ rad/s. Neglecting armature inductance, find the braking time for a speed reduction to $\omega_2 = 60$ rad/s regeneratively ($i$) by halving the armature applied voltage, ($ii$) by reducing the field-circuit resistance from 80 $\Omega$ to 40 $\Omega$. The field inductance is 4.0 H.

Writing $(r_a + L_a p) = z_a$ and $(r_f + L_f p) = z_f$, the general equations are

$$Gi_f = Gv_f/z_f \quad v_a = z_a i_a + G\omega_r(v_f/z_f) \quad M_e = Gi_f i_a = Jp\omega_r.$$

Substituting $i_a = (Jz_f/Gv_f)p\omega_r$ gives

$$\omega_r = \frac{Gv_a v_f}{Jz_a z_f} \cdot \frac{1}{p + (Gv_f)^2/Jz_a z_f^2} \tag{7.7}$$

($a$) $V_a = 100$ V: The field current is constant, so that $v_f/z_f$ becomes $I_f = V_f/r_f = 200/80 = 2.5$ A, and $GI_f = K_f = 0.80 \times 2.5 = 2.0$. With neglect of the armature inductance, $z_a = r_a$. Then

$$\omega_r = V_a \frac{K_f}{Jr_a} \cdot \frac{1}{p + (K_f^2/Jr_a)}$$

The initial conditions (p = 0) are $\omega_r = \omega_1 = 100$ rad/s, $i_a = 0$. The subsequent speed is readily found to be

$$\omega_r = 50 \,[1 + \exp(-2\,t)] \text{ rad /s}$$

This is plotted in Fig. 7.9(*i*) with the current and power. The braking time to $\omega_2 = 60$ rad/s is $t_b = 0.80$ s. The armature power-time integral is 12.8 kJ, equal to the reduction of k.e. The hatch-area is armature $I^2R$, the remainder being the regenerated energy, amounting to 8.0 kJ.

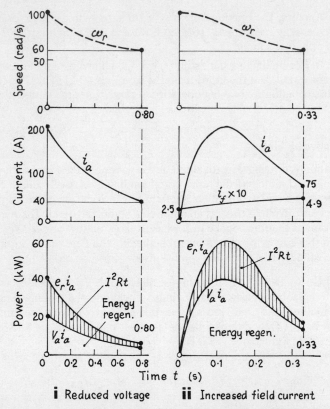

7.9 Shunt motor: regenerative braking characteristics: Example 7.6

(*b*) $r_f = 40 \,\Omega$. With 40 $\Omega$ switched out of the field circuit, $i_f$ rises exponentially from 2.5 A to 5.0 A with a time-constant $L_f/r_f = 0.10$ s, as indicated in Fig. 7.9(*ii*). At first, $e_r$ balances $V_a$ and both current and torque in the armature are zero. As $i_f$ rises, so does $e_r$, developing a current, a torque and a retardation. Eq. (7.7) is solved step-by-step with intervals $\Delta t = 10$ ms. The braking time is $t_b = 0.33$ s to the speed $\omega_2 = 60$ rad/s. Again the area under the $e_r i_a$ curve is 12.8 kJ, but the generated energy is now 9.3 kJ. The hatched area, 3.5 kJ, is the loss in the armature resistance.

## 7.6 MOTOR: SPEED CONTROL

The practical techniques of motor control are discussed in Chapter 8. The elements of two methods are considered here.

### Chopper control

The limited energy capacity of a secondary battery on a vehicle such as a milk float or a fork-lift truck demands an economic method of speed and torque control. Conventional methods include series-parallel connection of battery sections or of field windings, with intermediate adjustment and starting by means of a series rheostat. A more economical arrangement, which has the advantage of extending distance run or work done between battery charges, is voltage control by *chopping*. The essentials are shown in Fig. 7.10. The thyristor Th is used in an on/off switching cycle, either a constant-frequency

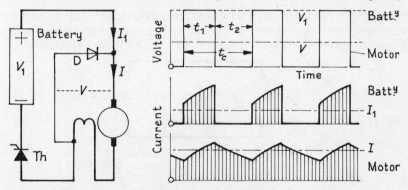

7.10 Chopper control

repetition at, say, 100 Hz with variable on-time, or by 2 ms pulses of variable number per second between 20 and 500. During the conducting time, a rectangular pulse of battery voltage is applied to the motor, the current of which rises at a rate determined by the inductance, the resistance and the change of rotational e.m.f. During the off-time the motor current continues to flow through the flywheel diode D, discharging some of its field energy. The mean current corresponds approximately to the mean voltage and speed. The battery output is the product of its terminal voltage $V_1$ and mean current $I_1$; the motor input is the product $VI$ of its mean voltage and current; both means are based on the on/off time ratio. Ideally the two mean products are equal (except for the volt-drop in motor resistance) giving control with no external loss. The system operates as a sequence of repeated transients.

Consider a separately excited motor with its speed controlled by chopping the armature voltage $V$, the chopping cycle covering the period $t_c$ comprising an 'on' or drive interval $t_1$ during which $V_1$ is applied, and an 'off' or quench interval $t_2 = t_c - t_1$ during which the applied voltage is zero but the armature current is continued through the flywheel diode. Assuming both the chopper thyristor and the diode to be ideal, then the operating modes have the following conditions:

*On*:   $V_1 = K_f\omega_r + (r_a + L_a p)i_a$   $K_f i_a = M_l + M_f + M_d + Jp\omega_r$

The armature current rises from a value determined by that at the end of the previous 'off' period, accelerating the load and storing kinetic energy.

*Off*:   $L_a p i_a = -(K_f\omega_r + r_a i_a)$   $Jp\omega_r = -(M_l + M_f + M_d)$

If the machine is running at high speed with low torque, the current may fall to zero before the end of the switching cycle.

Evaluation of the transients is tedious. Mellitt and Rashid [20] describe a computer method using linearized characteristics. Dubey and Shepherd [21] give simplified methods, one for cases where the armature current falls to zero during the 'off' periods.

### Ward-Leonard control

This system of speed control gives a wide range in both directions of rotation to a motor required, as in mine-winding or rolling-mill service, to drive and reverse large and varying loads. The motor M, Fig. 7.11, has a separately

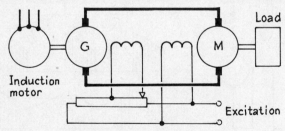

7.11 Ward-Leonard control

excited field, set to full-field for lower speeds but weakened for upper speed ranges. The armature of M is fed from that of the generator G, which is driven by an induction motor. Control of the field of G enables the armature of M to be supplied with a widely adjustable armature voltage and to run at a variable speed. In rolling-mill duty, the shaft of G may carry a flywheel to absorb and equalize its load peaks, the system then being termed Ward-Leonard-Ilgner. Traditionally the control of the field of G was by means of a shaft-mounted exciter. Later both rotating-machine and magnetic amplifiers were substituted. In modern sets both the generator and its driving motor are replaced by a static rectifier/inverter equipment.

EXAMPLE 7.7: The Ward-Leonard system in Fig. 7.11 has the following parameters.

*Generator*   Field resistance, $r_f = 100\ \Omega$, inductance $L_f = 60$ H. Armature resistance $r_{ag} = 0.35\ \Omega$, inductance negligible, rotational inductance $G_g = 1.0$ H, speed $\omega_g = 100$ elec. rad/s.

*Motor* (2-pole). Field current $I_{fm} = 2.0$ A. Armature resistance $r_a = 0.65\ \Omega$, inductance negligible, rotational inductance $G_m$ = 0.50 H, coupled inertia $J$ = 1.5 kg-m$^2$.

Obtain an expression for the motor armature speed $\omega_r$ and current $i_a$ following the application of 40 V to the generator field winding with the motor initially at rest and unloaded. Neglect rotational loss.

The behaviour is described by the rotational e.m.f.s $e_{rg} = G_g i_f \omega_g = 100\, i_f$ for the generator and $e_r = G_m I_{fm} \omega_r = \omega_r$ for the motor, together with the common armature current $i_a = (e_{rg} - e_r)/(r_{ag} + r_a) = (100\, i_f - \omega_r)$, and the torque $G_m I_{fm} i_a = Jp\omega_r$, whence $i_a$ = 1.5 $p\omega_r$. Combining the two expressions for $i_a$ gives 100 $i_f$ = (1 + 1.5p)$\omega_r$, where $i_f = v_f/(r_f + L_f p)$. Taking transforms

$$\omega_r(s) = \frac{44.4}{s(s + 1.67)\,(s + 0.67)} = \frac{40}{s} + \frac{27}{s + 1.67} - \frac{67}{s + 0.67}$$

$$\omega_r(t) = 40 + 27 \exp(-1.67\, t) - 67 \exp(-0.67\, t) \text{ rad/s}$$

$$i_a(t) = 67\, [\exp(-0.67\, t) - \exp(-1.67\, t)]$$

The current is zero at $t$ = 0, peaks to 22 A at $t$ = 1 sec and then sinks towards zero. The acceleration is a maximum at $t$ = 1 sec, and the speed is eventually 40 rad/s, reaching 38 rad/s in about 4.5 sec.

## 7.7 GENERATOR: SELF-EXCITATION

The conditions for the self-excitation of a shunt generator were discussed in Sect. 6.5. The problem of the *transient* build-up process is now considered. Let a shunt generator driven at constant speed have its armature terminals on open circuit. Any current $i$ in the field winding must arise as a generated current $-i$ from the armature. The voltage equation of the closed armature-field circuit, which has a total resistance $r$ and inductance $L$, is $(r + Lp)i - G\omega_r i = 0$, where $\omega_r$ is the speed at which the generator is driven. The current $i$ is, and remains, zero. But suppose that there is an initial current $i_0$ to represent the remanent flux $L_f i_0$ in the magnetic circuit: the transforms of the voltage equation and of the armature-field circulating current are

$$L_f i_0 = (-G\omega_r + r + Ls)I(s) \qquad I(s) = i_0/[s + (r - G\omega_r)/L]$$

where $L_a$ is taken as negligible compared with $L_f$. If $r$ is greater than $G\omega_r$, the current decays to zero; if $r = G\omega_r$ it remains constant; and if $r$ is less than $G\omega_r$ the current rises exponentially without limit. Thus there is a critical resistance $r_c$ which, if exceeded, inhibits self-excitation. There is also, for a given $r$, a critical speed $\omega_c$ below which the machine will not self-excite. In an actual machine, rise of field current develops magnetic-circuit saturation, reducing $G$ and stabilizing the field current.

A linear analysis is impossible. One approach is to linearize the field inductance/current relation as in Example 11.5, another, to use an iterative computation based on the actual magnetization characteristic. Even so, there may be unaccounted eddy-current reactions in poles and yokes, particularly if they are not laminated.

EXAMPLE 7.8: A 4-pole shunt generator with 1500 turns/pole has a steady-state open-circuit characteristic at $\omega_1 = 100$ rad/s as given in Fig. 7.12($a$). For

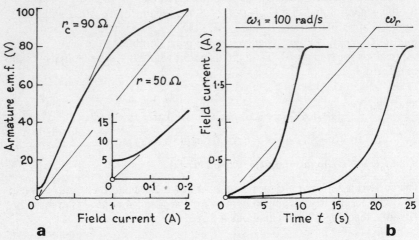

**a**                                                      **b**

7.12 Shunt generator: self-excitation: Example 7.8

a field current of 2.0 A the flux/pole is 22.5 mWb. Obtain the self-excitation (field-current/time) relation for a field-circuit resistance $r = 50 \Omega$ (including the armature) when the field circuit is closed ($a$) with the machine driven at constant speed $\omega_1 = 100$ rad/s, ($b$) with the machine at rest and the speed subsequently increased uniformly to 100 rad/s in 20 s (i.e., $\omega_r = 5t$).

($a$) *Constant Speed* $\omega_1 = 100$ rad/s. The o.c.c. in Fig. 7.12($a$) shows that $r_c = 90 \Omega$ and that $E_r = 100$ V for a field current $i = 2.0$ A. The corresponding field linkage is $\Psi = 22.5 \times 10^{-3} \times 1500 \times 4 = 135$ Wb–t: hence $\Psi = 1.35 \, e_r$. The voltage equation of the field circuit is $e_r = ri + \mathrm{d}\Psi/\mathrm{d}t$, whence $\mathrm{d}t = \mathrm{d}\Psi/(e_r - ri)$. This is evaluated in steps of current using finite differences:

| $i$ | $e_r$ | $\Psi$ | $\Delta\Psi$ | $e_r - ri$ | mean | $\Delta t$ | $t$ |
|---|---|---|---|---|---|---|---|
| 0 | 5.0 | 6.7 | | 5.0 | | | 0 |
| | | | 1.4 | | 4.25 | 1.4/4.25 = 0.33 | |
| 0.05 | 6.0 | 8.1 | | 3.5 | | | 0.33 |
| | | | 4.0 | | 3.75 | 4.0/3.75 = 1.07 | |
| 0.10 | 9.0 | 12.1 | | 4.0 | | | 1.40 |

Full excitation is reached in about 10 s, the field current settling to 2.0 A and the open-circuit e.m.f. to $E_r$ = 100 V, Fig. 7.12(*b*).

(*b*) *Rising Speed* $\omega_r$ = 5*t*. For a speed $\omega_r$ lower than $\omega_1$, the critical resistance is $r'_c = r_c(\omega_r/100)$: this becomes 50 $\Omega$ at $\omega_r$ = 55 rad/s, *t* = 11 s. For the first 11 s self-excitation is inhibited. Initially, therefore, the current rises from zero to $i = e'_r/50$, where $e'_r = e_r(\omega_r/100)$ and $e_r$ is the e.m.f. from the full-speed saturation curve. Thereafter the current rises more steeply as the speed approaches normal. The solution in Fig. 7.12(*b*) demands an iterative technique because both speed and e.m.f. rise during a time-interval $\Delta t$, the former being a function of *t* and the latter of both *t* and *i*.

*Series Generator*  Self-excitation is not possible on open circuit. If the generator is connected to a load, the terminal voltage will reach a value determined by the combined generator and load resistances, the speed and the magnetization characteristic. There are, as with the shunt machine, a critical resistance and a critical speed. Further, armature reaction cannot now be neglected.

*Rapid Field-reversal:*  When a shunt or separately excited field circuit has to be reversed, the stored magnetic energy can be dissipated in a quenching resistor, or a split-field arrangement used. A method described by Barney [19] is available if the field winding is fed from a 3-ph supply through a controlled bridge rectifier.

## 7.8 GENERATOR: LOADING

Transient conditions occur in a loaded generator when the operating conditions of speed, load or field current change. A linear analysis can be used to estimate the subsequent behaviour of the machine if saturation, commutation and armature-reaction phenomena can be ignored.

EXAMPLE 7.9: A separately excited generator driven at constant speed has a field circuit of resistance $r_f$ = 100 $\Omega$ and inductance $L_f$ = 40 H. The armature, in which the rotational e.m.f. is $Gi_f\omega_r$ where $G\omega_r$ = 50, is connected to an inductive load such that the armature-circuit parameters are $R$ = 20 $\Omega$, $L$ = 10 H. (*a*) Find the armature current following the sudden application of 200 V to the field-circuit terminals. (*b*) After steady-state conditions have been reached in (*a*), the load resistance is suddenly increased by 5 $\Omega$; determine the load current thereafter.

(*a*) The armature current is $i_a = G\omega_r i_f/(R + Lp)$ where $i_f = v_f/(r_f + L_fp)$. Taking transforms

$$I_a(s) = G\omega_r V_f(s)/(r_f + L_fp)(R + Lp) = 25/(s + 2.5)(s + 2)$$

$$= 5/s - 25/(s + 2) + 20/(s + 2.5)$$

$$i_a(t) = 5 - 25\exp(-2.0\,t) + 20\exp(-2.5\,t)$$

(*b*) The steady-state current in (*a*) is $i_a = I_0 = 5$ A. The change of load resistance $\Delta R$ results in the introduction into the armature circuit of a step-function voltage $-I_0 \Delta R$, superposing a current $i$ given by
$I(s) = (-I_0 \Delta R)/s(R + sL) = 2.5/s(s + 2.5)$; whence $i(t) = -[1 - \exp(-2.5\ t)]$. The required armature current is

$$i_a = I_0 + i = 4 + \exp(-2.5\ t).$$

In the foregoing the current is treated for convenience as positive.

The effect of compole and compensating windings, and approximately of armature reaction, can be introduced by adding a q-axis field winding in series with the armature, the behaviour then being expressed by eq. (7.2).

### Cyclic change of speed

Internal-combustion-engine drives produce pulsating torques, and small cyclic changes of speed are inevitable. The results of these (and other) small perturbations can be dealt with by expressions in eq. (7.4).

EXAMPLE 7.10: A 2-pole generator/diesel-engine set has an inertia $J = 12.0$ kg-m$^2$. The field current is constant, giving $GI_f = K_f = 3.81$ V per rad/s. The resistance of the armature is $r_a = 0.030$ $\Omega$, but its inductance and armature-reaction effect are negligible. The armature is connected to a pure resistance load $R = 0.97$ $\Omega$, delivering a current $I_a = 400$ A. Calculate (*a*) the mean speed $\omega_1$ and engine torque $M_m$, disregarding mechanical loss; (*b*) the current and speed fluctuations if the engine torque varies sinusoidally by $\pm 10\%$ at a frequency $f = 2.73$ Hz.

(*a*) *Steady State*  Given $I_a = 400$ A, then directly

$$V_a = 400 \times 0.97 = 388 \text{ V} \qquad E_r = 400(0.97 + 0.03) = 400 \text{ V}$$

$$\omega_1 = 400/3.81 = 105 \text{ rad/s} \qquad n_1 = 103/2\pi = 16.4 \text{ r/s (984 r/min)}$$

$$E_r I_a = 400 \times 400 = 160 \text{ kW} \qquad M_m = 160/105 = 1.52 \text{ kN-m}$$

(*b*) *Fluctuation* Writing $M_m = K_f I_a = \omega_1 K_f^2/(R + r_a)$, then for small changes $\Delta M, \Delta i_a$ and $\Delta \omega$

$$\Delta M = [K_f^2/(R + r_a)]\ \Delta \omega + Jp(\Delta \omega)$$

$$\frac{\Delta \omega}{\omega_1} = \frac{\Delta M}{M_m} \cdot \frac{1}{1 + [J(R + r_a)\omega_1/M_m]\,p}$$

For sinusoidal variation we write $p = j2\pi f = jm$ to obtain the phasor expression

$$\frac{\Delta \omega}{\omega_1} = \frac{\Delta M}{M_m} \cdot \frac{1}{1 + jm(J\omega_1/M_m)} = 0.1\ \frac{1}{1 + j14} = 0.07\ \angle{-86°}$$

The speed fluctuation is therefore $\pm 0.07 \times 105 = \pm 7.2$ rad/s (or 1.15 r/s).

The corresponding current fluctuation is $\pm 0.07 I_a$ in the absence of appreciable armature inductance.

## 7.9 GENERATOR: REGULATION

Field-winding inductance delays the response of a generator to the voltage control devices when there is a sudden change of load. Two examples are considered.

EXAMPLE 7.11: A generator on no load and driven at constant speed has its field winding F excited from a fully compensated amplidyne, Fig. 7.13. The control field C is derived from the difference $v_c = v - V_r$ of the generator armature and reference voltages. Obtain the overall transfer function relating armature voltage $v$ to control-field voltage $v_c$.

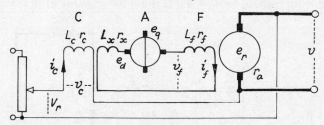

7.13 Separately excited generator: voltage regulator

With the symbols indicated in the diagram, and writing $z$ for $(r + Lp)$, then with $K$ representing the e.m.f. per unit field current, the t.f. is obtained as follows.

*Generator* $v = K_f i_f = K_f(v_f/z_f)$, whence $v/v_f = K_f/z_f$.

*Amplidyne* $v_c = i_c z_c \qquad e_q = K_c i_c = i_q z_a$

$\qquad e_d = K_a i_q = (K_a K_c/z_a z_c)v_c$

also $\quad e_d = v_f + i_f(z_a + z_x) = v_f[1 + (z_a + z_x)/z_f]$

whence $\quad v_f/v_c = K_a K_c z_f/[z_a z_c(z_f + z_a + z_x)]$

*Overall T.F.* The product $(v/v_f)(v_f/v_c)$ gives

$$v/v_c = K_a K_c K_f/[z_a z_c(z_f + z_a + z_x)]$$

Substituting $z_a = (r_a + L_a p) = r_a(1 + p\tau_a)$ etc. where $\tau_a = L_a/r_a$ etc., and with $(z_f + z_a + z_x) = R(1 + p\tau)$ where $R = r_f + r_a + r_x$ and $L = L_f + L_a + L_x$, the overall t.f. becomes

$$\frac{v}{v_c} = \frac{K_a K_c K_f}{r_a r_c R} \cdot \frac{1}{(1 + p\tau_a)(1 + p\tau_c)(1 + p\tau)}$$

EXAMPLE 7.12: A constant-speed generator of armature resistance $r_a$ and rotational e.m.f. $e_r = K_f i_f$ supplies a current $i$ at voltage $v$ to a load resistance $R$. A voltage-divider across the armature terminals provides a voltage $v_i = kv$ to an amplifier in series with a reference voltage $V_r$ as in Fig. 7.14. The error voltage $v_e = v_i - V_r$ is amplified to give the output voltage $v_o = Av_e$, which supplies the generator field winding with a current $i_f$.

(*a*) Show that, in the steady state with connector S open, the generator voltage is $v = (m/k)v_e$ where $m = (kAK_f/r_f)\,[R/(R + r_a)]$ ; and with S closed, $v_e = V_r/(m - 1)$.

(*b*) Given that $V_r = 12$ V, $A = -100$, $K_f = 80$ V/A, $r_a = 0.25$ $\Omega$, $r_f = 50$ $\Omega$, $L_f = 10$ H and $k = 0.10$, calculate the steady-state values of $v$, $v_e$ and $i$ for a load resistance (i) $R = 1.0$ $\Omega$, (ii) $R = 4.0$ $\Omega$.

(*c*) The generator is running in the steady state with $R_1 = 1.0$ $\Omega$; at $t = 0$ the load is suddenly changed to $R_2 = 4.0$ $\Omega$. Find the response of the terminal voltage $v$.

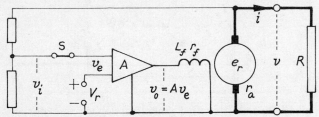

7.14 Generator voltage control: Example 7.12

(*a*) In the steady state with connector S open and an amplifier input voltage $v_e$, the field current is $i_f = v_o/r_f = Av_e/r_f$, and the rotational e.m.f. is $e_r = K_f i_f = v(R + r_a)/R$, where $v = v_i/k$. Hence

$$\frac{v_i}{v_e} = m = \frac{kK_fA}{r_f} \cdot \frac{R}{R + r_a}$$

When S is closed, $v_e = v_i - V_r = mv_e - V_r = -V_r(1 - m)$.

(*b*) (i) $R_1 = 1.0$ $\Omega$                                       (ii) $R_2 = 4.0$ $\Omega$

$$m_1 = -\frac{0.1 \times 80 \times 100}{50}\,\frac{1}{1.25} = -12.9 \qquad m_2 = -15.1$$

| | | | | |
|---|---|---|---|---|
| $v_{e1}$ | $= -12/(1 + 12.9)$ | $= -0.86$ V | $v_{e2}$ | $= -0.75$ V |
| $v_{i1}$ | $= (-12.9)(-0.86)$ | $= 11.1$ V | $v_{i2}$ | $= 11.2$ V |
| $v_1$ | $= 11.1/0.1$ | $= 111$ V | $v_2$ | $= 112$ V |
| $i_1$ | $= 111/1.0$ | $= 111$ A | $i_2$ | $= 28$ A |

(*c*) The steady-state conditions in (*b*)(i) give $e_r = 137.6$ V and $i_f = 137.6/80 = 1.72$ A. When at $t = 0$ the load resistance becomes $R_2 = 4.0$ $\Omega$, the armature current falls instantaneously to $137.6/4.24 = 32.5$ A (ignoring armature inductance) and the terminal voltage rises to $32.5 \times 4.0 = 130$ V because $i_f$ and therefore $e_r$ cannot change instantaneously. During the transient, $v_o = Av_e = A(v_i - V_r) = A(kv - V_r)$. Substituting $e_r \cdot R_2/(R_2 + r_a)$ for $v$, where $e_r = K_f i_f$, gives $Akv = m_2 r_f i_f$, whence

$$[r_f(1 - m_2) + L_f p] \, i_f = -AV_r$$

representing an exponential with a time-constant $\tau_f = L_f/r_f(1 - m_2) = 12.4$ ms. Thus the terminal voltage $v$ drops from 130 V down to the new steady-state value $v_2 = 112$ V exponentially with the time-constant of 12.4 ms.


## 7.10 GENERATOR: SHORT CIRCUIT

First we consider an ideal case of a separately excited machine with constant field current, an armature brush position exactly on the q-axis, constant speed, no armature reaction, no compoles nor compensating windings, no saturation, no reaction from armature coils undergoing commutation, and an armature with the constant parameters $r_a$ and $L_a$. If the armature, initially on open circuit and developing a rotational e.m.f. $E_r$, has its terminals short-circuited at $t = 0$, the process is represented analytically by the application of a cancelling step voltage $-E_r$. The resulting transient current is superposed on any existing armature current (in this case, zero). At $t = 0$ the armature transient current begins to rise at a rate $-E_r/L_a$ and settles eventually to $-E_r/r_a$, the steady-state short-circuit current. The two extreme conditions are linked through the exponential relation

$$i_a = -(E_r/r_a) \, [1 - \exp(-t/\tau_a)] \tag{7.8}$$

where $\tau_a = L_a/r_a$ is the armature-circuit time-constant. The field current remains constant and unaffected.

The short-circuit of a *practical* machine is very different. The condition is a severe one, especially in a large generator. Saturation occurs and eddy-currents are induced; compoles and compensating windings couple the armature to the field system magnetically; strong armature reaction disturbs the field axis and commutation is difficult; the precise position of the effective brush axis is crucial. Compole saturation inhibits the production of the very high commuting flux, the development of which may also be delayed by eddy-currents unless the pole cores are laminated.

Some of these effects — though not saturation — can be incorporated into a linear theory. The representative parameters are very difficult to evaluate, and they may have to be inferred from the oscillograms derived from test. Nevertheless, a closer approximation to reality can be achieved, giving insight into the complex conditions to which a short circuit can lead.

Consider eqs. (7.2), where $r_a$ and $L_a$ refer to the complete armature circuit including q-axis compole and compensating windings, and apply them to the case of a separately excited generator with no series field winding. It might be presumed that both $x$ and $L_{fs}$ are zero. But armature reaction on the field flux and brush-shift resulting from under-commutation introduce, to some approximation, the effect of a series field. In view of the dominance of the armature current, the term $L_{fs}p i_f$ is omitted; and for simplicity the generator speed $\omega_r$ is considered as invariable during the brief duration of the transient currents, enabling $G\omega_r$ to be replaced by a constant $K$. The transients are superposed on any current that exists at the instant of short circuit. For the main field winding this is $I_f$, while for the armature on no load it is $I_a = 0$. Eqs. (7.2) can now be written for the transient component currents $i_f$ and $i_a$:

$$0 = (r_f + L_f p)i_f + L_{fs}\, p\, i_a \quad = r_f(1 + \tau_f p)i_f + L_{fs}\, p\, i_a$$

$$-E_r = (r_a + L_a p)i_a + K(i_f + x i_a) \quad = R_a(1 + \tau_a p)i_a + K i_f$$

where $\tau_f = L_f/r_f$ is the field time-constant, $R_a = (r_a + Kx)$ and $\tau_a = L_a/R_a$ is the effective armature time-constant. The field-circuit equation is used to express $i_f$ in terms of $i_a$, whence the armature equation becomes

$$i_a = -\frac{E_r}{R_a} \cdot \frac{(1 + \tau_f p)}{(1 + \tau_f p)(1 + \tau_a\, p) - (KL_{fs}/R_a r_f)p}$$

The denominator is $[1 + (\tau_f + \tau_a - C)p + \tau_f \tau_a p^2]$, where $C = KL_{fs}/R_a r_f$. If we write $D = C/\tau_f = KL_{fs}/R_a L_f$ and define two new time-constants $T_f = \tau_f(1 - D)$ and $T_a = \tau_a/(1 - D)$, then the product

$$(1 + T_f p)(1 + T_a\, p) = 1 + (\tau_f + \tau_a - C)p + \tau_f \tau_a\, p^2$$

is a close approximation to the denominator, remembering that $\tau_a$ is small compared with $\tau_f$. The transforms of the armature and field transient currents for a sudden armature terminal short-circuit are therefore

$$I_a(s) = -\frac{E_r}{R_a} \cdot \frac{(1 + s\tau_f)}{s(1 + sT_f)(1 + sT_a)}$$

$$I_f(s) = \frac{V_f}{r_f} \cdot \frac{L_{fs}}{R_a} \cdot \frac{1}{(1 + sT_f)(1 + sT_a)}$$

Inverse transformation gives for the transient armature current

$$i_a(t) = -\frac{E_r}{R_a}\left[1 - \frac{T_f - \tau_f}{T_f - T_a}\exp(-t/T_f) - \frac{T_a - \tau_f}{T_a - T_f}\exp(-t/T_a)\right]$$

If $T_a$ can be neglected in comparison with $T_f$ or $\tau_f$, the armature current reduces to

$$i_a(t) = -\frac{E_r}{R_a}\left[1 + \frac{1}{1 - D}\{D\exp(-t/T_f) - \exp(-t/T_a)\}\right] \qquad (7.9)$$

where $D = KL_{fs}/L_f R_a$. The quantity $R_a(1 - D)$ is termed the *transient resistance*. The armature short-circuit current rises very rapidly at first (or instantaneously if $L_a = 0$), attaining a peak and then sinking more slowly to a steady short-circuit value $E_r/R_a$. The transient field current, again assuming $T_a$ to be small, is

$$i_f(t) = \frac{V_f}{r_f} \cdot \frac{L_{fs}}{R_a} \cdot \frac{1}{T_f - T_a} \left[ \exp(-t/T_f) - \exp(-t/T_a) \right] \qquad (7.10)$$

which reaches a peak and then decays to zero. The total field current is $I_f + i_f(t)$.

## Shunt generator

An armature terminal short circuit reduces the field as well as the armature voltage to zero. The solution can be obtained in much the same way as for the separately excited generator, giving $i_f = 0$ and $i_a = 0$ for the final steady state. In practice an armature-field circulating current will persist because of the effect of remanence. This can be taken into account if necessary by providing the field with an additional fictitious winding in which a small current $i_0$ is circulated, as noted in Sect. 7.7.

EXAMPLE 7.13: A 240 V generator has a separately excited main field taking 6.0 A at 240 V. the parameters are:

Field: $r_f = 40.0\ \Omega; L_f = 10.0\ H; L_{fs} = 18.7\ mH; x = 1.9 \times 10^{-3}$.

Armature: $r_a = 0.05\ \Omega; L_a = 0.10\ mH; K = G\omega_r = 40.0$.

Estimate the field and armature currents following an armature terminal short circuit (a) assuming the machine to be ideal, (b) taking account of armature/ field coupling and armature reaction. The generator is initially unloaded, and the speed remains constant.

(a) *Ideal* The field current remains unaffected at $I_f = 6.0$ A. The armature time-constant is $L_a/r_a = 2.0$ ms. The armature current is given (in kA and ms) by eq. (7.8),

$$i_a(t) = -4.8\ [1 - \exp(-t/2)]$$

and is plotted in Fig. 7.15(a).

(b) *Practical* The modified parameters must first be calculated:
$R_a = r_a + Kx = 0.125\ \Omega; D = KL_{fs}/R_aL_f = 0.60; (1 - D) = 0.40; \tau_f = L_f/r_f$
$= 250$ ms; $T_f = \tau_f(1 - D) = 100$ ms; $\tau_a = 2.0$ ms; $T_a = \tau_a/(1 - D) = 5.0$ ms.
Then eq. (7.9) and (7.10), with $t$ in millisec., give the transient current components

$$i_a(t) = -1920\ [1 - 1.5\{0.6\exp(-t/100) - \exp(-t/5)\}]$$

$$i_f(t) = 9.5\ [\exp(-t/100) - \exp(-t/5)]$$

whence the total field current is $i_f(t) + 6.0$ A. In Fig. 7.15 the armature current is reversed for convenience of plotting. Comparison of (a) and (b) demonstrates the considerable influence of armature-field magnetic coupling and of armature reaction. While the peak armature currents do not greatly differ, the final steady-state current in (*b*) is markedly reduced; and there is a surge in the field current to oppose the rise in field linkage that the armature current would generate, in accordance with the constant-linkage theorem.

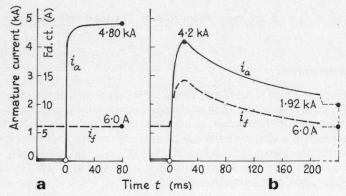

7.15 Separately excited generator: short circuit: Example 7.13

## 7.11 GENERALIZED MACHINE ANALYSIS

In the foregoing, transient phenomena have been treated in an *ad hoc* manner by solving equations set up for each case on the basis of eq. (7.1), with additions as necessary to suit the problem in hand. But each case could be considered as deriving from a comprehensive 'generalized' machine in which (i) the rotor is provided (as in Fig. 4.16) with brush sets forming windings D and Q that magnetize respectively along the d- and q-axes; and (ii) the stator carries both d- and q-axis exciting windings. Fig. 7.16 shows such an arrangement in two diagrammatic forms for a basic 2-pole machine. The d-axis stator

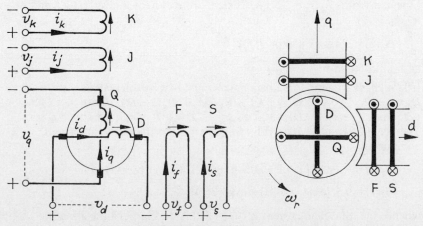

7.16 Generalized machine

coil F could represent a separate or shunt field winding, coil S a series field winding. The q-axis stator coils might include a compole winding J and a compensating winding K. Only those windings relevant to a particular case require to be included. Thus a small series motor could be represented by coils S and Q, a larger one by S, Q and J, with the addition of K if it were compensated.

The generalized 2-axis model was developed primarily for a.c. machines [Refs. 1, 18], in which all windings are assumed to be capable of developing sine-distributed gap fluxes. This is not the case, even approximately, in the d.c. machine, but the concept is a useful one for machines with several windings and elaborate interconnections. The generalized machine model is adaptable to conventional heteropolar structures with saliency on *one side only* of the airgap, but not to double-saliency machines such as stepper motors, and in this respect it is not, in fact, completely general.

Stator windings are grouped on one or other axis, a positive current in any winding developing an m.m.f. in the appropriate positive direction. As the d- and q-axis groups are effectively in space quadrature, there is *ideally* no magnetic coupling between them. Coils on the same axis, however, have mutual inductance. Coil F, for example, has a self inductance $L_f$ due to self-produced flux, and also mutual inductances $L_{fs}$ with S and $L_{fd}$ with D.

The commutator rotor winding has two pairs of brushes. Consider the q-axis pair: it gives rise to a rotor current pattern of fixed m.m.f. direction which can be represented by a current in a fixed coil Q. But the conductors forming Q rotate in the gap flux developed by currents in d-axis windings, and Q is therefore the seat of a rotational e.m.f. Similarly, the current pattern set up by a pair of d-axis brushes can be represented by a current in a fixed coil D which nevertheless has the property of generating an e.m.f. by rotation in the q-axis flux. Rotational e.m.f.s appear only in rotor windings, and the equivalent coils D and Q, with fixed m.m.f. directions but also motional e.m.f.s, are termed *quasi-* (or *pseudo-*) *stationary* windings. Thus Q has a self inductance $L_q$, mutual inductances $L_{qj}$ and $L_{qk}$ with q-axis windings J and K, and motional inductances $G_{qf}$, $G_{qs}$ and $G_{qd}$ with the d-axis windings. Correspondingly, D has a self inductance $L_d$, mutual inductances $L_{df}$ and $L_{ds}$, and rotational inductances $G_{dj}$, $G_{dk}$ and $G_{dq}$. The resistances of D and Q are $r_d = r_q = r_a$, the normal armature resistance between brushes.

*Conventions* Those set out in Sect. 1.5, Fig. 1.15, are 'natural' and convenient for machines with q-axis brushes only. The transient analysis of cross-field machines is preferably cast in the conventions of the generalized theory, in which the positive direction of the d-axis quantities is from left to right, giving symmetry. The diagrams in Fig. 7.16 are drawn in this convention. (The 'natural' convention is regained by assuming the current direction in d-axis stator windings to be reversed). The consequences are indicated in Fig. 7.17, where positive m.m.f. (indicated by an arrow) in each winding is that of positive current flowing through the winding from the $+$ to the $-$ terminal.

Consider ($a$), the q-axis rotor coil Q and stator coil J. A change of $i_q$ in Q induces the pulsational e.m.f. $e_{pj} = L_{jq} \, p \, i_q$ in J, and similarly a change of $i_j$

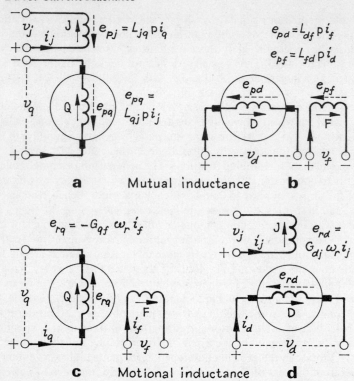

**a**     Mutual inductance     **b**

**c**     Motional inductance     **d**

7.17 Mutual and motional inductances

induces $e_{pq} = L_{qj} \, p \, i_j$ in Q. These pulsational e.m.f.s of mutual inductance oppose the applied terminal voltages $v_j$ and $v_q$, which must therefore contain balancing components.

Diagram ($b$) for d-axis coils D and F shows a comparable relationship. Additional stator axis windings are treated in the same way to account for pulsational e.m.f.s appearing in them as a result of mutual inductances.

Now consider the motional e.m.f.s. In ($d$) the d-axis rotor winding D contains turns that rotate in the q-axis flux of J at a speed $\omega_r$. The rotational e.m.f. $e_{rd}$ in D is $G_{dj} \, \omega_r \, i_j$; it opposes $v_d$ so that for steady-state conditions $v_d = r_a i_d + e_{rd} = r_a i_a + G_{dj} \, \omega_r \, i_j$. But in ($c$), for the q-axis rotor winding Q and the d-axis field winding F, the m.m.f. convention results in the rotational e.m.f. $e_{rq}$ aiding the applied voltage. This means that $v_q + e_{rq} = r_a i_q$ in the steady state, i.e. $v_q = r_a i_q - G_{qf} \, \omega_r \, i_f$.

### Six-coil machine

Returning to the generalized machine of Fig. 7.16, let a voltage $v_f$ be applied to winding F, and let a positive current $i_f$ flow in it. The volt drop in resistance is $r_f i_f$ and a d-axis flux $L_f i_f$ is produced. If there are also positive currents $i_s$ and $i_d$ respectively in S and D, the flux in F is increased by

$L_{fs}i_s + L_{fd}i_d$. Should all three currents be changing at a rate $di/dt = pi$, then the voltage $v_f$ applied to F is made up of the components

$$v_f = r_f i_f + L_f p i_f + L_{fs} p i_s + L_{fd} p i_d$$

A corresponding equation can be written for the voltage $v_s$ applied to S, and likewise for the voltages applied to the q-axis stator windings J and K.

The motional e.m.f. between the q-axis rotor brushes results from rotation in the d-axis flux, to which F, S and D contribute. If Q carries the positive current $i_q$, its applied voltage $v_q$ must have the components

$$v_q = r_a i_q + L_q p i_q \qquad \text{resistance and self inductance}$$
$$+ L_{qj} p i_j + L_{qk} p i_k \qquad \text{mutual inductance}$$
$$-(G_{qf} i_f + G_{qs} i_s + G_{qd} i_d)\omega_r \qquad \text{motional inductance}$$

A similar equation can be written for the d-axis rotor coil D, except that the change of orientation leads to positive coefficients of rotational inductance G.

*Matrix Equation* The six voltage equations can be assembled in a composite matrix equation, Fig. 7.18. For a given machine, the rows and columns of

|  |  |  | Stator d-axis | | Stator q-axis | | Armature | |  |
|---|---|---|---|---|---|---|---|---|---|
|  |  |  | F | S | J | K | D | Q |  |
| F | $v_f$ | $=$ | $r_f + L_f p$ | $L_{fs} p$ |  |  | $L_{fd} p$ |  | $\cdot$ $i_f$ |
| S | $v_s$ |  | $L_{sf} p$ | $r_s + L_s p$ |  |  | $L_{sd} p$ |  | $i_s$ |
| J | $v_j$ |  |  |  | $r_j + L_j p$ | $L_{jk} p$ |  | $L_{jq} p$ | $i_j$ |
| K | $v_k$ |  |  |  | $L_{kj} p$ | $r_k + L_k p$ |  | $L_{kq} p$ | $i_k$ |
| D | $v_d$ |  | $L_{df} p$ | $L_{ds} p$ | $G_{dj} \omega_r$ | $G_{dk} \omega_r$ | $r_d + L_d p$ | $G_{dq} \omega_r$ | $i_d$ |
| Q | $v_q$ |  | $-G_{qf} \omega_r$ | $-G_{qs} \omega_r$ | $L_{qj} p$ | $L_{qk} p$ | $-G_{qd} \omega_r$ | $r_q + L_q p$ | $i_q$ |

7.18 Matrix equation for generalized machine

absent windings are omitted, and constraints (such as equal voltages applied to windings in parallel or zero to a short-circuited winding, and equal currents in windings connected in series) taken into account. The matrix equation is then solved in any appropriate way.

*Torque* The instantaneous torque developed by a rotor axis winding carrying a current $i$ is the converted power $p_e = e_r i$ divided by the speed $\omega_r$, where $e_r$ is the summation of component rotational e.m.f.s. For winding D

$$e_{rd} = (G_{dj} i_j + G_{dk} i_k + G_{dq} i_q)\omega_r$$

The $G\omega_r$ terms appear explicitly in row D, columns J, K and Q of the matrix equation. The d-axis rotor torque is therefore

$$M_{ed} = (G_{dj}i_j + G_{dk}i_k + G_{dq}i_q)i_d$$

Similarly, rotor coil Q contributes the torque

$$M_{eq} = -(G_{qf}i_f + G_{qs}i_s + G_{qd}i_d)i_q$$

and the total output torque is $M_e = M_{ed} + M_{eq}$.

As an example, consider a simple separately excited motor with a d-axis field winding F and one pair of brushes set in the q-axis to form the armature winding into an equivalent rotor coil Q. Only the F and Q rows and columns are required, giving

$$\begin{bmatrix} v_f \\ v_q \end{bmatrix} = \begin{bmatrix} r_f + L_f p & \\ -G_{qf}\omega_r & r_a + L_q p \end{bmatrix} \cdot \begin{bmatrix} i_f \\ i_q \end{bmatrix}$$

that is, $v_f = (r_f + L_f p)i_f$ and $v_q = (r_a + L_q p)i_q - G_{qf}\omega_r i_f$. The torque is $M_{eq} = -G_{qf}i_f i_q$. The torque is negative because it is an output quantity. If the field voltage and current were reversed to give the 'natural' convention, the result would be eq. (7.1).

*Parameters*  The parameters of resistance and inductance in Fig. 7.17 can be calculated from design data, though only with some difficulty in the case of self and mutual inductances. They can, however, be derived from tests. Such tests made with a.c. are not suitable, as disturbing effects of core and eddy-current loss are introduced. A ballistic method using a d.c. inductance bridge has been described by Jones [17]. In elementary magnetic-field analysis it is often assumed that the mutual inductance $L_{12}$ of a coil 1 to a coil 2 is the same as $L_{21}$ of coil 2 to coil 1. This is true only if the coils are situated in air or in a medium of constant permeability, and in a practical machine such reciprocity may not be valid. Such inductance coefficients as $L_{fd}$ and $L_{df}$ may not always be taken as identical.

Field-winding resistances are obtained by simple calculation or by voltage/current measurements, and are constant, apart from variation with temperature, if during transients the eddy-current effects are minimal. Armature resistances measured across appropriate commutator sectors are similarly constant, but if measured across the armature terminals will include the brush-contact volt drop which is roughly independent of current. It is usually adequate to adopt an overall averaged value.

Motional inductances vary considerably with magnetic-circuit saturation, so that a value may have to be chosen appropriate to the level of flux density. A $G$-curve is determined by calculation, or by an open-circuit magnetization characteristic relating e.m.f. to field current at a known speed, whence $G = E_r/\omega_r I_f$.

Parametric variation makes a linear analysis inaccurate, particularly over a wide operating range. For transient analysis there is, however, no simple alternative.

EXAMPLE 7.14: A 2-pole commutator machine has two d-axis field windings F and N, and both d- and q-axis brush-pairs. Write down the primitive matrix equation in terms of the currents and voltages in Fig. 7.19(a). The machine is

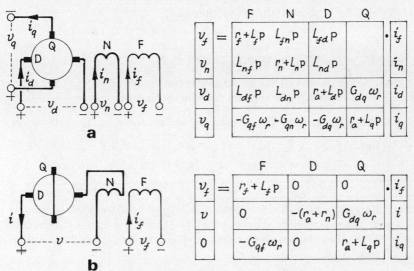

$$\begin{bmatrix} v_f \\ v_n \\ v_d \\ v_q \end{bmatrix} = \begin{bmatrix} r_f + L_f p & L_{fn} p & L_{fd} p & \\ L_{nf} p & r_n + L_n p & L_{nd} p & \\ L_{df} p & L_{dn} p & r_a + L_d p & G_{dq} \omega_r \\ -G_{qf} \omega_r & -G_{qn} \omega_r & -G_{dq} \omega_r & r_a + L_q p \end{bmatrix} \cdot \begin{bmatrix} i_f \\ i_n \\ i_d \\ i_q \end{bmatrix}$$

**a**

$$\begin{bmatrix} v_f \\ v \\ 0 \end{bmatrix} = \begin{bmatrix} r_f + L_f p & 0 & 0 \\ 0 & -(r_a + r_n) & G_{dq} \omega_r \\ -G_{qf} \omega_r & 0 & r_a + L_q p \end{bmatrix} \cdot \begin{bmatrix} i_f \\ i \\ i_q \end{bmatrix}$$

**b**

7.19 Cross-field machine: Example 7.14

now connected as in (*b*), with N in series with D to provide complete compensation, F supplied with a control voltage, and the rotor d-axis circuit providing a terminal voltage $v$ and an output current $i$, the q-axis brushes being short-circuited. Obtain the transfer function relating output voltage to field voltage with a load resistor $R$ across the output terminals.

(*a*) *Primitive* The matrix equation is written down from Fig. 7.18 by eliminating rows and columns J and K, renaming S as the compensating winding N.

(*b*) *Amplidyne* The connections in (*b*) are those of a fully compensated amplidyne. An approximate solution is found by introducing a few simplifications. The rotor d-axis current is $i_d = -i$ (because the machine is a generator), and reversal of the winding N makes $i_n = -i_d = i$. The terminal voltage is $v = v_d - v_n$, and across the short-circuited winding Q the applied voltage is $v_q = 0$. With full compensation and with negligible leakage, $L_n = L_d$ and $L_{fn} = L_{fd}$. The primitive matrix equation becomes that in (*b*). The field circuit equation is reduced to $v_f = (r_f + L_f p)i_f$, as the complete compensation means that D and N together produce no resultant flux. The output voltage $v$ is the rotational e.m.f. $e_{rd} = G_{dq} \omega_r i_q$ reduced by the volt drop in the resistance of D and N in series. The current $i_q$ is the rotational e.m.f. $e_{rq}$ divided by the operational impedance, i.e.

$$i_q = i_f \cdot G_{qf} \omega_r / (r_a + L_q p) = v_f \cdot G_{qf} \omega_r / (r_f + L_f p)(r_a + L_q p)$$

Then

$$v = Ri = -(r_a + r_n)i + G_{dq}\omega_r i_q$$

from which the required transfer function is

$$\frac{v(t)}{v_f(t)} = \frac{G_{dq}G_{qf}\omega_r^2 R}{(r_a + r_n + R)(r_a + L_q p)(r_f + L_f p)} \qquad (7.11)$$

In the steady state (p = 0) the inductance terms vanish, and if $(r_a + r_n)$ is small compared with $R$, the transfer function becomes that in eq. (6.16).

EXAMPLE 7.15: An amplidyne driven at constant speed has the parameters

Control field:      Q-axis:    $r_a$ = 4.0 Ω    D-axis:    $r_n = 0$

$r_f$ = 20 Ω              $L_q$ = 1.0 H              $L_d = 0$

$L_f$ = 2.0 H            $G_{qf}\omega_r$ = 240 V            $G_{dq}\omega_r$ = 120 V

Nominal steady-state load is $R$ = 20 Ω at voltage $V$ = 200 V. Find (i) the steady-state power amplification, (ii) the open-circuit voltage amplification, and (iii) the output current for a control voltage step of 1 V with the system initially on no load.

(i) *Power Amplification*  For $V$ = 200 V, then $I$ = 10 A, $V_d = V + Ir_a$ = 240 V, $I_q = V_d/G_{dq}\omega_r$ = 2.0 A, $V_q = I_q r_a$ = 8.0 V, $I_f = V_q/G_{qf}\omega_r$ = 8/240 = 0.033 A, $V_f = I_f r_f$ = 0.67 V. Thence

$$P/P_f = 200 \times 10/0.67 \times 0.033 = 90\ 000$$

(ii) *Voltage Amplification*  $V/V_f$ = 240/0.67 = 360

(iii) *Output Current Transient*  From eq. (7.11), $i(t)/v_f(t) = [v(t)/v_f(t)]/R$, whence

$$\frac{i(t)}{v_f(t)} = \frac{G_{dq}G_{qf}\omega_r^2}{(r_a + R)(r_a + L_q p)(r_f + L_f p)} = \frac{600}{(p + 4)(p + 10)}$$

In terms of Laplace transforms, with $V_f(s) = 1/s$

$$I(s) = 100/[s(s + 4) - s(s + 10)]$$

and by inverse transformation

$$i(t) = 15 - 25 \exp(-4t) + 10 \exp(-10t)$$

which indicates a comparatively slow response. Developments in power electronics have made cross-field machines obsolescent.

# 8 Industrial Motor Control: Conventional

## 8.1 REQUIREMENTS

Motors have to be started, run up to operating speed, and stopped in normal duty or in emergency. The frequency of start and stop is of importance, particularly with high-inertia loads. Speeds may have to be closely constant, or variable over a specific range. Braking and reversal may be a requirement: electric braking is more readily predictable, but mechanical braking may be needed if the motor is to be held at rest against a load torque.

   Motor duties vary widely, but can be classified in general terms as below. This chapter deals with motors operated from conventional d.c. supplies *not* involving rectification of an a.c. source.

### Starting

*Light Duty*  Starting unloaded, with load applied subsequently.

*Normal Duty*  Starting torque up to $1-1\frac{1}{2}$ times rated value.

*Soft Duty*  Gradual increase of load torque during starting to avoid damage to products or processes (e.g., wire-drawing, textiles).

*Heavy Duty*  High-inertia plant, high static friction, repetitive starting at frequent intervals.

### Speed control

*Normal Range*  Up to 2/1 or 3/1.

*Wide Range*  Between 10/1 and 100/1.

*Constant*  Accurately held speed at all loads, e.g. within 0.1% or less of a specified value at a given speed setting.

*Speed-drop*  Inherent reduction of speed with impulsive and fluctuating loads to utilize the stored kinetic energy of a shaft-mounted flywheel.

*Creep*  Creeping speeds for adjustment and setting up of work on the driven machine.

*Reversal*  Change (sometimes cyclic) in the direction of rotation.

189

## Braking

*Rheostatic* Dissipation of the stored kinetic energy in resistors, by dynamic or counter-current methods.

*Regenerative* Return of the major part of the stored energy to the supply.

## Load

Idealized torque/speed and power/speed characteristics for typical loads are shown in Fig. 8.1(*a*). These are

|   | Typical drive | Proportionality | |
|---|---|---|---|
|   |   | Torque | Power |
| A | Traction | $1/\omega_r$ | const. |
| B | Traction, lift, hoist | const. | $\omega_r$ |
| C | Machine-tool | $\omega_r$ | $\omega_r^2$ |
| D | Fan, centrifugal pump | $\omega_r^2$ | $\omega_r^3$ |

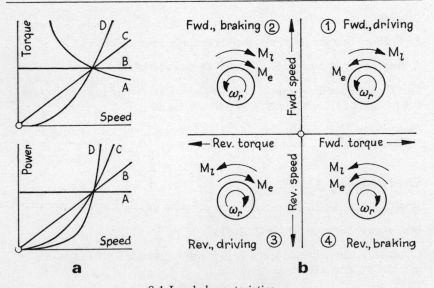

8.1 Load characteristics

All such load requirements can be met by d.c. machines with appropriate (and usually simple) conventional control equipment. The choice of a motor for a given duty requires a matching of the individual characteristics. A series motor, for example, can provide for both high-speed trains and for the lower speed and greater tractive effort of heavy freight trains; a reeler motor (Example 8.3) is suited to a separately excited controlled field.

The relations between electromagnetic torque $M_e$, load torque $M_l$ and speed $\omega_r$ for driving and braking in either direction are shown in the four-quadrant diagram in Fig. 8.1(*b*).

## 8.2 STARTING

Small and fractional-kilowatt motors are started direct-on-line. For larger machines the armature resistance is less than 0.01 p.u., so to prevent excessive temperature-rise and mechanical shock the voltage applied to the armature must initially be limited to that giving an initial starting current not exceeding $1\frac{1}{2}$–$2\frac{1}{2}$ rated full-load level. If a variable-voltage supply is available (as in some methods of speed control) the starting involves no additional equipment; but if the supply is at constant normal voltage it is necessary to absorb the excess by resistors in series with the armature. A liquid rheostat gives a smooth variation in resistance, but in almost all cases the robust and convenient metallic resistor is preferred, arranged in series sections that can be cut out successively by manual or automatic operation.

### Resistor grading

Analytical or graphical calculation can be used to determine the stepping necessary to ensure that the starting current varies between specified upper and lower current limits $I_1$ and $I_2$.

*Shunt and Separately Excited Motors* – Fig. 8.2. At switch-on the total armature-circuit resistance $R_1 = V/I_1$ is made up of the $n$ series sections

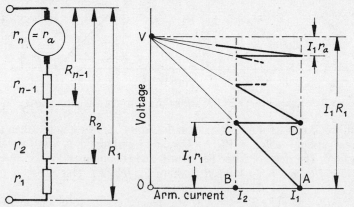

8.2 Shunt or separately excited motors: starter rheostat and e.m.f./ current diagram

$r_1, r_2 \ldots r_{n-1}$ and $r_n = r_a$, and the start is made in $n$ steps by cutting out the sections in sequence. Initially $I_1$ is OA. As the motor gains speed and develops a rotational e.m.f. $E_r$, the current falls along the voltage/current line AV until it reaches $I_2$ at C, when resistor section $r_1$ is cut out, the current jumping back to $I_1$ at D. If the change is instantaneous then $V - R_1 I_2 = E_r = V - R_2 I_1$, whence $R_2/R_1 = I_2/I_1 = k$, and $R_2 = kR_1$. Similarly $R_3 = kR_2$, and with $n$ steps $R_n = r_a = k^{n-1} R_1$. Thus

$$k = {}^{n-1}\sqrt{(r_a/R_1)} = {}^{n-1}\sqrt{(r_a I_1/V)} = {}^{n}\sqrt{(r_a I_2/V)}$$

The section resistances are $r_1 = (R_1 - R_2) = R_1(1 - k), r_2 = (R_2 - R_3)$ $= R_2(1 - k) = kR_1(1 - k) = kr_1 \ldots . r_n = r_a = k^{n-1} r_1$, so that both the total resistances $R$ and the section resistances $r$ are in geometrical progression.

The graph in Fig. 8.2 can be used directly for calculation of the resistance sections, fixing either $I_1$ or $I_2$ and fitting in the required number of steps by trial.

*Series Motor*, Fig. 8.3. A similar procedure with $I_2/I_1 = k$ can be adopted, but (i) with $r_a$ replaced by the total motor resistance $r = r_a + r_f$, and

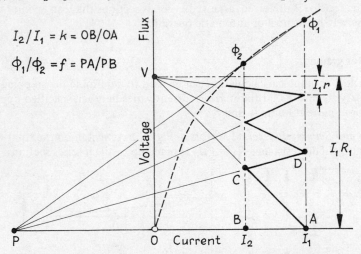

8.3 Series motor: e.m.f./current diagram for rheostat starter

(ii) with the flux $\Phi_1$ for $I_1$ greater than $\Phi_2$ for $I_2$. Let $\Phi_1/\Phi_2 = f$: then the switch-in current is still $I_1 = V/R_1 = OA$, and it reduces to $I_2 = OB$ along the line AV. The step condition between C and D combines $E_{r2} = V - I_2 R_1$ and $E_{r1} = V - I_1 R_2$, where $E_{r1}/E_{r2} = f$. This gives $R_2 = R_1 [kf - (f - 1)]$. The section resistances are $r_1 = R_1 - R_2 = R_1 f(1 - k), r_2 = kfr_1$ etc. Thus the steps form a geometrical progression but the total resistances do not.

The graphical method takes account of the flux difference by drawing lines such as CD from the pole P instead of horizontally. Again the fitting is by trial.

*Practical Starting Characteristics*  The methods above assume some conditions that do not obtain in practice. The instantaneous current jumps between $I_2$ and $I_1$ are opposed by the armature circuit inductance in the shunt motor, and by the considerably greater field inductance additionally in the series motor. Further, the inertia of the motor and its mechanical load may result in minor overshoot. A more probable voltage/current relation is shown in Fig. 7.4($b$).

## Manual starters

*Faceplate*  A spring-loaded contact arm, Fig. 8.4(a) is moved over resistor contacts to start the motor, and is held in the 'on' position by an electromagnet (uv) in series with the shunt field winding. The electromagnet releases the arm should the supply voltage fail, or should its coil be short-circuited by action of the over-current relay (oc). The faceplate starter is obsolete except for small motors up to about 5 kW.

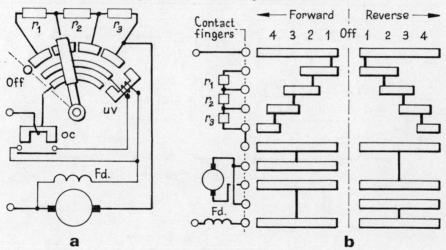

8.4 Manual starters

*Drum Controller*  Contact fingers under spring pressure bear on a set of contact bars mounted on a cylindrical drum, Fig. 8.4($b$), the required switching being made by rotating the drum manually. A magnetic 'blow-out' coil is incorporated to force the arcs away from the contacts and to aid their extinction. The manual effort needed for operation limits the application to motors of about 100 kW in light traction, hoist or haulage service, in which the starting resistors may be used also for speed control. Small drum controllers energizing separate electromagnetic contactors may be employed for large motors.

## Automatic starters

Modern practice for both small and large machines is to initiate starting by means of a push-button, the subsequent switching being effected by automatically controlled contactors, Fig. 8.5.

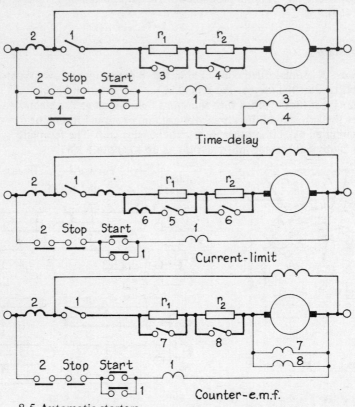

8.5 Automatic starters
Contactor coils and contacts: 1 main, 2 overcurrent,
3–4 time-limit, 5–6 current limit, 7–8 e.m.f.

*Time-delay Relays* With eddy-current discs or oil dashpots to control their operating time, these are set to operate at prefixed intervals after the start is initiated, the delays being timed so that an upper current limit is not exceeded with the heaviest prospective load. On light loads the starting time is likely to be excessive.

*Current-limit Relays* Each resistance step is cut out as soon as the current falls to the lower limit. The starting time thus depends on the load; but on heavy load the sequence is lost if the current fails to drop to its lower limit.

*Counter-e.m.f. Relays* These are connected across the armature and adjusted to operate at successively increasing values of the e.m.f. The starting time is thus load-dependent, but operation is subject to variation in the supply voltage. Further, the e.m.f. for which the contacts open is less than that at which they close, so that an attempt to re-start a motor before it has slowed down sufficiently for the e.m.f. to fall below the opening level will result in excessive motor current.

Most automatic starters embody additional control and safety features. The main contactor closes auxiliary *retaining contacts* in parallel with the 'start' button contacts so that the starting sequence continues after the button is released. An *inching* button without this feature may be provided. Several 'stop' buttons are placed strategically for *emergency* shut-down, and *interlocks* comprising contacts in series with the stop button are fitted to gates and safety guards to reduce hazard to operatives. *Signal lamps* are used to indicate the control state, including a 'trip' lamp to show when the machine has been disconnected by action of a protective relay.

Starter *resistors* are short-time rated, but with frequent starts and heavy duty their $I^2 R$ loss demands adequate ventilation.

## 8.3 SPEED CONTROL

From eq. (6.1) the speed of a motor is $\omega_r = E_r/K\Phi = (V_a - I_a r_a)/K\Phi$, whence the three basic methods of speed control are (i) field variation by adjustment of the main flux, (ii) armature-circuit resistance control by series rheostat, and (iii) variation of the applied armature voltage. Basic speed ranges available are shown in Fig. 8.6.

### Field control

The only means usually available is a reduction of the field current so that only speeds above the full-field value can be obtained, and the torque for a given armature current is reduced. The flux distortion due to armature reaction makes too weak a flux impracticable, and a reduction of field current below 70% of the maximum is unusual, unless compensating windings are provided.

*Shunt and Separately Excited Motors* The field current control in shunt machines is achieved by a series rheostat in the field circuit, comprising a simple slide-wire for small motors or an arc of studs over which a contact arm can be moved for larger machines. Up to 50 studs may sometimes be required for fine control of speed. As the field current is normally small, there is but little $I^2 R$ loss. For separately excited motors, the field current control method depends on the kind of source employed.

*Series Motor* A diverter resistor in parallel with the field winding reduces the proportion of the main motor current that passes through the field system. A preferable method is to provide tappings on each field coil, as the low resistance of a diverter may be affected by contact resistance, and under

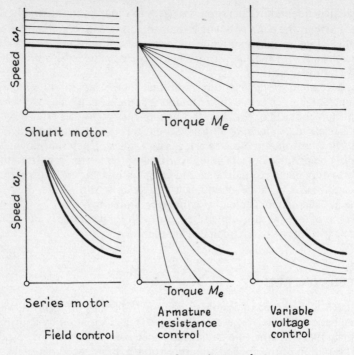

Shunt motor

Series motor

Field control

Armature resistance control

Variable voltage control

8.6 Ranges of speed control

transient conditions will divert an excessive proportion of the motor current because of the field-winding inductance. For the *steady state*, with a motor voltage $V$ and current $I$, let the diverter resistance be $R_d$ and carry a current $I_d$, so that the field current is $I_f = I - I_d$ in the field resistance $r_f$. Then $R_d I_d = r_f I_f$ , whence

$$I = I_f + I_d = I_f(1 + r_f/R_d)$$

$$E_r = V - I_f(r_a + r_f + r_a r_f/R_d)$$

(8.1)

For a given field current, $E_r$ can be found, and the speed $\omega_r$ determined from $\omega_r = \omega_m(E_r/E_m)$, where $E_m$ is the e.m.f. for the specified field current from a magnetization characteristic at speed $\omega_m$.

### Armature resistance control

Strictly this is a form of voltage control by increasing the IR volt drop. It has wide application, often using suitably rated starter resistors for the purpose; but it is wasteful.

*Shunt Motor: Series Resistor*   The speed/torque curves in Fig. 8.6 show that high torque can be obtained at low speed, but for low torque (high series resistance) the characteristic is so steep that stable creep speeds are not readily achieved.

Consider a motor with an armature supply voltage $V_a$ and a speed $\omega_0$ when running without added series resistance and taking a current $I_{a0}$. The economics of speed reduction depends on the load characteristics:

1. Load torque constant. For a lower speed $\omega_r$, the volt drop in the armature-circuit series resistance (neglecting $r_a$) is $V_a - E_r = V_a(\omega_0 - \omega_r)/\omega_0$. The $I^2 R$ loss is proportional to the speed drop and the efficiency is proportional to $\omega_r$. Prolonged operation at low speed is therefore wasteful.
2. Load torque proportional to (speed)$^2$. For a lower speed $\omega_r$ the armature current is $I_a = I_{a0}(\omega_r/\omega_0)^2$. The power loss in resistance is $V_a I_{a0}(1 - \omega_r/\omega_0)(\omega_r/\omega_0)^2$. This is a minimum for $\omega_r = \omega_0$ (2/3) and is only about 15% of the power input to the motor at normal speed. In such a case the simple series resistance control may be deemed economic.

*Shunt Motor: Parallel Resistor (Shunted Armature).* Stable creeping speeds can be achieved by the arrangement in Fig. 8.7, where the armature is in

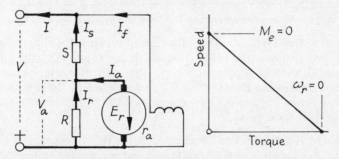

8.7 Shunt motor: shunted-armature control

series with a resistor $S$ and in parallel with a resistor $R$. With the indicated voltages and currents

$$V_a = E_r + I_a r_a \quad V = V_a + I_s S \quad I_s = (V - V_a)/S \quad I_r = V_a/R$$

The armature current is $I_a = I_s - I_r$, and writing $S/R = \alpha$ gives

$$I_a = [V - E_r(1 + \alpha)]/[S + r_a(1 + \alpha)]$$

For a given field current $I_f$, the rotational e.m.f. is $E_r = GI_f\omega_r = K\omega_r$ and the electrical torque is $M_e = KI_a$. Then

$$M_e = K[V - K(1 + \alpha)\omega_r]/[S + r_a(1 + \alpha)]$$

The torque is zero for a speed $\omega_r = V/K(1 + \alpha)$, the speed is zero for a torque $M_e = KV/[S + r_a(1 + \alpha)]$, and the torque/speed curve is a straight line between these values. The connection is suitable where creep speeds are required for occasional short periods only, because the input current is high and the efficiency low.

EXAMPLE 8.1: On normal full load, a 500 V shunt motor runs at 126 rad/s

(1200 r/min) and takes 25 kW. The field current is 3.0 A and the armature resistance is 0.40 $\Omega$. The shunted-armature connection is to be used to give a torque/speed characteristic falling from 600 r/min (62.8 rad/s) at zero torque to 300 r/min (31.4 rad/s) at full-load torque. Estimate (i) the resistors $R$ and $S$ required, (ii) the supply current and overall efficiency for the full-load torque condition.

At normal full load, $I = 50$ A, $I_a = 47$ A, $E_r = 481$ V, $K = 481/126 = 3.82$, $M_e = 3.82 \times 47 = 180$ N-m.

(i) For zero torque, $\omega_r = 62.8 = V/K(1 + \alpha) = 500/3.82(1 + \alpha)$, whence $(1 + \alpha) = 1 + S/R = 2.08$. Thus $S = 1.08 R$. For full-load torque

$$180 = 3.82 \, [500 - 3.82 \times 2.08 \times 31.4] \, / \, [S + 0.40 \times 2.08]$$

from which $S = 4.5 \, \Omega$ and $R = 4.2 \, \Omega$.

(ii) The speed is $\omega_r = 31.4$ rad/s. The armature voltage is $V_a = K\omega_r + I_a r_a = 139$ V, whence $I_r = 139/4.2 = 33$ A, $I_s = 33 + 47 = 80$ A, $I = 80 + 3 = 83$ A, and the efficiency is $180 \times 31.4/500 \times 83 = 0.14$ p.u.

*Series Motor: Series Resistor*  The torque/speed curves with series resistance are increasingly steep as the added resistance is raised, and the stalling torque is delimited. If the load torque exceeds this limit the rotor reverses and enters the 'reverse, braking' quadrant of Fig. 8.1, being driven by the overhauling load (e.g. a descending hoist). High reverse speeds may be reached if the series resistance is excessive.

*Series Motor: Parallel Resistor (Shunted Motor).*  This arrangement, Fig. 8.8(*a*), is virtually a voltage divider reducing the applied motor voltage

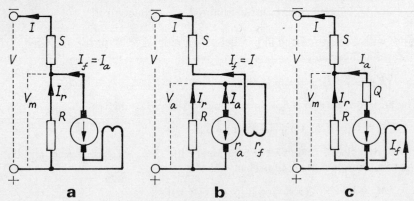

8.8 Series motor: resistance control

from $V$ to $V_m = V - IS$. Then

$$I_r = (V - IS)/R \quad I_a = I_f = I - I_r = I(1 + \alpha) - V/R$$

where $\alpha = S/R$, so that $I = (I_a + V/R)/(1 + \alpha)$. The rotational e.m.f. is

$$E_r = V - IS - r_m I_a = V/(1 + \alpha) - I_a [r_m + S/(1 + \alpha)]$$

where $r_m = r_a + r_f$. For a given $I_f$, the e.m.f. can be calculated and the torque/speed relation determined using a magnetization characteristic. For a given torque the speed is reduced, though to a minor extent only at light loads as the diversion of current through $R$ weakens the field.

*Series Motor: Parallel Resistor (Shunted Armature).* If as in Fig. 8.8(*b*) the armature only is shunted by $R$, the field current $I_f$ is the line current $I$, and low speeds can be obtained. The field current and the no-load speed can be limited by appropriate choice of $S$ and $R$. Let $S + r_f = S'$: then $V_m = V - S' I_f$ and $I_r = V_m/R$, whence

$$I_a = I_f - I_r = I_f(1 + S'/R) - V/R$$

$$E_r = V_m - r_a I_a = V(1 + r_a/R) - I_f[S' + r_a(1 + S'/R)]$$

If the load overhauls the motor, operation is in the second, 'forward braking' quadrant. The armature current reverses, decreasing $I_f$, and the torque has a maximum which, if exceeded by the load torque, may result in a dangerously high speed.

*Series Motor: Potentiometer Lowering* With crane and hoist drives it is desirable to have the facility of rapidly lowering the empty hook or container. The connection, Fig. 8.8(*c*), comprises a voltage divider, one branch of which includes the field winding and supplies the armature through a resistor $Q$. As the electrical torque has to augment the gravity effect, the armature connections must be reversed when changing from 'hoist' to 'lower' operation. Writing $R' = R + r_f$ and $Q' = Q + r_a$, then the voltage across the armature circuit is $R'I_f$ and across $S$ is $V - R'I_f$, whence

$$I = (V - R'I_f)/S \quad I_a = I_f - I = I_f(1 + R'/S) - V/S$$

$$E_r = -R'I_f - Q'I_a = V(Q'/S) - I_f(R' + Q' + R'Q'/S)$$

The torque/speed relation can be plotted, as in the following Example.

EXAMPLE 8.2: A series motor with armature and field resistances of 0.10 and 0.05 $\Omega$ respectively has a magnetization characteristic at 600 r/min given by

| Field current (A): | 25 | 50 | 75 | 100 | 125 | 150 |
|---|---|---|---|---|---|---|
| E.M.F. (V): | 57 | 106 | 135 | 155 | 167 | 180 |

The motor operates on a 230 V supply. Obtain the speed/torque characteristics for the following conditions:

A. Basic: normal connection as series motor.
B. Diverter: $R_d = 0.10 \ \Omega$ across field winding.
C. Series resistor: $S = 2.0 \ \Omega$.
D. Shunted motor, Fig. 8.8(*a*): $S = 3.0 \ \Omega$, $R = 4.0 \ \Omega$.
E. Shunted armature; Fig. 8.8(*b*): $S = 3.95 \ \Omega$, $R = 2.0 \ \Omega$.
F. Potentiometer lowering, Fig. 8.8(*c*): $S = 2.0 \ \Omega$, $R = 0.95 \ \Omega$, $Q = 0.5 \ \Omega$.

The evaluations are tabulated below. For each condition the speed is obtained from $\omega_r = \omega_m(E_r/E_m)$ where $E_m$ is obtained from the magnetization characteristic at $\omega_m = 62.8$ rad/s.

Magnetization curve at $\omega_m = 62.8$ rad/s:

| | | | | | | | |
|---|---|---|---|---|---|---|---|
| Field current $I_f$ | (A) | 25 | 50 | 75 | 100 | 125 | 150 |
| E.M.F. $E_m$ | (V) | 57 | 106 | 135 | 155 | 167 | 180 |

| | | | | | | | |
|---|---|---|---|---|---|---|---|
| A $E_r = 230 - 0.15\,I_f$ | (V) | 226 | 222 | 219 | 215 | 211 | 207 |
| $\omega_r$ | (rad/s) | 247 | 132 | 102 | 87 | 79 | 72 |
| $M_e$ | (N-m) | 23 | 84 | 161 | 247 | 324 | 430 |

| | | | | | | | |
|---|---|---|---|---|---|---|---|
| B $E_r = 230 - 0.20\,I_f$ | (V) | 225 | 220 | 215 | 210 | 205 | 200 |
| $\omega_r$ | (rad/s) | 248 | 130 | 100 | 85 | 77 | 70 |
| $I_a = 1.5\,I_f$ | (A) | 38 | 75 | 113 | 150 | 188 | 225 |
| $M_e$ | (N-m) | 34 | 127 | 243 | 371 | 500 | 643 |

| | | | | | | | |
|---|---|---|---|---|---|---|---|
| C $E_r = 230 - 2.15\,I_f$ | (V) | 176 | 122 | 69 | 15 | −39 | −92 |
| $\omega_r$ | (rad/s) | 194 | 72 | 32 | 6 | −16 | −32 |
| $M_e$ | (N-m) | 23 | 84 | 161 | 247 | 324 | 430 |

| | | | | | | | |
|---|---|---|---|---|---|---|---|
| D $E_r = 131 - 1.86\,I_f$ | (V) | 85 | 38 | −8 | −55 | −102 | −148 |
| $\omega_r$ | (rad/s) | 94 | 23 | −4 | −22 | −38 | −52 |
| $M_e$ | (N-m) | 23 | 84 | 161 | 247 | 324 | 430 |

| | | | | | | | |
|---|---|---|---|---|---|---|---|
| E $E_r = 242 - 4.2\,I_f$ | (V) | 137 | 32 | −73 | −178 | −283 | −388 |
| $I_a = 3.0\,I_f - 115$ | (A) | −40 | 35 | 110 | 185 | 260 | 335 |
| $\omega_r$ | (rad/s) | 151 | 19 | −34 | −72 | −106 | −135 |
| $M_e$ | (N-m) | −36 | 59 | 236 | 457 | 694 | 963 |

| | | | | | | | |
|---|---|---|---|---|---|---|---|
| F $E_r = 69 - 1.90\,I_f$ | (V) | 22 | −26 | −74 | −121 | −168 | −216 |
| $I_a = 1.5\,I_f - 115$ | (A) | −78 | −40 | −2 | 35 | 73 | 110 |
| $\omega_r$ | (rad/s) | 24 | −15 | −34 | −49 | −63 | −75 |
| $M_e$ | (N-m) | −71 | −69 | −4 | 86 | 195 | 317 |

The characteristic curves are plotted in Fig. 8.9, marked in accordance with the operating conditions A to F above.

**Voltage control**

A variable voltage for supplying a motor or its armature may be obtained from a separate d.c. generator, from a thyristor rectifier (Chapter 9) or by series-parallel connection.

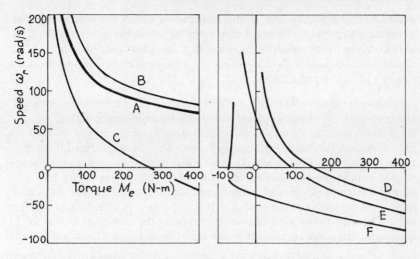

8.9 Series motor: speed control: Example 8.2

*Ward-Leonard System* The basic arrangement is given in Fig. 7.11. The generator G is driven by an induction motor, alternatively by a synchronous motor or occasionally by a prime-mover; and its field, provided by a controlled exciter or a rectifier, is variable from zero to a maximum in either polarity to feed the armature of the main motor M. With M fully excited, its output speed is consequently variable over a wide range in either direction up to maxima determined by the maximum output voltage of G. The range can be extended by weakening the field of M, yielding the rated torque/speed and power/speed relations in Fig. 8.10. Starting M requires no additional

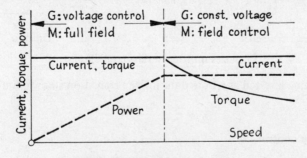

8.10 Ward-Leonard control

equipment, but if the starting duty is severe, G has to supply large currents when its field is weak, so that a compensating winding is necessary to limit the field distortion and to secure good commutation. The obvious drawback of the Ward-Leonard system is the capital cost of a set of three machines of comparable rating and the space that it occupies, but the system gives a wider and more controllable range of speeds than alternative methods, a facility of value in such applications as the accurate 'decking' of mine winders.

*Ward-Leonard-Ilgner System*  If M is subject to a widely fluctuating load (as in rolling-mill drives), a flywheel may be mounted on the generator shaft, with G driven by an induction motor with a drooping speed/torque characteristic to enable the kinetic energy stored in the flywheel to provide a measure of load equalization.

*Series-parallel Connection*  This method is restricted to traction equipments where at least two motors operate in parallel for full-speed running. Reconnecting the motors in series reduces the voltage per motor without the use of resistors (although further speed control could be obtained by series resistors and/or by field diverters). With a pair of motors, each can, by series-parallel control, be operated on one-half and full line voltage. With four motors a series/series-parallel/parallel connection gives additionally a one-quarter of the line voltage per motor. It is important to effect transition from one connection to the next without interrupting the current and to avoid transients. The *short-circuit* and *bridge* transitions, Fig. 8.11, are available,

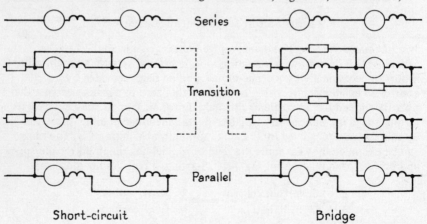

8.11 Series motors: series-parallel control

resistors being inserted as required during the transition steps to limit the currents.

### Closed-loop control

The addition to a motor system of sensors and amplifiers enables a wide variety of simple automatic controls to be devised. A typical case is outlined in the following Example.

EXAMPLE 8.3: A constant-tension reeler, Fig. 8.12, is fed with strip from a rolling mill at speed $u$ on to a reel of diameter varying between $d_0$ (empty) and $d_1$ (full). The armatures of generator G, tension-reference T and exciter E are driven at constant speed on a common shaft. E has two identical field windings, one carrying a constant current $I_k$, the other a current $i_j$ derived

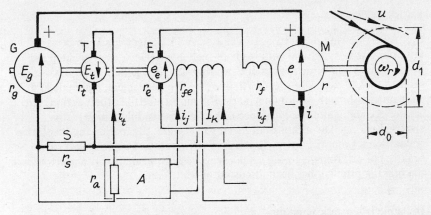

## 8.12 Strip-mill reeler control: Example 8.3

from an amplifier of gain $A$ and input proportional to the difference between the constant e.m.f. $e_t$ of T and the volt drop across a resistor $S$. The system parameters are

Feed: tension 45 kN, $u = 10.0$ m/s, $d_0 = 0.38$ m, $d_1 = 0.91$ m.
Armature resistances: $r_g = 0.030, r = 0.025, r_t = r_e = 1.0\ \Omega$.
Field resistances: $r_f = 4.0, r_{fe} = 50\ \Omega$.
E.M.F.s: $E_g = 800$ V, $e = 0.95\ \omega_r i_f, E_t = 21.6$ V, $e_e = 104(I_k - i_j)$.
Amplifier: $A = 500$ V/A, $r_a = 5.0\ \Omega$. Resistor: $r_s = 0.035\ \Omega$.

As the motor torque is proportional and its speed inversely proportional to reel diameter, show that the conversion power is substantially constant, given that the motor armature current with the reel empty is $i = 603$ A and assuming the system to be linear.
The main-loop resistance is $r_g + r_s + r = r_1$; the volt drop across $S$ is $r_s i$; the amplifier input current is $i_t = (E_t - r_s i)/(r_t + r_a)$; the net field current of E is $(I_k - i_j)$ where $i_j = Ai_t/r_{fe}$, whence $e_e = K(I_k - i_j)$; the motor field current is $i_f = e_e/(r_f + r_a)$. Combining these expressions yields

$$e = \omega_r GK \frac{[I_k - A(E_t - r_s i)/(r_f + r_a)]}{(r_e + r_f)} = \omega_r\, 19.2\ [I_k - 36.0 + 0.058i]$$

where $G = 0.95$ and $K = 104$. For $d_0 = 0.38$ m, $\omega_r = \omega_0 = 2u/d_0 = 52.6$ rad/s, $i = 603$ A and $e = E_g - r_1 i = 746$ V. The equation above then gives $I_k = 1.8$ A, whence the speed relation is

$$i = 588(\omega_r + 1.18)/(\omega_r + 0.08)$$

For $d_1 = 0.91$ m and $\omega_r = \omega_1 = 22.0$ rad/s, the motor current is $i = 617$ A and the e.m.f. is $e = 745$ V. For the limiting reel diameters the conversion powers $ei$ are 450 and 459 kW, giving gross torques proportional to reel diameter within 2%.

## 8.4 BRAKING

Controlled slowing or stopping of a motor and its driven load can be as important as starting, and with loads of high gravitational energy, such as hoists, cranes and vehicles on a downward grade, it is necessary to brake the motor to avoid excessive speed. Friction, electromechanical, magnetic and eddy-current brakes do not depend on the motor itself for their action, but greater economy and the avoidance of brake-wear and dust may justify electric braking. The motor is arranged to generate, dissipating the kinetic or gravitational potential energy in resistors (*dynamic* and *counter-current* braking) or returning it to the supply (*regenerative* braking). The dynamics of the braking problem has been discussed in Sect. 7.5.

### Dynamic (rheostatic) braking

In this simple method for bringing a motor almost to standstill, the motor (or its armature) is disconnected from the supply and immediately re-connected across a resistor, normally all or part of the starting rheostat. The system energy is dissipated in the armature-circuit resistance.

The most important parameter is the braking time $t_b$ during which the energy has to be dissipated. Fast braking demands a low circuit resistance, the minimum being of the order of three times the resistance of the armature. The mean brake power over a duty cycle may be small, but both the brake resistor and the motor armature must be capable of dealing with the peak power, the former because of temperature-rise and the latter because of commutation. The braking time that can be achieved depends naturally on the system inertia $J$ and the friction and load torques, and on the motor power rating $P$. Considering only the motor itself, its inertia is a function of the armature length $l$, the diameter $D$ and the radius of gyration. For a range of motors of fixed ratio $D/l$, the inertia is $J = k_1 D^5$; for a given normal speed $P = k_2 D^3$, whence $J = k_3 P^{5/3}$. In general the armature resistance is inversely proportional to $P$, so that the minimum braking time, eq. (7.6), for a total brake-circuit resistance $R = 3r_a$ and a 90% speed drop, is $t_b = kP^{2/3}$. A typical value of $k$ (in $s/W^{2/3}$) is 0.002.

*Shunt Motor*   The field circuit may be left connected to the supply as in Fig. 7.7(*a*), but braking fails if there is a supply fault. With the field left connected across the armature, Fig. 7.7(*b*), the braking torque is initially the same but falls more rapidly with speed, and collapses when the speed has fallen below the critical value for self-excitation.

EXAMPLE 8.4:   A 250 V shunt motor, with field and armature resistances $r_f = 50 \ \Omega$ and $r_a = 0.10 \ \Omega$, has a magnetization characteristic at 1000 r/min ($\omega_m = 105$ rad/s) given by

| Field current $I_f$   (A) | 0.5 | 1 | 2 | 3 | 4 | 5 |
|---|---|---|---|---|---|---|
| E.M.F. $E_m$   (V) | 42 | 83 | 150 | 183 | 203 | 215 |

The load and friction torque is $M_l = (25 + 1.0\omega_r)$ N-m and the system inertia is $J = 0.80$ kg-m$^2$. The motor is dynamically braked with the field circuit across the armature terminals and a braking resistor $R_e = 2.0\ \Omega$ connected across the armature. Estimate (i) the steady-state speed/braking-torque relation for the motor, and (ii) the braking time to standstill of the system. Ignore inductance effects and armature reaction.

(i) *Speed/torque* If the current in the braking resistor is $I$, the steady-state operating conditions are represented by

$$V_m = R_e I = r_f I_f = E_r - r_a I_a$$

$$I_a = I + I_f = I_f(1 + r_f/R_e) = 26 I_f$$

$$E_r = r_f I_f + r_a I_a = I_f(r_a + r_f + r_a r_f/R_e) = 53 I_f$$

$$M_e = E_r I_a/\omega_r = -0.25\ E_m I_f$$

Using the data from the magnetization characteristic, the speed and motor brake torque values can be tabulated:

| $I_f$ | (A) | 0.5 | 1 | 2 | 3 | 4 | 5 |
|---|---|---|---|---|---|---|---|
| $I_a$ | (A) | 13 | 26 | 52 | 78 | 104 | 130 |
| $E_r$ | (V) | 27 | 53 | 106 | 159 | 212 | 265 |
| $\omega_r$ | (rad/s) | 67 | 66 | 73 | 90 | 108 | 128 |
| $M_e$ | (N-m) | −5 | −21 | −75 | −137 | −203 | −269 |

The critical field resistance at $\omega_m = 105$ rad/s is $42/0.5 = 84\ \Omega$, and it becomes equal to the field resistance at $\omega_r = 105(50/84) = 62$ rad/s. Below this speed the motor develops negligible braking torque, as indicated by the curve $M_e$ in Fig. 8.13.

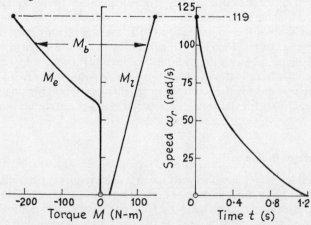

8.13 Shunt motor: dynamic braking: Example 8.4

(ii) *Braking Time*  By trial, the normal running condition on 250 V supply is found to be $I_f$ = 5.0 A, $E_r$ = 243 V, $I_a$ = 70 A, $\omega_r$ = 119 rad/s. On instantaneous switching to dynamic braking, the flux and speed are momentarily unchanged, whence $E_r$ = 243 V, $I_a$ = $-243/(2.0 + 0.1) = -116$ A approximately, and $M_e$ = 243($-116$)/119 = $-237$ N-m. This is augmented to $-(237 + 25 + 119) = -381$ N-m by the load torque, and represents the initial braking condition. The load torque is plotted in Fig. 8.13, enabling the total braking torque for any speed to be obtained. The braking time is found by the method of Fig. 7.3, or by calculation in steps $\Delta\omega_r$ of speed and $\Delta t$ of time, where $\Delta\omega_r/\Delta t = M_b/J$ and $M_b = M_e + M_l$:

| Speed $\omega_r$ | $M_e$ | $M_l$ | $M_b$ | Mean $M_b$ | $\Delta\omega_r/\Delta t$ | $\Delta t$ | $t$ |
|---|---|---|---|---|---|---|---|
| 119 | 237 | 144 | 381 | | | | 0 |
| 110 | 210 | 135 | 345 | 363 | 454 | 0.020 | 0.020 |
| 100 | 175 | 125 | 300 | 322 | 402 | 0.025 | 0.045 |
| . . . | | | | | | | . . . |
| 20 | 0 | 45 | 45 | | | | 0.725 |
| 10 | 0 | 35 | 35 | 40 | 50 | 0.200 | 0.925 |
| 0 | 0 | 25 | 25 | 30 | 37 | 0.270 | 1.20 |

The braking time is therefore 1.2 s, the machine being brought to rest by the load and friction torque. It must be noted that inductances have been neglected: the field inductance will, in practice delay the collapse of the main flux and the armature inductance will impede sudden changes of armature current.

*Series Motor*  The same principle may be applied, except that it is rarely practicable to excite the field winding from an auxiliary source and it is necessary for braking to reverse either the field or the armature connections to ensure build-up of the armature e.m.f. The braking resistor must be such that $(R_e + r_a + r_f)$ is less than the critical resistance for the speed at which braking is initiated. As with the shunt motor, braking torque is lost after the critical speed has been reached. In traction service with two or more motors it may be convenient for them to feed in parallel a common braking resistor; if so, the conditions for the parallel operation of series generators must be observed.

## Counter-current braking

A strong braking effect down to zero speed is achieved by maintaining the supply voltage $V_a$ to the armature but reversing its connections. The effective armature-circuit voltage is $V_a + E_r$, and initially this approaches $2V_a$, so that a limiting resistor must be switched into the armature circuit.

If the starting rheostat is used for this purpose, the initial braking current will be about twice the starting current. Considerable power is drawn from the supply during braking and this, as well as the system kinetic energy, has to be dissipated in the armature and the limiting resistor. The supply must be disconnected at zero speed (unless it is intended that the motor should run up in the reverse direction), using a current or speed directional relay. The large initial current and the consequent high mechanical stress limit the application of counter-current braking ('plugging') to small machines.

## Regenerative braking

This method, in which most of the braking energy is returned to the supply, is used for slowing or stopping a machine especially where the duty cycle demands this action frequently, but its most useful application is in holding a descending load of high potential energy at a constant speed. Significant saving in overall energy consumption can be made at the expense of some complication, but it is essential that the supply system be capable of accepting the regenerated energy without excessive change of voltage. The condition for regeneration is that the rotational e.m.f. exceeds the applied voltage to reverse the current and to change the mode from motoring to generating. This is effected by increasing the motor field current or the armature speed or, occasionally, reducing the supply voltage. With a supply of 'constant' voltage the range of speed in which regeneration is possible is limited.

*Shunt Motor*   Regeneration is automatic if the rotational e.m.f. exceeds the supply voltage, operating conditions passing smoothly from quadrant 1 to quadrant 3 in Fig. 8.1.

*Series Motor*   Regenerative braking of industrial series motors is uncommon, but on d.c. traction systems where electric locomotives haul heavy trains and the route profile has long stretches of downhill track, regenerative equipment economizes electrical energy demand and reduces the wear of mechanical brakes. The regenerative braking must be mechanically stable (i.e. the braking effort must increase with the train speed) and electrically stable (i.e. insensitive to line-voltage variation and capable of a wide controlled speed range). Plain series motors, almost universal for d.c. traction duty, are unsuitable: they must be compounded. Fig. 8.14($a$) shows the main series field winding S carrying both the regenerated current $I$ and an auxiliary current $I_x$ furnished by an exciter. Using the symbols in ($b$), then for the motor mode ($I$ positive)

$$E_r = G\omega_r(I_x + I) = V - r_a I - r_s(I_x + I)$$

and the traction machine operates as a cumulatively compounded motor. For a given $I_x$ there is a balancing speed $\omega_0$ for which $E_r = V$ and $I = 0$, i.e. $G\omega_0 I_x = V - r_s I_x$. For higher speeds the machine generates ($I$ negative) for which

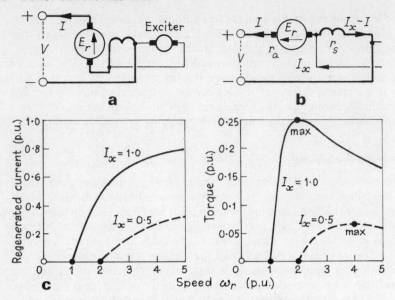

**c**

8.14 Series motor: regenerative braking

$$E_r = G\omega_r(I_x - I) = V + r_a I + r_s(I_x - I)$$

showing that $I_x$ must be greater than $I$ for the generator mode (a condition that gives the essential reversal of the field). The current/speed and torque/speed characteristics can be found to adequate approximation if both magnetic linearity and negligible resistance are assumed: then
$E_r = G\omega_0 I_x = V = G\omega_r (I_x - I)$, whence the regenerated current and braking torque become

$$I = I_x(\omega_r - \omega_0)/\omega_r \qquad M_e = VI_x(\omega_r - \omega_0)/\omega_r^2$$

These characteristics are plotted in (*c*) for $I_x$ values of 1.0 and 0.5 p.u., assuming the line voltage to be constant at $V = 1.0$ p.u. The regenerated current $I$ rises with speed above the balancing level $\omega_0$, becoming asymptotic to $I_x$. The braking torque has a maximum of $\frac{1}{4} VI_x/\omega_0$ at a speed $\omega_r = 2\omega_0$ for which $I = \frac{1}{2}I_x$. Thus if regeneration is set to begin at a train speed of, say, 30 km/h, then the maximum braking effort occurs at 60 km/h. Should the downhill gradient be such as further to accelerate the train, $I$ rises but $M_e$ falls, and train control can be maintained only by application of mechanical braking. The system is therefore electrically but not mechanically stable.

Industrial series motors that have, as in crane operation, to deal with over-hauling loads, are provided with mains-fed shunt field windings which limit the motor speed for regeneration in the same way as for normal shunt machines.

## 8.5 PROTECTION

A generator or a motor requires adequate protection against hazards arising in the machine and its control gear, in the supply system to which it is connected, in its attached prime mover or mechanical load, and in its environment. Apart from such electrical faults as insulation failure, breakdown of a machine may occur through overload, overspeed (due, for example, to a field failure), flooding, blockage of ventilation ducts, bearing failure, dry coil-riser joints, commutator metal fatigue, short-circuited sectors, seized brushes, shaft currents and vibration. The electrical result will usually be a condition of overcurrent or overvoltage.

### Generators

Protection against excessive current is normally provided by a conventional circuit-breaker, set to trip at 20–30% overcurrent, a mechanical or thermal time-lag being incorporated to give an inverse time/current characteristic with a minimum tripping time of 150–500 ms. Where the cost is justified, embedded temperature sensors are installed in likely hot-spots in windings and bearings to give an alarm signal or to initiate tripping.

Generator short-circuit currents have a rate of rise limited by armature inductance and a prospective final value limited by armature resistance. These parameters depend on the rating, speed and voltage of the generator: inductance lies between 1 H and a few millihenrys, and resistance between 0.01 and 0.1 p.u. for large and medium ratings respectively. To prevent commutator flashover it is desirable that a short-circuit current be interrupted well before its prospective level is reached. High-speed circuit-breakers with contacts of complex construction, and special features to achieve rapid flux change in the magnetic tripping system, can be used for large machines to give tripping within 10–15 ms or less.

Generators operating in parallel require reverse-power relays to prevent a machine from motoring should a field or prime-mover fault occur.

Internal faults, due usually to insulation breakdown, are cleared by reverse-power protection, or (for an isolated machine) by earth-leakage protection.

### Motors

Motors are subject to a wider range of faults than are generators, as they are often less effectively maintained and may have to operate in hazardous environments. Basic protection is provided by standard circuit-breaker or contactor gear, tripped by overcurrent, thermal or other relays at 20–30% overcurrent.

Supply-system faults result in a brief or sustained interruption. Protective devices must trip the main switchgear so that the full starting procedure is imposed when the supply is restored. Although a motor may momentarily feed energy back into a system short circuit and trip on overcurrent, a voltage-failure (undervoltage) relay is essential.

Overspeed, usually the result of a field failure, can quickly destroy a motor and its mechanical load. Protection is provided by a current relay in the field circuit, sometimes with back-up by a mechanical overspeed device. In some cases the overspeed protection initiates the application of a mechanical brake.

Faults in the driven load, other than overload, include stalling, overspeed due to an overhauling load, and excessive load fluctuation. Stalling normally results in a very large overcurrent, but if the machine has current limitation by voltage reduction, the current may sink below a value sufficient to operate the normal overcurrent protection and, with the motor at rest, flow through the same commutator sectors and risers continuously. This may cause a burn-out, so that a low-set current-operated relay associated with a speed relay (operating if the speed falls below a preset minimum) may be required).

A sudden stalling of the load may impose a severe mechanical shock to the transmission even if the supply is tripped. A 'weak link' is sometimes interposed to prevent damage to the shaft and keys.

# 9 Industrial Motor Control: Electronic

## 9.1 SOLID-STATE SUPPLY AND CONTROL SYSTEMS

As d.c. mains supplies are rare, power for industrial d.c. motors is normally obtained from an a.c. power network through rectifiers in order to retain the speed-control advantages offered by the d.c. machine. Even with Ward-Leonard control, the traditional a.c.-motor/d.c.-generator set is reduced to a controlled thyristor converter, and the d.c. main-motor field derived from a separate rectifier. The outstanding advantage of the thyristor is its controllability, which makes feasible a high-performance drive. With the armature voltage as the speed signal, a 20/1 speed range with a regulation within 2% of maximum speed is readily obtained; with a shaft-mounted tachogenerator these figures may be improved to 100/1 and 0.1%. Still higher precision is achieved by digital techniques, correcting the basic analogue system to give long-term speed holding to 0.01%.

Modern d.c. motors are designed for rectified supply. A completely laminated magnetic circuit provides partial suppression of the core-loss increment, and improves both commutation and the response of the machine to control commands. For a given shaft height the output of the d.c. motor is greater than that of an a.c. machine, and its overload capability is markedly superior: a small permanent-magnet motor, for example, might sustain a 5 p.u. overload. Such performance, together with wide and accurate thyristor speed control, has opened to the d.c. machine a field of application in which it can outdo its a.c. rivals.

Fractional-kilowatt 1-ph/d.c. drives are well established for hand-tools, food-mixers and washing machines. 'Packaged' drives for industrial applications up to 500 kW are readily available, and custom-built equipments can be designed for ratings of 10 MW or more. For drives up to 5 kW, 1-ph converters are employed. Three-phase half-controlled bridges with three diodes and three thyristors are in use for motors up to 150 kW where regenerative braking is not a requirement. Fully-controlled 6-pulse bridges with six thyristors are applicable up to 1 MW, and 12- or 24-pulse converters' for largest ratings.

The d.c. series motor has for many years been applied in electric traction to locomotives, multiple-unit rapid-transit trains and tramways. With a 1-ph a.c. contact-line feed, the tap-changing transformers on the railway rolling stock can control the driving-motor voltage through small robust diodes. The control by a thyristor converter, however, can be 'stepless', permitting maximum exploitation of the available adhesion without wheelslip, a considerable advantage on rapid-transit trains. Conventional diesel-electric locomotives have engine-driven d.c. generators, but the trend is to substitute a.c. generators and solid-state rectifiers to deliver d.c. to the traction motors.

## 9.2 SOLID-STATE DEVICES

The rectifiers here concerned are based on the properties of p- and n-type materials in conjunction.

### Diode

An *ideal* diode offers zero impedance to current flow in one direction and infinite impedance in the other. A *practical* solid-state diode has a very low forward resistance (giving a 0.5–1.5 V drop) and a reverse current of a few milliamperes when blocking several hundred volts, Fig. 9.1(*a*). It can operate

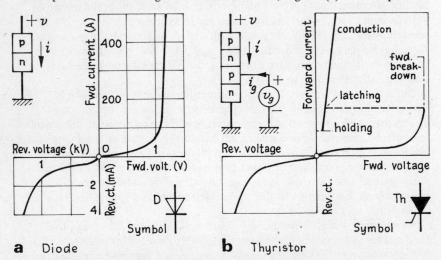

**a**   Diode          **b**   Thyristor

9.1 Semiconductor diode and thyristor

with a mean forward current of about 100 A/cm$^2$ of active area. The voltage rating is determined by the reverse characteristic in. terms of the breakdown voltage, at which the reverse current is such that the power loss reaches a predetermined value related to the temperature-rise. Thermal dissipation is augmented by means of a *heat-sink* thermal emitter. High-power diodes comprise a number of units connected in series-parallel, with selection for parallel current-sharing and a resistor chain for proper series voltage division. Protection is required against voltage surges, overcurrents and excessive temperature-rise.

## Thyristor

This resembles the diode except that a four-layer structure is used. The core is a wafer of n-type silicon with two p-type layers diffused into its surfaces. The cathode connection forms a p-n junction, with the trigger electrode attached to the p-layer. The anode connection is made to the heat-sink base, which provides a rigid platform for the whole device.

The static current/voltage characteristic, Fig. 9.1(*b*), is similar to that of the diode in the reverse direction. The forward characteristic is the same up to the *turn-on voltage*, at which avalanche multiplication begins and the current rapidly increases to a value determined by the supply voltage and the load impedance. This is the conducting mode. The thyristor remains in this mode until the forward current is reduced below a *holding current* level, when it reverts to the forward blocking state. Increasing the gate (trigger) current reduces both the forward breakdown voltage and the holding current. The thyristor can be made to 'fire' (i.e. to conduct) at a given forward voltage by control of the gate current.

*Turn-on* When a trigger pulse is applied to the gate of a thyristor having a positive anode, there is a brief delay, after which the device switches rapidly into a conducting mode. Typical delay and rise times are 0.2 and 0.1 μs, but these are affected by the character of the trigger pulse, which must have a duration suited to the type of d.c. load. The thyristor remains conducting after the pulse has ceased, provided that the load current has risen above the level of the *latching current* (typically twice the holding current).

*Turn-off* A conducting thyristor can be switched off only by reducing the load current below the minimum holding current (typically 10 mA) and be reverse-biased for a minimum turn-off time (15–50 μs). The rate of change of the anode voltage after turn-off must be limited, for if too great the thyristor will re-enter a conducting mode.

*Commutation* This term refers to the means by which turn-off is achieved. In rectifying a.c., the supply voltage periodically reverses, providing the requisite reverse-bias. The conduction current decays and turns off by *natural* or *line* commutation. But when the thyristor is required to interrupt the current from a d.c. source, *forced* commutation is necessary. One method, Fig. 9.2,

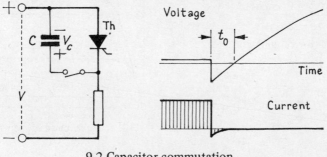

9.2 Capacitor commutation

employs a charged capacitor. Initially the thyristor is conducting and has a small forward voltage drop. Capacitor $C$, pre-charged to voltage $V_c$ and isolated, is then closed across the thyristor Th to impose a reverse bias; the thyristor current is quenched, and $C$ discharges through the load. $V_c$ falls to zero and then rises with reversed polarity towards the supply-voltage level. Provided that $t_0$ is greater than the turn-off time, thyristor conduction stops. If the load is a motor, it must be parallelled by a 'flywheel' diode to provide a path for the inductively prolonged load current.

*Gating*  For proper triggering, the gate pulse must lie between specified upper and lower voltage and current limits. The trigger signal is applied to the gate electrode of small area, the concentration giving rise to appreciable power-loss density. The turn-on time is related to the rate at which conversion of the material into the conducting state can be propagated, and the rise-time of the triggering pulse must be less than this. It is usual to apply a gating signal comprising a short pulse train, e.g. of 30 $\mu$s pulses of a high repetition frequency.

*Protection*  A thyristor can be fired spuriously by an excessive rate of change $dv/dt$ in the anode voltage. System transient voltages of 'spike' waveform have to be suppressed by input filters. The turn-on and turn-off of inductive loads can produce such waveforms, and so can voltage surges resulting from fuse blowing. Thermal protection is essential, for semiconductor devices have very short heating time-constants; and if by excessive temperature the reverse blocking property is lost, the device will be destroyed. Protection against overcurrent (due e.g. to a short circuit) may be provided by fast-acting fuses. The necessary protective features are therefore (i) temperature limitation to prevent thermal runaway, (ii) overcurrent limitation, and (iii) suppression of impulsive transient voltages and currents.

### Transistor

Transistors have been developed in ratings suitable for motor control. Basically the device is a three-layer p-n-p (or n-p-n) structure, Fig. 9.3(*a*).

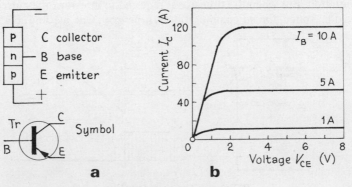

**a**                    **b**

9.3 Power transistor

In normal operation the emitter/base junction is forward biased and the collector/base reversed biased. The collector current is just less by about 2% than the emitter current, and can be controlled by the base current, as indicated by the static characteristics in (*b*). Except for miniature motors, transistors are used in a switching (on/off) mode. With output typically 100 A and a forward volt drop of 1–2 V, a substantial heat-sink is needed. In the 'off' state there is a 'leakage' current that rises with temperature.

Operating conditions are strongly affected by the time of transition between the two states, and the waveform of the base-current drive. For *turn-on,* the base-current waveform gives a rise ideally in less than 1 $\mu$s to give a switching delay of less than 3 $\mu$s: slower rates increase the switching losses and limit the rating capability. For *turn-off* the base current must be reversed, the nonconducting state being attained in about 10 $\mu$s. Such very rapid rates of change of current and voltage make design of the circuit layout critical.

The technology of semiconductor diodes, thyristors and transistors for use in power engineering is discussed in Ref. [22].

**Motor control**

A d.c. motor fed from an a.c. supply through diodes will give a performance close to that on a conventional d.c. supply. Prime interest, however, lies in the combination of rectification with speed control, for which purpose diodes are replaced by thyristors or transistors. The control advantage is not achieved without complication, but the adaptability to speed control is impressive.

For d.c. motors operated from secondary batteries, solid-state equipment can be used to give a maximum economical exploitation of the limited battery capacity in the control of torque and speed.

Series field windings are normal for fractional-kilowatt machines and for traction motors; separate excitation from a separate rectifier may be used for industrial machines. Advantages in size and efficiency are obtained in small and miniature motors by the adoption of p.m. excitation.

Three basic methods of solid-state control are listed below.

*Phase Control* Current is permitted to flow from an a.c. supply to the motor for a controlled fraction of each positive half-period, the pattern being cyclically repeated. The method is applicable to all motor ratings.

*Integral-cycle Control* Current is permitted to flow from an a.c. supply for a number of complete periods, then quenched for a further number, the sequence being continuously repeated. Control is effected by adjusting the ratio of the 'on' and 'off' durations. This method is suitable for fractional-kilowatt machines.

*Chopper Control* A motor is rapidly and repeatedly connected to and disconnected from a d.c. supply (such as a battery), control being imposed by varying the on/off time ratio. Chopping is employed in both railway and battery-vehicle traction.

## 9.3 PHASE CONTROL

Some basic concepts are introduced by first considering uncontrolled diode rectification. In the half-wave circuit of Fig. 9.4(*a*), a load is connected to a

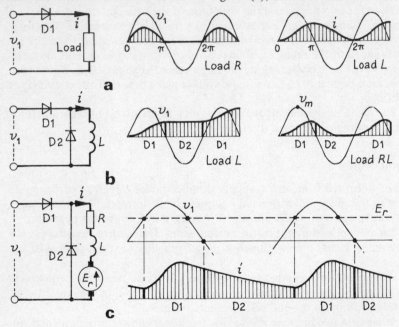

9.4 Diode rectification

sinusoidal a.c. source of voltage $v_1 = v_m \sin\omega t$ through an ideal diode D1. Let the load be a pure resistance $R$: then the voltage and current waveforms of the load comprise the positive half-periods of $v_1$ and of $i = v_1/R$. In the negative half-periods the diode blocks the voltage and the current is zero. The mean load voltage is $V_d = (1/\pi)v_m$. With $R$ replaced by a pure inductor $L$, the current in any positive half-period grows at the rate $di/dt = v_1/L$, and so rises throughout the half-period to a peak $i_m = 2v_m/\omega L$, the inductor storing a magnetic energy $\frac{1}{2}Li_m^2$. During the negative half-period this energy is returned to the supply by a falling (but still forward) current: the diode continues to conduct because the negative voltage $v_1$ is balanced at every instant by the e.m.f. induced in $L$. Thus the mean output voltage is $V_d = 0$.

If, as in Fig. 9.4(*b*), $L$ is shunted by a second ('freewheel') diode D2, a path is provided for the load current during the negative half-period, so that $L$ retains its stored energy with a constant current. In the following positive half-period the supply voltage raises the current level by a further

$2v_m/\omega L$, and so on. Current is carried by D1 and D2 alternately, and the mean load voltage is $V_d = (1/\pi)v_m$. The loss of mean output voltage has been avoided.

A practical inductive load has resistance. Fig. 9.4(*b*) shows the effect of a series-inductive-resistive load *RL*; all, or some, of the magnetic energy is dissipated in *R*. In the former case the current ceases at some instant in a negative half-period; while in the latter some current is still flowing when the succeeding positive half-period begins, and eventually there is a continuous conduction between upper and lower current limits.

If the load is a motor armature generating a rotational e.m.f. $E_r$ proportional to speed and field flux, conditions (*c*) apply. Here $E_r$ is assumed for simplicity to be constant. Current starts in D1 when $v_1$ equals $E_r$ and is rising, and continues until $v_1$ falls to zero at the end of the positive half-period. Thereafter, current is maintained through D2 until again $v_1 = E_r$. A condition of continuous conduction is illustrated, the armature-circuit inductance being presumed large enough for this. If the mean load current is $I_d$, then the mean voltage across the armature circuit is $V_d = E_r + RI_d$. In all the cases in Fig. 9.4, diode commutation takes place naturally when the p.d. across D1 becomes negative.

In Fig. 9.5, D1 is replaced by thyristor Th, and we consider only the case

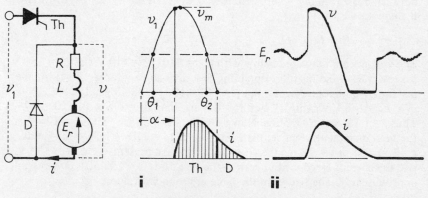

9.5 Thyristor control

of the motor load, with the same assumptions as before. Conduction can be initiated only in the positive half-period between time-angles $\theta_1$ and $\theta_2$ ($= \pi - \theta_1$), where $\theta_1 = \arcsin(E_r/v_m)$. For example, if $E_r = 0.5\ v_m$, then $\theta_1 = \pi/6$ rad (30°) and $\theta_2 = 5\pi/6$ rad (150°). Phase-controlled rectification is obtained by delaying the triggering instant of Th by a time-angle $\alpha$ (measured from the start of a positive half-period). On firing, Th applies to the motor the net instantaneous voltage $v_m \sin \alpha - E_r$, providing a current *i* which is continued beyond $\theta_2$ by the circuit inductance, switching to the freewheel diode D after the applied-voltage zero by natural commutation. Fig. 9.5 shows (i) a typical idealized waveform, and (ii) corresponding oscillograms of the motor circuit voltage *v* and the current *i*. The voltage waveform gives evidence of commutator ripple in $E_r$; it follows the input voltage $v_1$ down to zero, remains substantially at zero while current flows in D, then rises to $E_r$ until the next triggering instant. The motor current is usually discontinuous, but may become continuous for minimum delay angle and a low level of rotational e.m.f.

Single-phase half-wave converters draw a d.c. component from the a.c. supply and are therefore suitable only for fractional-kilowatt motors. Full-wave 1-ph bridge converters are preferred for ratings up to 5–10 kW. For industrial machines of higher rating a 3-ph supply is necessary, but of necessity the 1-ph supply must be used for a.c. rail traction.

EXAMPLE 9.1: The separately excited motor of Fig. 9.5 is supplied through a thyristor from a 240 V 50 Hz supply, and the armature-circuit parameters are $R = 1\ \Omega$, $L = 50$ mH. For a given load condition and a delay angle $\alpha = 75°$, the mean rotational e.m.f. is $E_r = 150$ V. Neglecting the forward voltage drops in Th and D, find the armature current waveform and mean value, and the mean converted power.
The armature-circuit voltage equations to be solved are

$$v_m \sin\theta - E_r = Ri + L(di/dt) \tag{i}$$

between $\theta = 75°$ and $\theta = 180°$. Thereafter $v_1$ is cut off by the thyristor and the equation is

$$-E_r = Ri + L(di/dt) \tag{ii}$$

The voltages are tabulated below with $v_m = 240\sqrt{2} = 340$ V and $E_r = 150$ V for intervals of 0.5 ms (corresponding to a time-angle interval of 9°). At the instant $t = 0$ of triggering, $i = 0$, $v_1 = 340 \sin 75° = 328$ V, and $v_1 - E_r = 178$ V $= L(di/dt)$. For d$t$ = 0.5 ms, then d$i$ = 178 × 0.5/50 = 1.78 A. Thus $i$ = 1.78 A at $t$ = 0.5 ms. Now $v_1 - E_r - Ri = 328 - 150 - 2 = 176$ V and the next increment of current is 1.76 A, giving $i$ = 1.78 + 1.76 = 3.54 A at $t$ = 1 ms. After t = 5.8 ms the only circuit voltage present is $-E_r = -150$ V, and the current is reduced to zero at the rate d$i$/d$t = -150/50 = 3$ A/ms, approximately. The figures in the table are rounded values.

| Time $t$ | (ms): | 0 | 0.5 | 1.0 | ... | 4.0 | 4.5 | 5.0 | ... | 8.0 | 8.5 | 9.0 |
|---|---|---|---|---|---|---|---|---|---|---|---|---|
| Angle $\theta$ | (deg): | 75 | 84 | 93 | | 147 | 156 | 164 | | 219 | 228 | 237 |
| $v_1$ | (V): | 328 | 337 | 340 | | 184 | 139 | 88 | | 0 | 0 | 0 |
| $v_1 - E_r - Ri$ | (V): | 178 | 185 | 186 | | 23 | −22 | −64 | | −153 | −152 | −150 |
| d$i$ | (A); | 1.8 | 1.8 | | | 0.2 | −0.2 | | | −1.5 | −1.5 | |
| Current $i$ | (A): | 0 | 1.8 | 3.6 | | 11.0 | 11.2 | 11.0 | | 3.2 | 1.7 | 0.2 |

The current waveform is that in Fig. 9.5(ii).
The mean current is found to be 2.93 A, and the converted power is therefore $P = 150 \times 2.93 = 440$ W.

## Converter arrangements

The most common are listed below. Except for (1), all are bridge circuits, and only for (7) is a mains transformer essential (provided that in the other cases the supply voltage has appropriate level). The symbol $q$ refers to the number of pulses of supply current that the converter delivers in each cycle.

### Single-phase

(1) *Single Thyristor, q* = 1 (half-wave). Motor current and torque discontinuous. For hand-tools, small domestic machines, etc.

(2) *Diode Bridge and Thyristor, q* = 2 (full-wave). Current discontinuity and torque fluctuation less than in (1). For industrial drives up to 5 kW.

(3) *Fully-controlled Bridge, q* = 2 (full wave). For industrial drives up to 10 kW where regeneration is required, and for railway traction service up to 10 MW.

(4) *Half-controlled Bridge, q* = 2 (full-wave). For industrial drives up to 10 kW and traction service up to 10 MW.

### Three-phase

(5) *Fully-controlled Bridge, q* = 6. Low p.f. at large delay angles; no supply-side harmonics below 5th; regeneration possible. For industrial drives of 10 kW−1MW.

(6) *Half-controlled Bridge, q* = 6. For industrial drives as in (5) not requiring regeneration.

(7) *Double Bridge, q* = 12. No supply-side harmonics below 11th. For paper-making, rolling-mill and mine-winder drives of 1 MW upward.

(8) *Anti-parallel Bridges, q* = 6 or 12. For drives requiring rapid reversal and regeneration, 100 kW−10 MW.

A simplified circuit for these converter arrangements (other than (1) and (2)) is given in Fig. 9.6. The 1-ph or 3-ph supply provides an output voltage $v_d$ from the converter to a motor of armature-circuit resistance $r_a$ and inductance $L_a$ (which may sometimes be augmented by a series inductor). The motor generates a rotational e.m.f. $e_r$. Supply-network impedance and the forward (conducting) volt drop of the converter are ignored. Then

$$v_d = r_a i_d + L_a p i_d + e_r$$

If $L_a$ is large, $i_d$ for a given operating condition is constant to a first approximation, and is consequently derived from thyristor currents of 'rectangular block' waveform (i). The thyristors are fired in a sequence that gives an input supply-line current (ii) comprising positive and negative blocks:

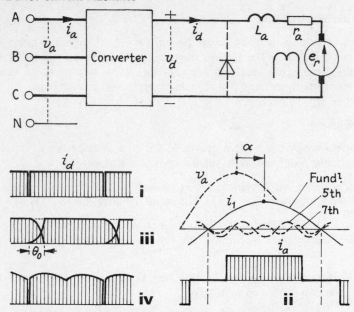

9.6 Simplified converter circuit and current waveforms

in converter configuration (5), for example, these have an angular time
duration of $2\pi/3$ rad, separated by intervals of $\pi/3$ rad. Fourier analysis of
this waveform yields a fundamental sine with harmonics of orders 5, 7, 11,
13..., and if the r.m.s. value of the fundamental is $I_1$, that of a harmonic of
order $n$ is $I_1/n$. The a.c. line currents consequently have a considerable
harmonic content, as indicated in (ii) for those of orders 5th and 7th.
Further, if firing is delayed by a time angle $\alpha$, the fundamental current
component $i_1$ lags by $\alpha$ on the line-to-neutral voltage $v_a$, giving the effect of a
power factor less than unity and presenting to the a.c. supply network a
lagging reactive load. Finally, if the a.c. supply network has appreciable
inductance, instantaneous transfer of current from one rectifier to the next is
inhibited, the current change-over involving parallel conduction by two
rectifiers during an *overlap* angle $\theta_0$, as indicated in (iii).

If $L_a$ is not large enough for the d.c. output current $i_d$ to be strictly
constant, it will carry a *ripple* (iv), introducing harmonics into the d.c. load.
Thus besides thermal limitation, the problems that arise from semiconductor
control include the introduction of harmonics into the a.c. network and a
demand from it of lagging reactive power. Fluctuation of converter d.c.
output increases the motor $I^2 R$ and core losses, and complicates the motor
commutation conditions. Initially, however, it is sufficient to describe the
*basic* behaviour of most of the converter arrangements on a 'rectangular-
block' and constant direct-current assumption, with the effects of harmonics,
power factor and overlap phenomena considered separately.

### 9.4 PHASE CONTROL: SINGLE-PHASE SUPPLY

#### (1) Single thyristor

Fig. 9.7 shows a simple system for a *constant*-field (e.g. permanent-magnet) motor. With the machine at rest and a supply voltage $v_1 = v_m \sin \omega t$

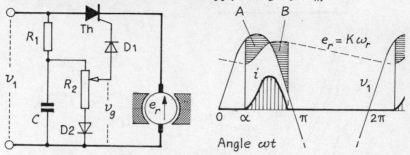

9.7 Single-thyristor control

beginning a positive half-period, capacitor $C$ is charged and a fraction of its voltage is supplied to the gate of thyristor Th, turning it on after a delay angle $\alpha$ depending on the setting of $R_2$. Towards the end of this half-period, the current falls below holding level, so that Th turns off until fired in the succeeding positive half-period. The motor current is discontinuous, but the pulse of torque starts the motor, which generates an e.m.f. $e_r$. Were the armature-circuit inductance zero, the current would cease after the balance point $v_d = e_r$; with a finite inductance, however, the current is maintained until the voltage-time integral, area $B$ in Fig. 9.7, matches the area $A$ (neglecting resistance). Triggering occurs regularly in successive positive half-periods when $v_g$ exceeds $e_r$. The effect of $C$ is to give a positive slope to the gate voltage for delay angles greater than $90°$, so providing stable firing at low motor speeds. Diode D1 isolates the gate circuit from Th while it is conducting, and D2 eliminates the loss in the resistors during negative half-periods.

If the motor has a *series* field, $e_r$ during zero-current conditions results from remanence only, so that a lower gate voltage is required. A speed drop resulting from a rise in load torque is thus compensated by earlier triggering. The inverse speed/torque characteristic of the conventionally fed series motor is modified to a roughly constant speed for any given control setting.

The control method is cheap, but the highly pulsating current increases the motor $I^2 R$ and core losses, the a.c. supply has a large harmonic-current content and it has to provide a d.c. component. These drawbacks can be tolerated for low motor ratings.

#### (2) Diode bridge and single thyristor

The diode bridge, shown in Fig. 9.8 for a separately excited machine, makes use of full-wave operation and eliminates the d.c. component from the a.c. supply. Assuming for simplicity that the rotational e.m.f. $E_r$ is constant, the

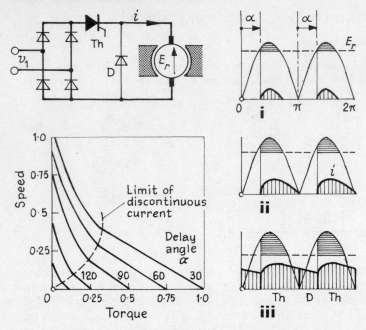

9.8 Diode bridge and thyristor control

action falls into three distinguishable operating modes as the duration of the current pulse increases.

(i) *Low torque, high speed* $E_r$ is high and the mean current is low. Turn-off occurs at an angle less than $\pi$ rad, the freewheel diode D being inactive.

(ii) *Intermediate torque and speed* $E_r$ is lower than in (i). A limiting condition occurs when the current ceases at the end of a positive half-period and D remains inactive.

(iii) *High torque, low speed* There is a further reduction in $E_r$, and D may maintain the current long enough to make it continuous. The speed/torque characteristics in Fig. 9.8 show two distinct regions. For continuous flow the speed falls with rise of torque in a manner resembling that for a conventionally fed machine. For discontinuous flow the slopes are steeper. An analysis of these modes is given by Holmes and Merrett [23].

### (3) Fully-controlled bridge

This configuration is widely used for low-power separately excited industrial motors and for high-power traction service with series motors, Fig. 9.9. It is convenient to assume that the 1-ph supply, of instantaneous voltage $v_1$ and peak $v_m$, is formed from two antiphase sources respectively of $v_a = +\frac{1}{2}v_m \sin \omega t$ and $v_b = -\frac{1}{2}v_m \sin \omega t$ between terminals A and B and a

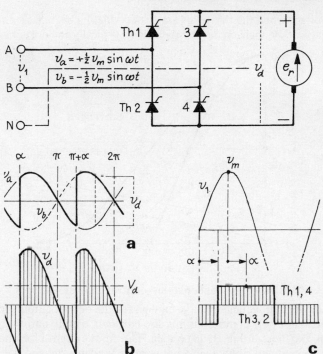

9.9 Single-phase fully-controlled bridge converter

midpoint ('neutral') N. Then $v_1 = (v_a - v_b)$ is the vertical intercept between the graphs of the two half-voltage sinusoids in ($a$).

During the positive half-period (A positive and B negative with respect to N), input current starts when Th1 and Th4 are fired at a delay angle $\alpha$ provided that $v_1$ exceeds $e_r$; it flows from A to the load through Th1 and returns to B through Th4. The output voltage is $v_d = v_1 = (v_a - v_b)$ as indicated in Fig. 9.9($a$). Beyond the time-angle $\omega t = \pi$, terminal A becomes negative, but the motor circuit inductance sustains the current through Th1 and Th4. At $\omega t = (\pi + \alpha)$, Th3 and Th2 are fired, and current transfers by natural commutation from Th1/Th4 to Th3/Th2. The input voltage $v_1$ is negative but the polarity of the output voltage $v_d$ is maintained as the result of commutation, its waveform being indicated in ($b$).

If the load inductance maintains the output current without discontinuity, the mean output voltage $V_d$, found by integration over a repetition period, is

$$V_d = (2/\pi)\, v_m \cos \alpha = (2\sqrt{2}/\pi)\, V_1 \cos \alpha \qquad (9.1)$$

where $V_1$ is the r.m.s. value of the supply voltage. In the waveforms shown in ($a$) the delay angle is $\alpha = \pi/3$ rad ($60°$), for which $V_d = (2/\pi)\, v_m \times 0.5$, i.e. 0.5 p.u. of that for zero delay.

Eq. (9.1) is plotted in Fig. 9.10, showing that the output voltage falls to a mean of zero for $\alpha = \pi/2$ rad ($90°$). But if the load current is *discontinuous*, the load voltage when the thyristors are not conducting becomes $e_r$, and the mean-voltage/delay-angle characteristic follows typically the dotted line.

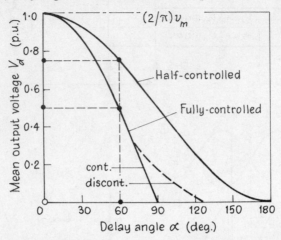

9.10 Single-phase converter, control characteristics

When the load inductance is large enough for the output current to be considered as not only continuous but also *constant*, supply-current conditions like those in Fig. 9.9(*c*) obtain. The input current block has a harmonic content and a phase lag $\alpha$ between its fundamental and the supply voltage, giving an inherent lagging power factor. The presence of supply-network inductance (together with the leakage inductance of the transformer, if used) produces *overlap,* as discussed in Sect. 9.3. The angle $\theta_0$ of overlap may not exceed two or three degrees if the converter is connected directly to the a.c. supply, but may be $10°$ or more on full load where a transformer is employed. During overlap the conducting thyristors are forced to have the same voltage, so lowering the mean output voltage, and increasing the effective phase lag.

**(4) Half-controlled bridge**

Two thyristors can be replaced by less expensive diodes if regeneration is not required, using one of the configurations in Fig. 9.11. Consider that in which the diodes are connected to the negative load terminal. Conduction takes place from $\alpha$ to $\pi$ through Th1 and D4. At $\pi$ rad, B becomes positive and the current commutates naturally from D4 to D2: this short-circuits the load terminals and gives zero output voltage as in (*a*), but provides a 'freewheel' path for the motor current. At $(\pi + \alpha)$ rad, Th3 is triggered and current transfers to it, to return to A through D2. The mean output voltage is

$$V_d = (1/\pi)\, v_m (1 + \cos \alpha) = (\sqrt{2}/\pi)\, V_1 (1 + \cos \alpha) \tag{9.2}$$

This relation, Fig. 9.10, shows that the mean output voltage is the same as that for the fully-controlled bridge for $\alpha = 0$, but is higher at other delay angles.

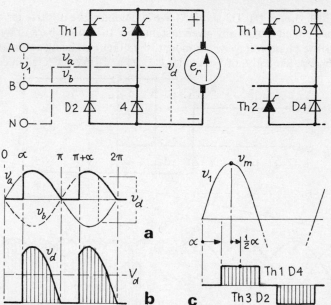

9.11 Single-phase half-controlled bridge converter

The voltage waveforms in Fig. 9.11 are for $\alpha = \pi/3 = 60°$, for which $V_d = (1/\pi) v_m (1 + 0.5)$, i.e. 0.75 p.u. of that for zero delay.

Assuming that the load inductance imposes a constant output current, an input current block, Fig. 9.11(c), begins at the delay angle $\alpha$ and ends at $\pi$ rad, giving a conduction angle $(\pi - \alpha)$. The effective phase angle is a lag of $\frac{1}{2}\alpha$, so that a power factor higher than that in the fully-controlled bridge results, especially for the larger delay angles required for low motor speeds. In the intervals of zero input current, the motor current is maintained through the 'freewheel' path.

The voltage and current waveforms are the same in both of the configurations in Fig. 9.11, but in the alternative connection the freewheel path is through D3 and D4, which are directly across the load. The thyristors are relieved of the burden of additional current, but as the two gate circuits are of necessity at different potentials, the control circuitry may be more complicated.

## 9.5 PHASE CONTROL: THREE-PHASE SUPPLY

Consider an ideal 3-ph supply with line terminals ABC and line-to-neutral voltages

$$v_a = v_m \sin \omega t \qquad v_b = v_m \sin(\omega t - 2\pi/3) \qquad v_c = v_m \sin(\omega t - 4\pi/3)$$

and let it be connected to a 3-ph rectifier bridge of ideal *diodes* as in

Fig. 9.12(*a*). Diodes D1, D3 and D5 have in common the positive terminal of the load circuit, but at any given instant only that diode backed by the most positive phase voltage will conduct, the other two being automatically reverse-biased. Similarly, of the diodes D2, D4 and D6 connected in common

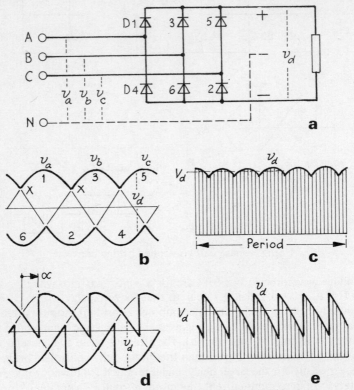

9.12 Three-phase converter

to the negative load terminal, only that with the most negative phase voltage can conduct. Current commutation between successively conducting diodes takes place at the 'crossover' points X where successive phase voltages are instantaneously equal. The phase with the reducing voltage quenches, while that with the increasing voltage takes over conduction.

The voltage to neutral of the positive load terminal is therefore a succession of sinewave tops, and similarly for the negative load terminal, giving a resultant direct output voltage $v_d$ formed by the intercept at any instant between the upper and lower graphed lines in (*b*), yielding the output waveform (*c*). This can be taken as a mean $V_d$ with a superposed ripple corresponding to the succession of $q = 6$ pulses per period of the a.c. supply.

If the diodes are replaced by *thyristors*, gated to conduct at the crossover points X, the result is unaffected. But it is now possible to delay firing by an angle $\alpha$ (measured from the natural commutation instants). That thyristor connected to the phase with the highest voltage is inhibited until it receives a gate pulse. The waveform (*b*) is modified to that in (*d*), yielding an output load voltage (*e*), which has a reduced mean value $V_d$ and considerably greater ripple. Varying the angle $\alpha$ of delay thus gives control of the mean output voltage. It also makes possible a regeneration mode, in which the direction of power flow can be reversed, as discussed in Sect. 9.6.

## (5) Fully-controlled bridge

In Fig. 9.13, Th1 conducts during part of the positive half-period of $v_a$. At first, current flows from A to the positive side of the load circuit and returns

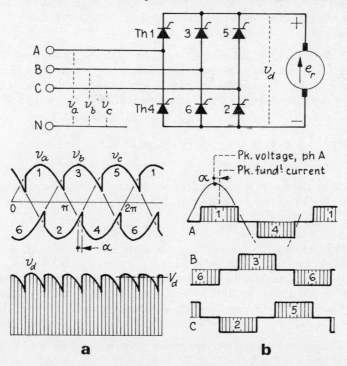

9.13 Three-phase fully-controlled bridge converter

to B through Th6. After Th2 is fired, the return current is through Th2 to C. Thereafter a corresponding sequence occurs for the remainder of a period. With zero delay, commutation takes place naturally at the crossover instants, but it can be delayed by a time-angle $\alpha$: The conduction pattern in (*a*) is drawn for $\alpha = 15°$, yielding the resulting output voltage $v_d$ with $q = 6$ pulses per period. The mean output voltage is

$$V_d = (3\sqrt{3}/2\pi)\, v_m \cos \alpha = (3\sqrt{6}/2\pi)\, V_1 \cos \alpha \qquad (9.3)$$

in which $V_1 = v_m/\sqrt{2}$ is the r.m.s. phase voltage. For $\alpha = 15°$, then $V_d = 0.80\, v_m = 1.13\, V_1$.

Motor inductance is such that discontinuous load current is not likely except for large delay angles, and the assumption of constant load current is usually adequate. The corresponding currents for phases A, B and C are shown in (*b*). The current shift by delayed triggering displaces the fundamental components by $\alpha$ from the phase voltages to give the effect of a power factor $\cos \alpha$, which is lowered still further if overlap is appreciable.

EXAMPLE 9.2: A separately excited motor fed from a 3-ph 6-pulse fully controlled converter develops full-load torque at 1500 r/min when the delay angle is zero, the armature taking 50 A at 400 V (mean) and having a resistance of 0.50 $\Omega$. Estimate the r.m.s. voltage per phase of the a.c. supply, and the range of delay-angle control required to give speeds between 1500 and 750 r/min at full-load torque.

*Phase Voltage* Consider the voltage waveform $v_d$ for zero delay angle in Fig. 9.12(*c*). It comprises a symmetrical series of cosine wave tops, each extending between time-angles $-\pi/m$ and $+\pi/m$ on each side of the peak $v_m$, where $m$ is the number of phases. The mean rectified voltage is

$$V_d = \frac{1}{2\pi/m} \int_{-\pi/m}^{+\pi/m} v_m \cos \omega t \cdot \mathrm{d}(\omega t) = \left[ \frac{m}{\pi} \sin \frac{\pi}{m} \right] v_m$$

With a delay-angle $\alpha$, as in Fig. 9.12(*e*), the integration is taken between the limits $[-(\pi/m) + \alpha]$ and $[+(\pi/m) + \alpha]$, now giving

$$V_d = (m/\pi) \sin (\pi/m) \, v_m \cos \alpha.$$

For $m = 3$, then $V_d = (3\sqrt{3}/2\pi) v_m \cos \alpha$, as given in eq. (9.3), and with $V_d = 400$ V the phase voltage required is 484 V (max) or 342 V (r.m.s.). These values are based on the assumptions that the supply voltage waveform is not distorted, that commutation takes place instantaneously without overlap, and that the thyristor forward volt drop (typically $1\frac{1}{2}$ V) can be ignored.

*Control Range* For 1500 r/min, the armature resistance drop is 25 V so that the rotational e.m.f. is $400 - 25 = 375$ V. For 750 r/min this e.m.f. is $375/2 = 188$ V, and the mean output voltage of the converter is $188 + 25 = 213$ V. Then $213 = 400 \cos \alpha$, giving $\alpha = 58°$. Thus $\alpha$ must be variable between zero and $58°$.

*Phase Current* For an output current $I_d$, the r.m.s. phase current is $I_1 = I_d/\sqrt{m}$. For the machine concerned, $I_1 = 50/\sqrt{3} = 29$ A.

## (6) Half-controlled bridge

For the configuration in Fig. 9.14 the waveforms are shown in (*a*), and the mean load voltage is

$$V_d = (3\sqrt{3}/2\pi) v_m(1 + \cos \alpha) = (3\sqrt{6}/2\pi) V_1(1 + \cos \alpha) \tag{9.4}$$

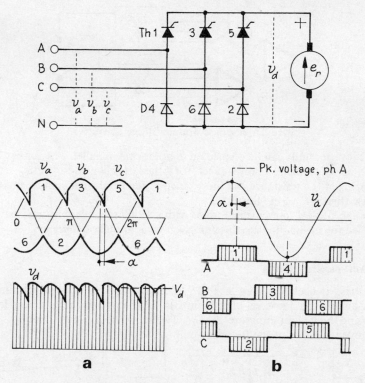

9.14 Three-phase half-controlled bridge converter

The phase currents to the positive side of the load are delayed by time-angle $\alpha$, but those from the negative side are not. If $\alpha > 60°$ the thyristor and diode of each phase conduct simultaneously; e.g., Th1 and D4 in phase A both carry current for an interval $(\alpha - 60°)$. There will be no resultant phase current for such intervals, but the thyristor and diode concerned will provide a freewheel path for the motor current.

Assuming a high value of load inductance, the phase currents are as indicated in (*b*). The effective angle of lag of the fundamental phase current is reduced to $\frac{1}{2}\alpha$.

## (7) Double bridge

With large motors the 3-ph 6-pulse connections (5) or (6) may produce unacceptable input and output harmonics. A 6-ph 12-pulse configuration with (ideally) no harmonic of order below the 11th is achieved with the aid of a transformer with two sets of secondary windings, Fig. 9.15, connected respectively in star and delta and each feeding a separate 3-ph 6-pulse converter. As there is a 30° phase displacement between the two sets of secondary voltage phasors, the combination gives a 12-pulse output.

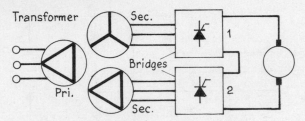

9.15 Three-phase double-bridge converter

The converter units can be connected in series or in parallel. The former has the advantage of *sequence* control (Sect. 9.7); the latter demands careful balancing of the resistance and inductance of the parallelled circuits and a precise triggering of the thyristors.

In exceptional cases a transformer with a 12-ph secondary output may be justified for the harmonic reduction given by a 24-pulse converter.

## (8) Anti-parallel bridges

This 'back-to-back' connection, Fig. 9.16, is adopted for motors requiring rapid and repeated reversal of direction; it eliminates power contactors for reversing the field or armature circuits. Two converters, 1 and 2, are employed in either of the following configurations.

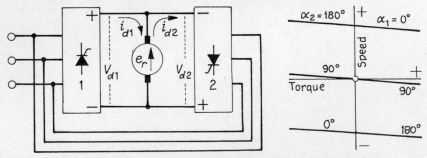

9.16 Three-phase anti-parallel bridge converter

(i) Gating is arranged so that both converters fire simultaneously with delay angles such that $\alpha_1 + \alpha_2 = 180°$. A circulating current, limited by an inductor to upwards of 5% of the rated current, flows continuously so that either converter can take up the load on demand by adjustment of the delay angles. Change-over from one converter to the other is virtually instantaneous, but there is a circulating-current loss.

(ii) With converter 2 blocked, the delay angle of 1 is increased to about $150°$, so reversing its voltage. Load current, maintained by stored energy in the mechanical system, produces a braking effect. At zero speed, 2 is triggered at a large delay angle which is then reduced to raise the motor speed in the reverse direction. The method involves a 'dead' time in reversal of 5–10 ms.

## 9.6 PHASE CONTROL: REGENERATION

Reversible speed control with regenerative braking in one direction is applied
to cranes, hoists, paper mills etc., and with regeneration in both directions to
traction, rolling-mill and comparable drives; these controls can be achieved
with fully-controlled bridge circuits. Reversal of direction of power flow
demands reversal of either current or voltage. As conduction through
thyristors is by nature unidirectional, then the motor armature connections
or the field current must be reversed, or a separate thyristor bridge for each
direction provided. Then during regeneration, the converter unit must be so
controlled that current flows only during the negative half-periods of the
supply voltage.

*Anti-parallel Bridges* To obtain drive in either direction, with regeneration, the
two bridges in Fig. 9.16 must so operate that one rectifies while the other
inverts, these modes being interchanged as the machine reverses. The delay
angles $\alpha_1$ and $\alpha_2$ must therefore be interrelated so that $\alpha_2 = (\pi - \alpha_1)$. As a
result, when the speed-reference is adjusted for reversal, the inverting bridge
produces conditions for regeneration. The relations of the four quadrants of
operation in Fig. 8.1(*b*) are obtained with the *idealized* delay angles indicated
in Fig. 9.16. The actual delay angles are not zero or 180° for maximum
speeds in practice: the minimum delay is greater than 0° in order to secure
reliable commutation of the thyristors, and the maximum delay is less than
180°. In Fig. 9.16 the reversal process is applied to the armature; in the case
of large machine systems for which the mechanical time-constant is greater
than that of the field circuit, the latter may instead be operated by its own
anti-parallel bridge to secure field and direction reversal without conventional
switching.

*Single Converter* Of the arrangements listed in Sect. 9.3, the fully-controlled
bridges (3) and (5) can be employed for regenerative braking. Consider the
1-ph bridge configuration in Fig. 9.17. The waveform for the *motor* mode in
(*a*) is the same as that in Fig. 9.9(*a*) except for the inclusion of the motor
e.m.f. $E_r$, taken for simplicity as constant. The vertically hatched region
represents the condition for which the bridge output voltage exceeds $E_r$ so
that forward conduction takes place with the machine receiving power as a
motor. In the horizontally hatched region the forward current is maintained
by the load inductance. For *regeneration* (*b*) the connections of the armature
are interchanged: the thyristor output voltages are now positive over almost
the whole of a period, as indicated by the vertically hatched region for $v_a$ in
(i). The current in the machine, produced by the summation of $v_a$ and $E_r$,
is limited only by the load impedance, and unless the thyristors are inhibited
from firing will rise to a level both excessive and destructive. As regeneration
is possible only if *positive* current is allowed to flow during the *negative* half-
periods of the a.c. supply, thyristor conduction must be confined thereto.
Suppose, as in (ii), that Th1 and Th4 have been triggered and are conducting
during the negative half-period of $v_a$. Following the negative peak, $v_a$ begins
to rise, increasing the forward voltage. At the same time, $v_b$ is falling and so

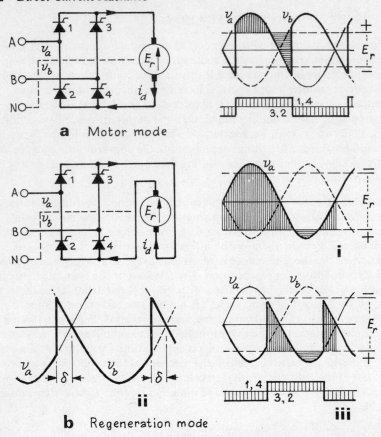

**a**  Motor mode

**b**  Regeneration mode

9.17 Regeneration

decreasing its forward voltage. Triggering of Th3 and Th2 to take over the load current must therefore take place before crossover (when $v_a$ and $v_b$ are momentarily equal) by an angle $\delta$ so that, despite overlap and thyristor turn-off time, commutation is complete before the instant of crossover; otherwise control is lost and the conditions shown in (i) are likely, with damaging results. The angle $\delta$ in (ii) is of the order $20°-30°$. The voltage and current waveforms in (iii) show that the supply current is substantially in phase opposition to the supply voltage, representing an output to the supply derived from the energy delivered by the machine as a generator.

Comparable conditions apply to the 3-ph bridge. With large minewinding or mill motors the time required to effect a change from one mode to the other may be of importance. When the mechanical time-constant is relatively long, it may be simpler to reverse the connections of the field circuit rather than that of the armature, using either conventional switches or an anti-parallel field converter. Reversal must occur accurately at zero current to avoid destructive induced e.m.f.s. Armature reversal involves a 'dead time' of

20–30 ms, but field reversal may take up to 3000 ms.

Regenerative braking can be combined with speed control in a sophisticated closed-loop system without significantly adding to the cost and complication. Typical of such a system would be the provision of a 500/1 continuously variable speed range, with a speed error of ± 0.3% for a supply voltage variation of ± 15% and a torque range of 10/1. The sequence of operations for a sudden reversal of the speed-control setting would then be:

(i) The thyristor trigger pulses are retarded to lower the current.

(ii) The motor reversing contactor operates.

(iii) The trigger pulses are advanced to give the maximum permissible regenerating current and braking torque.

(iv) The pulses are further advanced as the motor slows, to maintain regeneration.

(v) The triggering is advanced into the driving mode to accelerate the machine to the preselected reverse speed.

## 9.7 PHASE CONTROL: HARMONICS AND POWER FACTOR

Phase-controlled rectification introduces harmonic voltages and currents into both the input (a.c.) and output (d.c.) sides of the converter. The former distort the supply-system waveform and give the effect of a lagging power factor. The latter impair the performance of the motor.

### D.C. side

Neglecting overlap, a mean output voltage $V_d$ has a superposed ripple comprising harmonics of r.m.s. voltage $V_n$ given by

$$V_n = V_d \sqrt{2} \, (1 + n^2 \tan^2 \alpha)/(n^2 - 1)$$

for a $q$-pulse converter with a delay angle $\alpha$, where $n$ is an integral multiple of $q$. Voltage harmonics increase the core loss, and give rise to harmonic currents that increase the $I^2 R$ loss. The compole current ripple induces e.m.f.s of pulsation in armature coils undergoing commutation, and eddy-currents are induced in the compole structure, delaying its flux response.

A motor at full load and speed may have typically 10% greater armature $I^2 R$ loss than when run on a steady d.c. supply. Low-speed running (i.e., $\alpha$ large) intensifies the ripple, but the $I^2 R$ loss increment may be offset by a lower core loss. Compared with its rating on a conventional supply, a motor fed from a 6-pulse converter may have to be de-rated by 10% or more.

### A.C. side

Here the current-harmonic effects predominate. Analysis is complicated, but if the input-current waveform is assumed to comprise rectangular blocks, a number of general features can be discerned.

Current harmonics of orders $n = (aq \pm 1)$, where $a$ is an integer, occur in a $q$-pulse fully-controlled converter input with a zero delay angle, and their respective r.m.s. magnitudes $I_n$ in terms of the r.m.s. fundamental $I_1$ are $I_n = I_1/n$. Thus in a 3-ph 6-pulse converter the harmonic content is

| Harmonic order $n$: | 5 | 7 | 11 | 13 | 17 | 19 | 23 |
|---|---|---|---|---|---|---|---|
| Ratio $(I_n/I_1)$ : | 0.20 | 0.14 | 0.09 | 0.08 | 0.06 | 0.05 | 0.04 |

Neither triple $n$ ($n$ a multiple of 3) nor even-order harmonics appear.

Delayed firing in a fully-controlled converter shifts the timing of the rectangular blocks without changing their shape, so that the harmonic magnitudes are unaltered; but their phase position is changed. With half-controlled converters an adjustment of the delay angle varies the duration of a block, and its harmonics may now include those of even order. Current ripple on the d.c. side increases the 5th-order harmonic but slightly reduces others.

Divergence of the supply voltage from the pure sinusoid (e.g. as a result of rectification in the converter) introduces input-current harmonic content, with the further effects of energy-meter errors, interference in communication circuits and reaction on the load of other consumers (for example, an increase of loss in p.f.-correcting capacitor banks). Supply authorities impose limits on the rating of converters connected to the system [24]. As the impedance of the system at the point of connection of the converter is directly related to the fault level, a general requirement is that the converter rating shall not exceed 2% of the fault level for fully-controlled and 1% for half-controlled converters. Typical limits for various system voltages are

| System voltage | (kV): | 0.415 | 6.6 | 11 | 66 | 132 |
|---|---|---|---|---|---|---|
| Converter rating, 6-pulse | (MW): | 0.25 | 0.6 | 1.0 | 3.0 | – |
| 12-pulse | (MW): | 0.75 | 1.8 | 3.0 | 7.0 | 14 |

The harmonic magnitudes of two or more converters with the same loading and delay angles, and operated from the same supply, are arithmetically additive. With different loads and delay angles, however, the summation is more complex and may in fact result in a significant reduction in gross harmonic levels.

Although the 5th and 7th harmonics are eliminated in a 12-pulse converter, it is usual to suppress higher-order harmonics by fitting band-pass inductance-capacitance filters to the a.c. side, normally in parallel to assist in power-factor correction. The filters must be so designed as to avoid resonance with the supply-network inductance.

The rapid changes of current and voltage at the instants of commutation of thyristors, which may cause interference with radio communication in the kilohertz range, are suppressed by resistance-capacitance circuits connected across the thyristors or across the supply terminals.

## Power factor

The power factor of the load presented by a thyristor converter to an a.c. supply of sinusoidal phase voltage $V_1$ is defined as the ratio $P/S$ of the active and apparent power inputs. In r.m.s. terms, $P = V_1 I_1 \cos \phi$ per phase, where $I_1$ is the fundamental-frequency component of the phase current and $\phi$ is a phase-angle closely related to the delay-angle $\alpha$. The term $\cos \phi$ is the *displacement factor*. The apparent power is $S = V_1 I_a$, where $I_a = \sqrt{(I_1^2 + \Sigma I_n^2)}$ is the r.m.s. phase current, and includes the contribution of harmonic currents of order $n$. Writing $(I_1/I_a) = \cos \delta$ as the *distortion factor*, the overall power factor is

$$P/S = V_1 I_1 \cos \phi / V_1 I_a = \cos \phi \cdot \cos \delta$$

With the assumption that the input current waveform has a rectangular block shape, the distortion factor for a 3-ph 6-pulse fully-controlled converter is 0.955.

In practice the distortion factor is commonly ignored and the power factor is taken as $\cos \phi$, which has been shown to be either $\cos \alpha$ or $\cos \frac{1}{2}\alpha$. Let a *fully-controlled* converter deliver a constant load current and have block-form input currents of constant amplitude. The mean output voltage falls with increase of delay angle, the input voltage being unaffected. The apparent power $S$ is therefore constant, but its active and reactive components are $P = S \cos \alpha$ and $Q = S \sin \alpha$, as indicated in Fig. 9.18. With a *half-controlled*

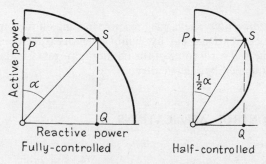

9.18 Active and reactive power

converter, however, both input current and apparent power are zero when the load voltage has been reduced by delay to zero. This gives the half-controlled configuration a significant advantage in power factor, although it has more severe harmonics and cannot regenerate.

It is possible, as described by Farrer and Andrew [25], to secure with full control the advantages of the half-controlled arrangement. By special triggering at appropriate instants, certain thyristors are made to operate as freewheel diodes to provide for the load current a path alternative to that of the supply lines. Another method is to connect a fully-controlled and a diode bridge in series, the two being fed from separate transformer windings with cophasal voltages. Control is effected by the thyristor bridge, the diode bridge delivering one-half voltage continuously. Above one-half output voltage the thyristor bridge is additive; for lower voltages it acts as an inverter to subtract its voltage from that of the diode bridge.

In *sequence control*, Fig. 9.19, two or more 1-ph or 3-ph thyristor bridges are connected in series. For high output voltage, bridge 1 is controlled, with 2 and 3 set to zero delay and therefore taking no reactive power. For the middle range of output voltage, 3 is set to maximum, 2 is controlled and 1 acts as a freewheel path. For low output voltages 3 is controlled, with 1 and 2 freewheeling.

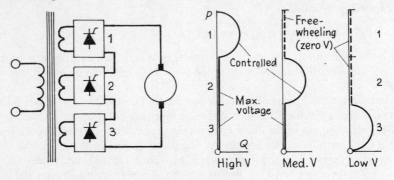

9.19 Sequence control

*Power-factor Correction*  The lagging reactive loading on the supply may be reduced in the conventional way by shunt capacitors at the converter input terminals. These will carry some of the harmonic currents, and may also form part of the harmonic filtering circuitry.

## 9.8 PHASE CONTROL: APPLICATIONS

### Industrial drives

In ratings up to about 100 kW, thyristor drives for industrial motors are engineered as standardized 'packages' with a number of options (close speed control, braking, inching, regeneration etc.) at choice for little extra cost. Maintenance is simplified by the reduction of mechanical contacts and by the provision of test points, diagnostic programs and replaceable modules. Larger-power equipments are individually engineered with attention to sophisticated control, p.f. correction and harmonic reduction. An approximate correlation between the d.c. motor armature voltage and that of the a.c. mains supply (direct or through a transformer) is

|  | 1-ph |  | three-phase |  |  |  |
|---|---|---|---|---|---|---|
| Supply voltage (r.m.s.) (V): | 240 | 415 | 380 | 415 | 440 | 500 |
| Motor voltage (mean) (V): | 170 | 290 | 440 | 480 | 500 | 570 |

*Single-phase* Converters for motors up to 5 or 10 kW are made for a wide
field of application, provision being made for a variety of input controls and,
where required, braking and reversal. For economy, a.c.-type relays and
contactors may be adopted; these necessitate phasing the thyristors back to
zero before the contactors are opened, and before they are reclosed, but the
contactors must still be capable of clearing a d.c. fault current. Automatic
speed references are built into the control equipment. Printed-circuit logic
functions such as the following may be provided: (i) local and remote start/
stop push-buttons; (ii) local and remote reversing; (iii) input inhibit (for
preventing an unwanted re-start); (iv) loss-of-field detection; (v) integrated-
circuit thyristor firing by 5 kHz pulse-trains for reliable triggering; (vi) speed
regulation operated by tachogenerator or armature-voltage feedback and
amplifier; (vii) dynamic braking; (viii) regenerative braking; (ix) analogue/
digital device to decode duty data (time-ramp, torque-taper, overload, field
take-over for constant-speed operation, IR compensator for armature-voltage
feedback). For constant-speed operation the motor field is adjusted, giving
economy of field power and response to the detection of field failure.
Separate excitation is the most common where speed control is needed, but
series motors are occasionally more suited to the drive. Current limiting is
necessary to permit the use of competitive thyristor ratings and for
overcurrent protection.

*Three-phase* Fig. 9.20 shows a typical manufacturer's range of 3-ph bridge

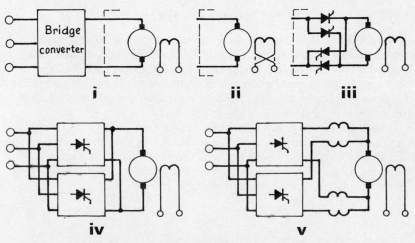

9.20 Range of phase-controlled industrial motors

converter configurations for industrial motors between 25 and 500 kW,
adaptable to a variety of applications. They are

(i) Single converter for non-reversible duty.

(ii) Single converter for reversal by interchanging field connections.

(iii) Single converter with armature reversal by thyristors (for which contactor gear may be substituted as an alternative.

(iv) Double converter with blocking control to suppress circulating currents; reversible.

(v) Double converter with circulating current limitation by inductor; reversible.

Field reversal (ii) takes a few seconds. Armature reversal (iii) can reverse the armature torque in 0.01 s (or, with contactors, 0.1 s). Double converters (iv) and (v) give stepless torque control through zero current. The several control modules follow the same general functions as for the 1-ph equipments. Speed sensing is normally by tachogenerator on the motor shaft.

The schematic diagram, Fig. 9.21, shows a speed-control arrangement with

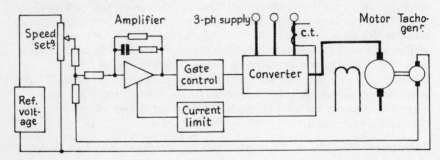

9.21 Speed-control system

a constant field current and a controlled armature voltage. The amplifier input is derived from a reference-voltage unit (a stabilized and smoothed rectified supply and a Zener diode) and a tachogenerator speed-voltage feedback. A current transformer and a current-limit unit provide protection at high current levels. The amplifier output controls the gate unit which in turn controls the delay angle of thyristor firing.

Criteria for speed-control accuracy are concerned, in both the steady and the transient states, with performance as affected by specified supply-system variations of voltage and frequency, with the variation of ambient temperature, and with the stability of the reference and feedback voltages. For the two latter, typical ranges of available accuracy are:

| Control | Error % | Reference source | Speed sensor |
|---------|---------|------------------|--------------|
| Digital | 0.0002 | crystal oscillator | digital pulse generator |
|         | 0.1    | precision direct voltage | precision d.c. tachogenerator |
| Analogue | 0.25  | stable direct voltage | high-quality d.c. tacho-generator |
|          | 1.0   | power supply | tachogenerator or armature voltage |

Digital pulse generators develop up to about 5000 pulses per revolution of the motor, constant in amplitude and independent of speed.

*Braking*  The methods described in Sect. 8.4 can be applied with appropriate modification to motors fed from phase-controlled thyristors. In *dynamic* braking the field excitation is maintained through its separate rectifier, the armature input is effectively disconnected by inhibiting the firing of the converter transistors, and a braking thyristor (connected permanently across the armature brushes in series with a braking resistor) is fired. The result is to set up a condition like that in Fig. 7.7(*a*). The process is rapid and does not require electromechanical contactors. With anti-parallel converters it is possible to provide *counter-current* braking (plugging). The requirements for *regenerative* braking are discussed in Sect. 9.6.

Typical industrial controls require accurate control of shaft speed, a wide speed range, controlled acceleration to a preset speed, controlled retardation to standstill and, for multi-section drives, an accurate speed-ratio. Solid-state control has a rapid response and very high gain, and in some cases these must be limited to avoid severe stress and oscillation where the mechanical natural frequency of the motor-load combination is low (e.g., 3–7 Hz). Variation of the a.c. supply voltage causes more disturbance to thyristor equipments than to conventional systems such as the Ward-Leonard, and may demand the introduction of additional amplifiers and control loops with sensors responsive to the armature current.

### Mine winders

Precise speed control is required to ensure accurate decking. Two winding speeds, for men (10 m/s) and materials (15–20 m/s), may be required. Peak powers of several megawatts are demanded, with motor speeds of 50–70 r/min when the machine directly drives the winding drum. Modern Ward-Leonard installations employ thyristor control for higher efficiency, smaller space, lower mass and economy in initial cost. On large units 12- or even 24-pulse converters are chosen to minimize supply-network harmonics. If two converters are necessary to cope with the load, sequence control can be applied to improve the power factor. Motor reversal is usually effected by reversing the field current using an anti-parallel converter; but some smaller equipments substitute armature reversal. With solid-state control, the equipment makes no contribution to the supply-system fault level, there is no 'creep' resulting from remanence and (unless high-speed fans are necessary for cooling) the station noise level is reduced.

### Paper-making machines

Accurate speed control is essential. At the pulp end the mix is very wet (99% moisture) and tender; at the reeling end it is strong and nearly dry (9%). Between these the paper passes through a series of rolls driven by up to a score of motors. To avoid tearing, all motors must run at carefully adjusted and interrelated speeds, but each must be individually controlled to allow for 'draw' as the paper becomes drier. Overall speeds must be adjustable for different paper qualities, and a creep speed (e.g. 0.1 m/s) for initial setting up. Thyristor control enables steady-state speeds to be held within 0.02% (less with digital methods), enabling paper speeds up to 10 m/s to be reached.

### Rolling-mills

Here the power fluctuations are severe. Fig. 9.22 shows typical speed, torque and power variations for one pass of a billet through the rolls of a mill. The

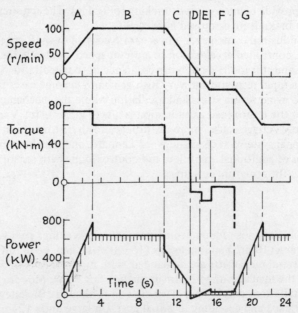

9.22 Rolling-mill drive characteristics

acceleration and retardation torques are each $M_a$ = 15 kN-m, the billet-squeezing (deformation) torque is $M_d$ = 55 kN-m, and the friction torque is

$M_f = 5$ kN-m. During the pass, the speed and torque demands are as follows:

A.  Billet enters mill at the no-load roll speed of 25 r/min, and the rolls are accelerated to 100 r/min: $\quad M_d + M_f + M_a = 75$ kN-m

B.  Rolling at 100 r/min: $\qquad\qquad\qquad M_d + M_f \quad = 60$ kN-m

C.  Retardation to 25 r/min: $\qquad\qquad M_d + M_f - M_a = 45$ kN-m

D.  Billet expelled and rolls stopped: $\qquad M_f - M_a \qquad = -10$ kN-m

E.  Rolls reversed and accelerated to no-load speed:
$$M_f + M_a \qquad = -20 \text{ kN-m}$$

F.  Rolls running at no-load speed: $\qquad M_f \qquad\qquad = - 5$ kN-m

G.  Billet re-enters rolls for re-pass.

To achieve maximum productivity a sophisticated, accurate and quick-response control system is needed. Thyristor-fed d.c. motors are replacing Ward-Leonard systems for billet-rolling, strip-mill and comparable drives around a steelworks.

Commutation presents difficult problems. The compole flux cannot accurately keep pace with rapid changes of armature-circuit current, whether due to load or to harmonic ripple, because of eddy currents induced in the liners by which the compole airgap is adjusted (and even more in the pole core if it is solid). A classic theory of compole flux-lag with magnetic liners was advanced by Rudenberg, but later work [26] has shown that high-resistance nonmagnetic liners considerably reduce the lag.

### Single-phase railway traction

Tap-changing transformers, mounted on the locomotives and feeding d.c. series motors through diodes, have been proved successful; but more recently phase-controlled thyristor equipments have been adopted for better torque control and elimination of contactor maintenance. Two or more converters are connected in series for sequence control, to give higher power factors over the whole voltage range. Improved performance can be obtained with separately excited field windings fed from an anti-parallel converter to obtain reversal without mechanical switching. Constant field current is maintained up to about one-half of full train speed, with field-weakening for higher speeds. Phase control of the armature current gives smoother acceleration, rapid compensation for the inevitable fluctuations of contact-wire voltage, and exploitation of the wheel-adhesion limits. Cost and complication usually inhibit the adoption of regenerative braking. As the contact-wire current level and its active length vary with locomotive location and loading, and with the traffic density, harmonic filtering is essential to limit interference by traction currents with railway signalling and communication circuits.

Phase control is also applied to rapid-transit ('suburban') motorcoach trains. A typical two-coach (motor-trailer) unit of overall length 50 m has a gross weight 80 t (adhesive 50 t), a continuous rating of 1000 kW corresponding to a tractive effort of 45 kN at 75 km/h, an initial acceleration of $1.0$ m/s$^2$, a maximum tractive effort of 125 kN and a maximum speed of 120 km/h. The main transformer has two 365 V secondary windings, respectively of ratings 750 and 600 kVA, one operating alone for high-speed running when the current demand is relatively low. The secondaries feed two half-controlled bridge rectifiers arranged for sequence control. All the equipment is carried under-frame. The four axle-mounted traction motors have compound excitation. The shunt windings provide field-weakening, and also field-strengthening in order to extend the full-load working range of the converters; the series excitation assists in proper current division between motors even when there is disparity between driving-wheel diameters resulting from wear. Each 510 V motor has the following ratings: 300 kW, 650 A, 1600 r/min (1-hour); 250 kW 540 A, 1650 r/min (cont.); maximum current, 1200 A; maximum speed, 2600 r/min. The speed regulation system relieves the driver of the onus of matching tractive effort and brake force to the changing traffic and route conditions: he has only to set the required speed on the master controller, phase-control of the converters then being automatic. This together with preset acceleration and braking rates (the latter with fixed-location application) gives short, accurate and reproducible inter-station journey times.

The electronic control system comprises (i) speed and braking control, and (ii) current-limiting and thyristor-firing control. System (i) receives the pre-selected speed in the form of a d.c. reference voltage, any change in which initiates acceleration or retardation at a fixed rate of $1.0$ m/s$^2$. The actual train speed is sensed by a pulse generator on a driving axle, the frequency being converted to a proportional direct voltage. The fixed-location brake is initiated by a pickup pulsed by a permanent magnet placed at a suitable position on the permanent way, for normal station stops; and other p.m.s for partial brake application are located at the top of steep down-gradients. Braking is electropneumatic. System (ii) imposes current-limiting and controls thyristor firing angle in conjunction with shunt-field adjustment.

## 9.9 INTEGRAL-CYCLE CONTROL

This, also termed *burst-firing*, is a simple speed control suitable for fractional-kilowatt d.c. motors fed from a rectified 1-ph source, which is switched on for a number of periods $c_1$ and off for a further number $c_2$ as indicated in Fig. 9.23($a$). The on/off sequence is cyclically repeated, in a time corresponding to $c = c_1 + c_2$. The switching function is provided by thyristors in either of the connections ($b$), 'freewheeling' being provided by the rectifier bridge diodes in (i) and by the diode D in (ii).

Let the r.m.s. value of the supply voltage be $V_1$. Then the mean direct output voltage over a full interval corresponding to $c$ is

$$V_d = (2\sqrt{2}/\pi)\, V_1(c_1/c)$$

Speed control is obtained by varying $c_1/c$. As both $c_1$ and $c$ must be integral, the speed control is coarse if $c$ is small; but if $c$ is large, the duration of $c_2$ may be such as to result in motor-current discontinuity. Smooth operation over an adequate speed range can generally be achieved with $c$ between 10 and 20, and $c_1/c_2$ less than unity.

An advantage of integral-cycle control is that thyristor triggering occurs at voltage zero, avoiding the abrupt changes of input current that, in phase control, cause radio interference. There is, however, a wide spectrum of harmonic frequencies. For a supply frequency $f$, the lowest harmonic has the order $n = f/c$. There is no component of frequency $f$. It has been shown [27] that the ripple factor exceeds unity for $c_1/c = 1/9$. Harmonics can be reduced by duplicating the thyristor/freewheel-diode/field-winding arrangements in ($b$) (ii), and phasing the two channels and their currents to mitigate torque and current fluctuation.

Motor performance, in terms of mean voltage, current, torque and speed, is virtually the same as when the supply is conventional, but motor losses are significantly greater.

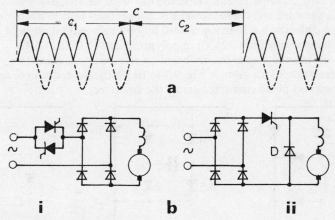

9.23 Integral-cycle control

## 9.10 CHOPPER CONTROL

A thyristor 'chopper' can be used to obtain a variable output voltage from a constant-voltage d.c. supply, such as a battery or a d.c. railway traction supply, to avoid the loss in speed-control resistors. The essential features are described in Sect. 7.6 and Fig. 7.10. In terms of the supply voltage $V_1$, and the mean values of the input current $I_1$, the output voltage $V$ and the output current $I$, and ignoring the small loss (2–3% of the throughput), then $VI = V_1 I_1$ and the chopper may be regarded as a variable-voltage 'd.c. transformer' providing from the input a higher output current at a lower voltage. Output control results from variation of the on-time/off-time ratio $t_1/t_2$. A fixed pulse frequency between 0.25 and 1 kHz is usually preferred, particularly for rail traction, because the inevitable harmonics on the input side have then a fixed frequency range and can more effectively be filtered.

The operating conditions are complex: Dubey and Shepherd [21] investigate them for series machines; Franklin *et al* [28] for separate excitation.

### Turn-off

The chopping frequency is limited by the thyristor characteristics. Turn-on is straightforward, but turn-off requires forced commutation. A previously charged capacitor controlled by an auxiliary thyristor is connected in parallel with the main thyristor and discharged into it in a direction opposing the main output current, so applying reverse bias. The commutating capacitor must be large enough to maintain the discharge current for longer than the turn-off time; for a 240 V 1.5 kW motor its capacitance might be of the order of 20 $\mu$F. This vital component has a severe duty. Large peak currents have to be discharged at high repetition rates within only two or three times the thyristor turn-off requirement. Dielectric loss must be low, heavy-duty terminals provided and the inductance of connections kept to a minimum. Paper as a dielectric has been superseded by polycarbonate (loss-angle 0.0005, temperature-limit 125°C) or polypropylene (0.0001, 85°C).

Turn-off circuits are commonly based on resonance. This and two alternative methods are given in Fig. 9.24: Th1 is the main thyristor and D1 is the freewheel diode connected across the armature.

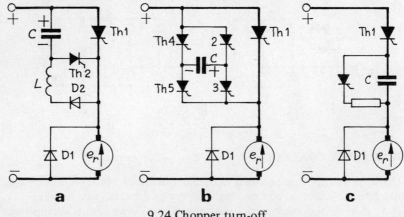

9.24 Chopper turn-off

*Resonance* (*a*) With the motor at rest, Th2 is fired to charge the commutating capacitor $C$ to a p.d. $V_1$ with the polarity indicated. When charging is complete the current ceases, and Th2 turns off naturally. To start the 'on' mode, Th1 is fired. Motor current begins and the machine rotates. Simultaneously $C$ discharges rapidly through Th1, diode D2 and inductor $L$ by means of an inherently oscillatory current of angular frequency $\omega = \sqrt{(1/LC)}$: but only the first half-cycle can flow because of the blocking property of D2. Thus $C$ is isolated with a p.d. of $-V_1$ (i.e. of polarity opposite to that shown in the diagram). The lower plate of $C$ is now at a potential $+2V_1$ and this condition persists throughout the duration of the 'on' mode. To quench the motor current, Th2 is triggered to reverse-bias Th1, and $C$ discharges through the motor. When Th1 has completed turn-off, $C$ is recharged to its original polarity ready to start the next 'on' mode. The resonance turn-off method may emit acoustic noise from its inductor, and precautions must be taken to avoid failure on overload should $C$ produce a discharge current insufficient to quench Th1.

*Bridged Thyristor* (*b*) Initially, Th2 and Th5 are fired to charge $C$ to the indicated polarity, after which they turn off naturally. To start the 'on' mode, Th1 is triggered and conducts current to the motor, the polarity of $C$ remaining unchanged with its right-hand plate at voltage $V_1$ with respect to the negative supply terminal. To initiate the 'off' condition, Th4 and Th3 are fired, raising the left-hand plate to $V_1$ and the r.h. to $2V_1$. This reverse-biases Th1 and it turns off. Discharge leaves the l.h. plate positive, and for the following 'on' mode the functions of Th4 and Th3 are taken over by Th2 and Th5. As with method (*a*), the use of a parallel commutating capacitor introduces the hazard of failure on overload; but the inductor $L$ with its cost and complication is avoided. McLellan [29] describes a modification that requires fewer additional thyristor elements.

*Series Capacitor* (*c*) This avoids commutation failure on overcurrent but passes only a succession of charging-current pulses, and is a method suitable only for very small motors.

### Braking

Basic circuits for three braking connections in Fig. 9.25 are applicable to either series or separately excited machines. Operation is subject to an upper

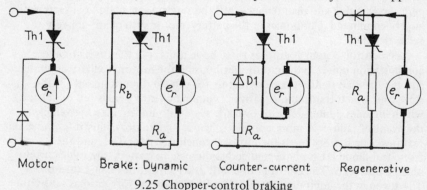

Motor     Brake: Dynamic     Counter-current     Regenerative

9.25 Chopper-control braking

current limit (u.c.l.) for the main thyristor Th1, for which the essential commutating circuit (not shown) has in each case to be provided. In *dynamic* braking, with Th1 conducting, stored kinetic energy is dissipated in the resistor $R_a$. As the braking current rises towards the u.c.l., Th1 is turned off and the current is transferred to a resistor $R_b$ of higher ohmic value.

Plugging, or *counter-current* braking, is not often required. First the armature connections must be interchanged; then the firing of Th1 results in a rapid rise of braking current and torque. Before the u.c.l. is reached Th1 is turned off, current now flowing through diode D1 and resistor $R_a$, the latter being chosen so that the current decreases during the 'off' mode. For *regenerative* braking, conditions during the 'on' mode of Th1 are similar to those in dynamic braking except that, for a series machine, the field connections must be interchanged. When the current approaches the u.c.l., Th1 is turned off so that, if the rotational e.m.f. exceeds $V_1$, current is returned to the supply during the 'off' mode. When the current has fallen to a lower current limit, Th1 is again triggered to initiate the 'on' mode. Thus variation in the receptivity of the supply transfers braking automatically between regenerative and dynamic conditions.

## 9.11 CHOPPER CONTROL: APPLICATIONS

### Battery vehicles

For fork-lift trucks, delivery vans and vehicles required to start and stop repeatedly, chopper control by replacing resistors makes maximum use of the limited battery capacity and may lengthen the duty between battery charging by regenerative braking. Objections to the use of battery road vehicles are concerned with the speed and range limitations imposed by the battery. In their favour are simplicity, quietness, economy and absence of atmospheric pollution — particularly in urban service where nose-to-tail traffic-jam conditions are increasingly common.

Typical battery voltages are 24 V for a 500 kg pedestrian-controlled van and 150 V for a 5-tonne refuse-collecting vehicle. The battery requires periodic charging and maintenance, but the servicing of switch contacts is largely eliminated. Harmonics in the battery/motor system are unlikely to cause trouble.

The control system comprises three basic units: (1) the thyristor and commutation unit, (2) the motor unit housing the motor auxiliary equipment and the off-load field-reverse contactors for vehicle direction, and (3) the drive unit which translates the driver's commands into signals fed to (1) and which includes a fault-detector override system to immobilize the vehicle in the event of failure. A representative scheme, Fig. 9.26(*a*), illustrates the main features. The battery feeds the motor through thyristor Th1, and the freewheel diode D1 is connected across the machine and a low-value current-sensing resistor $S$. The control-pulse unit deals with turn-on and turn-off of Th1 through the commutating circuit, and with the on/off ratio as set by the driver's 'accelerator' pedal. The voltage across $S$ blocks the gate pulses if the u.c.l. is exceeded.

The characteristic curves in (*b*) relate motor voltage and mean current for various positions 1–5 of the control pedal. The motor torque/current and speed/current relations in (*c*) indicate the limitation imposed by the u.c.l.; a

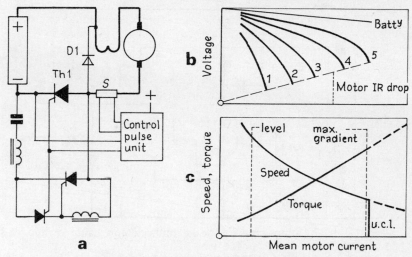

9.26 Chopper control for battery vehicle

fully loaded vehicle can start and accelerate up a gradient for which the current limit is not exceeded. For such conditions thyristor Th1 may be by-passed by a shunted contactor. Protection is afforded by overspeed and armature over-current relays, and by a temperature sensor in the main thyristor heat sink with a feedback to lower the u.c.l. as necessitated by the loading conditions.

*Transistor Chopper*  It is possible to drive a power transistor as a switch by arranging it to operate at either high or zero base current, so that the emitter-collector resistance is very low ('on') or very high ('off'). If the time of transition between these two modes is achieved rapidly by means of a rectangular pulse waveform, the power losses in the transistor are small. Transistor pulse controllers can be applied to light vehicles with low battery voltages, such as 24 V with motor currents up to about 200 A.

An assessment of the design and control features of chopper-controlled battery vehicles has been made by Mangan and Griffith [30].

### Railway traction

Multiple-unit rapid-transit trains having a d.c. supply to the contact-wire or rail, conventionally controlled by series rheostats, can alternatively be equipped for chopper control. The typical system shown in Fig. 9.27 has four series traction motors connected in permanent series pairs, each pair having its own chopper equipment. The choppers operate at a fixed frequency of 274 Hz and are connected in antiphase so that a common 548 Hz filter can be used to mitigate interference with track-circuit signalling. Design constraints make it necessary to introduce rheostat control at speeds below 4 km/h, the resistors being cut in and out by thyristors. At high speeds, field-

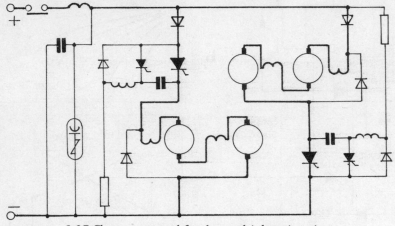

9.27 Chopper control for d.c. multiple-unit train

weakening (to 0.8 or 0.5 of full-field current) is employed to modify the torque/speed characteristics of the traction motors. Compared with conventional resistance control, energy saving up to 7% can be achieved. Regenerative braking could give further energy economy if the cost and complication were justified, but it is less important than with battery vehicles. On long-haul intercity service with locomotives, chopper control may be somewhat less efficient than conventional resistance control.

Input voltage fluctuations of ± 10% may occur in rail traction as a result of varying demands of trains on the system, and pantograph collectors are subject to bounce causing momentary interruptions. These are injurious to regenerative braking as they affect the receptivity of the supply to regenerated energy. The control system must be able to transfer from regenerative to dynamic braking within a few milliseconds.

Supply-side harmonics carried on the contact wires can cause serious interference with parallel running signal and communication circuits. Motor-circuit harmonics give rise to losses and may impair brush commutation. The 'fundamental' of the chopper waveform is the pulse frequency, and must be well above any neighbouring signal-circuit frequency. Further, both sides of the converter will be affected by the harmonics produced in the contact wire by 6- or 12- pulse substation rectifiers, and these may be more deleterious than those developed by chopping.

Shunt capacitors connected across the chopper input for smoothing can produce with the varying length of contact wire (which is inductive) a wide range of resonance frequency. The chopper pulse frequency must always exceed this resonance frequency by a factor of about 2, so that a fixed inductor has to be connected in series with the chopper input. To limit the size of the inductor a high pulse frequency is desirable; for a single chopper a frequency in the range 250–1000 Hz may be chosen, while two or more choppers in parallel, as in Fig. 9.27, can be phased to give a higher effective frequency in combination.

## 9.12 PROTECTION

Thyristors are vulnerable to both overcurrent and overvoltage. Careful choice of thyristor units and comprehensive protection are both essential, and cost-saving in the latter must be balanced against possible damage and the cost of 'down time' and repair.

### Overcurrent

Excessive currents may result from overload, d.c.-side short circuit and thyristor commutation failure. The small thermal capacity of thyristors is augmented by cooling fins and by a heat sink which, besides absorbing heat due to transient currents, can be cooled by natural or forced air or oil circulation. Even with a massive heat sink, the thermal time-constant of a thyristor assembly is no more than a few minutes, but that of the motor it controls may be several hours. Failure of a forced-cooling system must therefore cause immediate tripping. The rated load (or overload) should produce a temperature not exceeding about 150°C, but transient short-circuit currents may be permitted to give momentary temperature rises up to about 300°C. Excessive temperatures, if sustained, impair the thyristor characteristics and may result in breakdown at less than rated reverse voltage. It is essential to select thyristor units of adequate surge-current rating. This is based on a half-sinusoid current of 10 ms duration followed immediately by the application of 0.5–0.8 of the peak reverse voltage rating. The $I^2 t$ value (i.e., [r.m.s. current]$^2$-time) is about 10 for a small 4 A thyristor up to 80 000 for a 250 A unit.

Overcurrent protection can be provided by fuses, circuit-breakers and suppression of trigger pulses. A main requirement is that the $I^2 t$ value that results must be less than that of any thyristor concerned. Further, operation of the protective method should not produce a voltage transient in excess of about 1.5 times the rated peak voltage.

Overload and low-level fault currents can be cleared by a conventional circuit-breaker or contactor on the a.c. side set to trip at about 20% overload with a clearance time of 50–100 ms. For large installations the time can be reduced to 10–20 ms by high-speed circuit-breakers or specially designed fuses. Where short-circuit current levels would otherwise be too high, they may be reduced by inductors in the a.c. feed or, where a transformer is employed, designing it with a high leakage reactance of 0.12–0.15 p.u. A high-speed circuit-breaker on the d.c. side may be used to clear motor faults, and is essential in regenerative braking because a supply failure might result in a short circuit within the converter. Tripping the a.c. breaker should simultaneously intertrip the d.c. breaker.

Trigger-pulse suppression does not limit the first current peak, which is likely to be the most severe, but it prevents the escalation of a minor fault into a major breakdown.

Where rated current level requires two or more thyristor units to be parallelled, impractibility of precise matching makes it necessary to fit a small inductor in series with each unit to equalize the current division, and a high-speed fuse to operate in the event of the unit failing to commutate or breaking down on reversed voltage. The fuse will not prevent damage to the affected thyristor but will protect others in parallel with it. Fuse operation should not cause tripping of the main protection (although an alarm signal should operate), and if there are several units in a parallel group the reduction in capacity is small, and can probably be tolerated for a limited period.

With the double-converter configuration, protection must be provided against excessive circulating current resulting from a commutation or blocking failure in either converter.

To prevent the rate of increase of current exceeding that required by the delay and turn-on times of a thyristor, inductors are fitted to limit the rate to 100–300 A/$\mu$s.

### Overvoltage

Voltage transients, whether generated internally or externally, may damage a thyristor or cause spurious triggering. Both the peak voltage and its rate of rise are significant. Protection is normally provided by absorption of the surge energy in a capacitor and its subsequent dissipation in a resistor. The transient peak should be limited to 1.5–2.5 times the rated peak reverse voltage of the unit, and the rate of rise to about 400 V/$\mu$s.

Voltage transients in the supply system, due to lightning, switching or faults, are not readily predictable, but can be mitigated by the provision of spark-gap suppressors. A more common cause is the operation of the main contactor (or of other contactors in adjacent circuits), particularly if the contacts are worn or are subject to 'bounce'. Voltage rises up to 1000 V/$\mu$s may result. Protection is afforded by a filter capacitor across the supply, located as closely as is practicable to the converter terminals. To avoid resonance with the supply-system inductance a resistor is connected in series with the capacitor and a discharge resistor in parallel with it. The capacitor value is not critical: for a 415/240 V supply a 5 $\mu$F capacitor might be suitable. Filter circuits for harmonic suppression and capacitors for power-factor correction have also to be connected across the supply terminals.

As an alternative, suppression devices may be located across the d.c. terminals of the converter. A single device suffices for a 3-ph equipment, and it does not have to cope with mains-frequency current. A further alternative is a single filter circuit on the a.c. side, fed from a 3-ph diode rectifier.

Protection against internal transient voltages arising from the thyristor commutation process is almost invariably furnished by a series resistor-capacitor 'snubber' circuit connected across each thyristor unit, with values of 0.05–0.5 $\mu$F and 10–50 $\Omega$. This arrangement has been found to give near-complete protection against internal voltage 'spikes' and rates of rise.

When thyristors are connected in series to deal with high operating voltages the inevitable differences between their respective turn-off times result in the first unit to quench being subjected to the full voltage across the chain. To prevent reverse breakdown a capacitor is connected across each unit, a typical value being 0.7 pF per A of forward current rating.

## Layout

Fig. 9.28 for a 3-ph 6-pulse phase-controlled thyristor converter for direct connection to 415 V 50 Hz a.c. supply incorporates most of the protective devices mentioned above. The items are listed below.

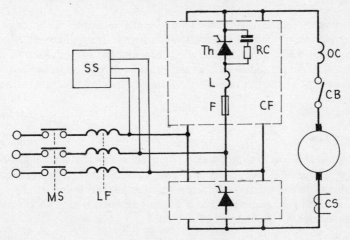

9.28 Protection scheme

*Supply*
MS   main circuit-breaker
LF   inductors for fault-current limitation
SS   line-surge suppressor

*Converter*
Th   Thyristor, voltage rating 1200 V normal, 1400 V peak
RC   resistance-capacitance combination to reduce commutation voltage transients and high ($dv/dt$) values when MS operates
L    inductor to limit rate of rise of thyristor current
F    fast-acting fuse
CF   thyristor cooling-fan failure relay

*Motor Armature*
OC   overcurrent relay to protect against overload or delayed start, acting on CB or on thyristor gate circuits to reduce voltage
CB   circuit-breaker (particularly for regenerative braking to protect against mains failure)
CS   current sensor for current-limit signal to control unit
Further devices normally applied include protection against field failure and interruption to the motor forced cooling fan.

# 10 Losses, Thermal Dissipation and Rating

## 10.1 LOSSES

In a d.c. machine, some or all of the losses listed below may occur. In general they may be roughly classified as *load-dependent* and *speed-dependent*, and in certain cases the total loss can be regarded as having a *fixed* and a *variable* component. All losses appear as heat and must be dissipated from the machine because of the limitation set on the working temperature of the insulation.

### Sources of loss

*Conduction*: $I^2 R$ loss in shunt, series, compole, compensating and armature windings, in shunt-field regulators and series-field diverters. The losses are predictable from the winding resistance and the current. In industrial machines these losses are referred to a winding temperature of 75°C for insulation classes A E B, and of 115°C for F H.

*Exciter*: For separately excited machines and for multi-machine installations sharing a common excitation system.

*Core*: In the armature teeth, the flux density is roughly constant (under steady-state conditions) and directed along the tooth axes while they are within the pole-arc, but are subject to sudden change as they enter or leave this region and also as they pass through the compole region. In the core below the teeth the flux density varies sinusoidally to a first approximation. Simple classical analyses do not apply to the armature as a whole: nevertheless, it is customary to assume that the armature iron is subject to a loss in hysteresis proportional to the rotational frequency $f = np$, and in eddy-current to $f^2$. The loss is normally calculated or measured for no-load conditions, while its design estimate is based unavoidably on empirical curves substantiated by tests on completed machines.

*Pole-face*: Local high-frequency flux pulsations occur as the teeth run past the poles (Sect. 4.10), producing a loss that may be mitigated by laminating the pole-shoes.

*Brush-contact*: The contact surface between brush and commutator has a nonlinear voltage/current characteristic. The combined effects of the semi-conducting film and the aerodynamic 'lift' can be taken as producing a constant volt-drop $v$ that is independent of current $I$, and a loss $vI$. Typically, $v = 1.0$ V per brush-arm for carbon/graphite brushes, 0.25 V for metal/graphite, and 0.1 V or less for small control machines.

*Brush-friction*: For a total brush-surface area $a_b$, applied at a pressure $f_b$ on a commutator running at a peripheral speed $u_c$, the brush-friction loss is $\mu_b f_b a_b u_c$. The coefficient of friction $\mu_b$ varies with the brush material, and may be in the range 0.1–0.3. Brush pressures in industrial machines are typically 12.5 kPa.

*Bearing-friction*: For a shaft of diameter $d_s$ in a journal bearing of length $l_s$, the brush-friction loss is calculable for a rubbing speed $u_s$ using an empirical expression of the form $k d_s l_s u_s$, where $k$ is typically 3000. The friction is much smaller for ball, roller and miniature bearings.

*Windage*: The air-drag on miniature armatures is a minor factor unless the speed is very high. It is, however, significant in large and high-speed machines, but is dependent upon too many intricate factors for ready assessment. In general it is proportional to the square of the armature peripheral speed. Fan loss may be proportional to a higher power of the speed.

*Load (Stray)*: This comprises eddy-current loss in armature conductors, coil/commutator/brush short-circuit loss in commutation, additional armature core loss arising from armature-reaction distortion of the airgap field, and pole-face loss (if not separately accounted). Load loss is difficult to assess, for it occurs in several parts of the machine in a way not susceptible to ready calculation.

*Rectifier*: In machines operated from a.c. supplies, and those having solid-state commutation.

## 10.2 EFFICIENCY

The *power* efficiency of a machine in steady-state operation is the ratio (output power/input power). As important, however, is the *energy* efficiency, the ratio (output energy/input energy) over a specified operating time: it is a function of the duty-cycle of the machine (Sect. 10.7) and of the manner in which the losses are affected by the load. It is the energy efficiency that determines the temperature-rise of a machine in service.

The power efficiency of a machine with an output $P_o$, a total loss $p$ and an input $P_i = (P_o + p)$ is

$$\eta = P_o/P_i = (P_i - p)/P_i = 1 - (p/P_i) \tag{10.1}$$

in which $(p/P_i)$ could be called the 'deficiency'.

Industrial machines rarely operate in the steady state. Small and miniature machines in control systems have even greater fluctuations of speed and load, and their efficiency has less significance than a number of other parameters, such as acceleration rate, linearity, and the electrical and mechanical time-constants.

The relation of power efficiency to load current for an industrial machine over its normal working range can be analysed if the losses can be expressed as functions of speed and current. The simple shunt and series machines are considered, to explain the usual shape of the efficiency/load graph.

*Shunt Machine*   For constant supply voltage, the speed and field flux can be taken as roughly constant. The losses can then be segregated into three components:

> Fixed: field and regulator $I^2 R$, no-load core, friction, windage.
> Proportional to current: brush-contact.
> Proportional to (current)$^2$ : conduction, stray (load).

Taking $I$ as the total input current in motor operation, the loss is $p = k_0 + k_1 I + k_2 I^2$ and the input is $P_i = VI$. The efficiency from eq. (10.1) is

$$\eta = 1 - [(K_0/I) + K_1 + K_2 I]$$

where $K_0 = k_0/V$, etc. The efficiency/current relation is plotted in Fig. 10.1($a$), the bracketted terms being deducted from unity level, the

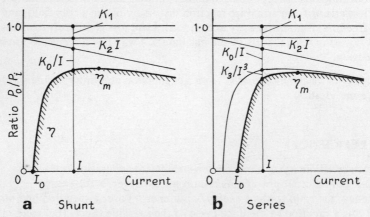

**a**   Shunt          **b**   Series

10.1 Losses and power efficiency

intercept below being the efficiency. The steep initial rise and the presence of a maximum-efficiency load are characteristic. Maximum efficiency occurs for $d\eta/dI = 0$, i.e. for $K_0 = K_2 I^2$ and the fixed loss equal to that proportional to $I^2$.

*Series Machine*   As both the speed $n$ and the current $I$ of a motor vary with load, only an approximate relation between losses and current can be found. Adopting the ideal expressions of eq. (2.15), we assume that the product $nI$ is a constant. Friction and windage are taken as proportional to $n^2$ and therefore to $1/I^2$. The core loss is proportional roughly to the squares of flux and frequency, i.e. to $(In)^2$ which is a constant. The loss components are then

Fixed: core.
Proportional to current: brush-contact.
Proportional to (current)$^2$ : conduction, stray (load).
Proportional to $1/$(current)$^2$ : friction, windage.

The total loss is consequently $p = k_0 + k_1 I + k_2 I^2 + k_3/I^2$, and the efficiency is

$$\eta = 1 - [(K_0/I) + K_1 + K_2 I + (K_3/I^3)]$$

with the results shown in Fig. 10.1($b$). The no-load current $I_0$ corresponds to maximum speed. The no-load loss prevents racing in small machines, but a large series motor must not be disconnected from its load when operating at normal rated voltage.

*Maximum Power* Fractional-kilowatt industrial motors, and permanent-magnet machines in control systems, commonly operate over the whole available range of current and torque from no load with current $I_0$ and speed $n_0$ to stalling (or 'short-circuit') current $I_{sc}$ and zero speed, At $I_0$ and $I_{sc}$ the power output is zero; between these limits there is a maximum output power $P_{om}$ at a current that is normally of the order of $0.5\, I_{sc}$. Fig. 10.2 shows the

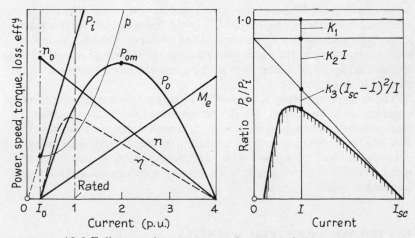

10.2 Full-range characteristics of constant-field motor

torque $M_e$, the speed $n$, the input and output powers $P_i$ and $P_o$, their difference $p$, and the efficiency, typical of a small p.m. motor. The curves are to a base of the per-unit armature current, where 1 p.u. represents rated load. If the field flux and the applied armature voltage are assumed to be constant, then the speed at a current $I$ is approximately $n = n_0 k(I_{sc} - I)$ where $I_{sc} = V/r_a$. The core and mechanical losses are roughly proportional to $n^2$, so that the loss components are

Proportional to $I$: brush-contact.
Proportional to $I^2$: armature conduction, stray (load).
Proportional to $(I_{sc} - I)^2$: core, friction, windage.

The efficiency is then

$$\eta = 1 - [K_1 + K_2 I + K_3 (I_{sc} - I)^2 / I]$$

## Declared efficiency

BS 269 specifies the assessment of efficiency on a basis of loss-summation in place of an actual load test. The loss schedule is

*Exciting-circuit Loss*: (a) shunt $I^2 R$, (b) field rheostat, (c) exciter.

*Fixed Loss*: (d) core, (e) bearing friction, (f) windage, (g) brush friction.

*Direct Load Loss*: (h) armature $I^2 R$, (j) series-winding $I^2 R$, (k) electrical brush loss.

*Stray Load Loss*: (l) in iron parts, (m) in conductors, (n) additional brush loss.

The efficiency is stated for the reference winding temperature (75 or 115°C in accordance with the insulation class). The stray load loss is taken to be proportional to the square of the current, and in view of the difficulty in its evaluation is assumed to be 1.0 or 0.5% of the 'basic output' for machines respectively without and with compensating windings. The basic output is that corresponding to the maximum rated current at the highest rated voltage for constant-speed machines. For variable-speed machines it depends on the method of speed control.

The declared efficiency is assigned by its maker to a machine for specified operating conditions. The $I^2 R$ losses are calculated from d.c. resistance measurements, and most of the remaining losses are inferred from a no-load test. The loss-summation method is valuable for machines of large rating, for not only may load testing be difficult and expensive, but also an efficiency based on the ratio of two nearly equal powers, one electrical and one mechanical, is liable to error.

## 10.3 TESTING: INDUSTRIAL MACHINES

The essential tests cover (i) the mechanical structure, (ii) magnetization characteristics, (iii) load characteristics, losses and efficiency, (iv) temperature rise and (v) commutation.

## Mechanical

The soundness of a machine must be established with attention to the bearings, shaft and rotor eccentricity, commutator surface condition, brush contact and pressure, vibration and noise.

## Magnetization

The machine, whether shunt or series, motor or generator, is driven at a constant speed $n_1$ by means of a pony motor (which for a generator may be the shaft-mounted exciter). The separate excitation is varied from zero. The armature is on open circuit, and the voltage at the brushes (corrected for brush-drop) is the rotational e.m.f. $E_r$. A plot of $E_r$ to a base of excitation current gives the *open-circuit magnetization characteristic*. The ordinates of $E_r$ hold for any other speed $n_2$ if multiplied by $n_2/n_1$. The characteristic for rising field current may not coincide with that for reducing current because of the effect of remanence. If the machine is a shunt generator, its ability to self-excite has to be tested.

Measurement of the power input to the machine represents the sum of the core, friction and windage losses. One method is to run the machine as a motor with varying separate excitation, the armature applied voltage being adjusted to maintain constant speed. After correction for the small armature $I^2 R$ loss and the brush-drop, the armature power input comprises a constant mechanical and a varying core loss. The latter is roughly proportional to the gap-flux density squared, and therefore to the square of the armature voltage. A plot of corrected loss to a base of (armature voltage)$^2$ approximates to a straight line. Extrapolated to zero excitation, the intercept is the friction and windage loss for the speed concerned.

## Load, loss and efficiency

Direct loading of small machines by friction brake, calibrated fan or eddy-current dynamometer is straightforward, and test rigs are commonly employed for this purpose. For larger machines, direct loading is economically possible by back-to-back testing if two identical machines are available. Alternatively a declared efficiency is evaluated by loss-summation.

## Loss-summation

*Shunt Machine* Core and mechanical losses can be obtained as indicated under 'Magnetization'. Winding resistances are measured by any suitable method (e.g. voltmeter/ammeter), it being essential to observe the winding temperature. In the case of an armature, current is fed through the winding at rest and voltage measurements taken between the brush terminals and also between the commutator sectors on which the brushes stand. The armature resistance $r_a$ and the brush-drop $v$ can then be found. For type-testing, the armature no-load power input at rated voltage and speed, corrected for $I^2 R$ and brush-contact losses, can be regarded as the 'fixed' loss. The direct load loss is then calculated for an assumed full-load armature current, and the estimated efficiency reduced by 1 or $\frac{1}{2}\%$ as a token of the stray load loss.

*Series Machine* There are no 'fixed' losses. Loss-summation is illustrated by the tests for series traction motors recommended in BS 173.

*Mechanical loss*: The machine is run on low voltage, adjusted to give range of speeds. The armature input (corrected) represents friction and windage on the assumption that the airgap flux is low enough for core loss to be negligible.

*No-load core and mechanical loss*: The machine is separately excited with a field current $I_f$. A low armature voltage is applied to give the same speed as if the machine were running on load at normal voltage with a current $I = I_f$. Then the net armature input is the sum of the core, friction and windage losses, except that the armature-reaction effect on the gap field distribution (and therefore on the core loss) is subnormal.

*No-load core loss*: This is the difference between the two foregoing loss measurements.

*Brush-contact loss*: Although the volt-drop $v$ is obtainable by direct measurement, BS 173 allows the assumption of 3 V or 2 V for carbon-graphite brushes, the former if the brushes are not stabilized.

*Direct load loss*: This is calculated as the $I^2 R$ loss in armature, field, compole and compensating windings at the appropriate temperature.

*Stray load loss*: A specified addition of 0.5, 0.28 and 0.2 p.u. of the core loss is made for 2.0, 1.0 and 0.25 p.u. of the rated output.

*Retardation*   If a machine running at normal speed is switched off, it slows down at an angular rate $\alpha = d\omega_r/dt$ proportional to the retarding torque $M_d$ and inversely proportional to the inertia $J$ of the rotating parts: thus $d\omega_r/dt = M_d/J = p_d/J\omega_r$, where $p_d = M_d\omega_r$ is the retardation power. By open-circuiting both armature and field, $p_d$ represents the mechanical loss; and if only the armature is switched, the core loss is included, making possible a separation of core from mechanical loss. The slope at A of the retardation speed/time curve, Fig. 10.3, is given by arctan $(AC/CB)$ = arctan $(DC/AC)$. A separate determination of the actual loss for a given speed serves to calibrate the curve.

EXAMPLE 10.1: When cold, a 10 kW 240 V shunt motor runs at 1490 r/min and takes 2.65 A on no load from a constant-voltage d.c. supply. The armature and field resistances, measured at 20°C, are respectively 0.40 and 144 $\Omega$. The winding insulation is class B, and the brush volt-drop is 1.0 V per brush. Estimate for full load (i) the declared efficiency, (ii) the speed.

*No-load*: The field current is $240/144 = 1.67$ A, whence the armature current is $2.65 - 1.67 = 0.98$ A. The rotational e.m.f. is $E_{r0} = 240 - (0.98 \times 0.40) - 2.0 = 237.6$ V. As the armature $I^2 R$ loss is negligible, the fixed loss (core, friction and windage) is $237.6 \times 0.98 = 233$ W.

*Full-load*: The $I^2 R$ losses are based on a winding temperature of 75°C, so that the resistances must be increased by the factor $(235 + 75)/(235 + 20)$ = 1.216 to become respectively 175 and 0.486 $\Omega$. The field current is therefore $240/175 = 1.37$ A. The full-load armature current must be

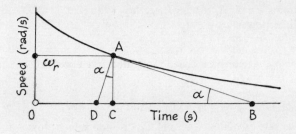

10.3 Retardation test

estimated and then checked. Assuming the efficiency to be 0.85, the input to the machine is $10/0.85 = 11.76$ kW, corresponding to a line current $11.76/0.24 = 49.0$ A, of which the armature takes 47.6 A. The losses (in W) can now be summed:

| | |
|---|---:|
| Fixed | 233 |
| Field, $240 \times 1.37$ | 329 |
| Brush-contact, $47.6 \times 2.0$ | 95 |
| Armature $I^2R$, $47.6^2 \times 0.486$ | 1101 |
| Total loss | 1758 |
| Output | 10 000 |
| Input | 11 758 |

The calculated efficiency is 0.85, and the declared efficiency is $0.85 - 0.01 = 0.84$ p.u.

The field currents are 1.67 A (cold) and 1.37 A (hot), a ratio of 0.82. The corresponding pole-flux ratio is estimated to be about 0.90. The full-load armature e.m.f. is $E_{r1} = 240 - (47.6 \times 0.486) - 2 = 214.9$ V, whence the full-load speed is

$$n_1 = 1490 \,(214.9/237.6)\,(1/0.90) = 1497 \text{ r/min.}$$

EXAMPLE 10.2: A series motor with armature and field resistances of 0.20 and 0.15 $\Omega$ respectively and a brush-drop of 2.0 V gave the following data on a no-load test with separate excitation and a constant speed of 2000 r/min.

| | | | |
|---|---|---:|---:|
| Field current, $I_f$ | (A): | 60 | 120 |
| Voltage across armature brushes, $V_a$ | (V): | 314 | 459 |
| Armature current, $I_a$ | (A): | 11 | 12 |

Estimate the terminal voltage for the motor to run at 2000 r/min and loaded to take 120 A. Evaluate the loss components, assuming the gap density $B$ to be proportional to current, and the hysteresis and eddy-current components $p_h$ and $p_d$ of the core loss $p_i$ both to be proportional to $B^2$. Disregard armature-reaction effects.

Extending the Table:

$$\text{Rotational e.m.f., } E_r = V_a - I_a r_a - v \quad \text{(V):} \qquad 310 \qquad 455$$
$$\text{Armature power, } p_a = E_r I_a \qquad\qquad \text{(W):} \qquad 3410 \qquad 5460$$

The power $p_a$ is the mechanical loss $p_m$ plus the two components of the core loss $p_i = p_h + p_d$. At constant speed (and therefore constant armature frequency)

$$p_a = p_m + p_h + p_d = p_m + (anB^2 + bn^2B^2) = p_m + cB^2 = p_m + kE_r^2$$

From the two sets of values of $p_a$ and $E_r$, the component losses at 120 A are $p_m = 1630$ W, $p_i = 3830$ W. The loss-summation (in W) is then

| | |
|---|---|
| $I^2R$, $120^2(0.20 + 0.15)$ | 5040 |
| Brush-contact, $120 \times 2$ | 240 |
| Core | 3830 |
| Friction and windage | 1630 |
| Total loss | 10 740 |

For the specified load, the terminal voltage required is

$$V = E_r + I(r_a + r_f) + v = 455 + 120 \times 0.35 + 2 = 500 \text{ V}$$

The electrical input is $500 \times 120 \times 10^{-3} = 60.0$ kW, the output is $60.0 - 10.74 = 49.3$ kW, and the efficiency is $49.3/60.0 = 0.82$ p.u.

## Back-to-back tests

Where two identical machines are available, they can be load tested if mechanically coupled, one machine as a motor driving the other as a generator which, in turn, feeds back power to the motoring machine. The power deficiency may be furnished electrically (from a mains supply or a booster, or by unbalancing the machines), or mechanically (by a drive motor coupled to the common shaft), or by a combination of both. In the following, the various elements are designated

> AM, AG   armature of motoring, generating machine
> FM, FG   shunt field windings        D drive motor
> SM, SG   series field windings       B series booster

Corresponding subscripts are used for currents and voltages, $V_1$ and $I_1$ for the mains supply, and $V_m$, $V_g$ for the motor and generator terminal voltages.

These differential tests give close estimates of losses and efficiency, and serve particularly for heat-run and commutation measurements. They demand appropriate test routines. Generally the set is started with the generating machine inactive.

*Shunt Machines,* Fig. 10.4. In its original form (*a*), the test machines are driven by a calibrated auxiliary motor D to furnish the total mechanical loss, and a booster B is included in the closed armature-circuit loop to provide the field and armature $I^2R$ losses. The test machines operate normally, but the

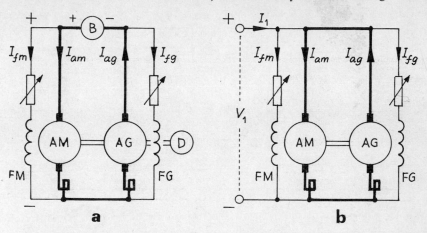

10.4 Shunt machines: back-to-back tests

use of two auxiliaries is inconvenient. Where, as in (b), a d.c. supply of rated voltage $V_1$ is available, both D and B can be dispensed with; but the loadings differ, and the loss measurements are less precise because the armature-current circulation requires $I_{fm}$ to be reduced to raise the speed of the set, and $I_{fg}$ to be increased to overcome the armature resistance volt-drops. The core losses therefore differ from normal, particularly if the test machines are small. The core, mechanical and $I^2R$ losses are furnished by a current $I_1$ from the d.c. supply, and on the assumption of equal loss division,

$$p_i + p_f = \tfrac{1}{2}[V_1 I_1 - (I_{am}^2 r_{am} + I_{ag}^2 r_{ag})] = p_0$$

for each machine. The efficiencies are evaluated from $V_1$ and the measured currents as follows:

| | *Motor* | *Generator* |
|---|---|---|
| Power input | $P_{im} = V_1(I_{am} + I_{fm})$ | $P_{ig} = P_{og} + p_g$ |
| Power loss | $p_m = p_0 + I_{am}^2 r_{am} + V_1 I_{fm}$ | $p_g = p_0 + I_{ag}^2 r_{ag} + V_1 I_{fg}$ |
| Power output | $P_{om} = P_{im} - p_m$ | $P_{og} = V_1 I_{ag}$ |
| Efficiency | $\eta_m = 1 - (p_m/P_{im})$ | $\eta_g = 1 - (p_g/P_{ig})$ |

In an alternative modification, the booster B is reintroduced to provide the armature $I^2R$ losses, leaving the d.c. supply to furnish the mechanical and field losses. In this case the fields do not have to be weakened or strengthened.

*Series Machines*, Fig. 10.5. Method (a) employs a booster B to provide the $I^2R$ loss and a calibrated drive motor D for the core, mechanical and stray load losses which sum to $p_s$. If no drive motor is available, then B provides all the losses. Method (b) requires a normal voltage d.c. supply, the fields SG and SM being connected to take the same current $I_m$. The booster B (which may alternatively be connected in the motor branch) deals approximately with the

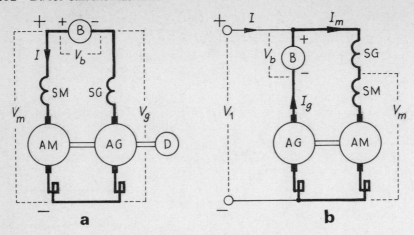

10.5 Series machines: back-to-back tests

$I^2 R$ losses. The input $V_1 I_1$ accounts for the remainder. It may be necessary to adjust the calculated efficiency by the difference $v(I_m - I_g) = v I_1$ for the brush-contact losses. The efficiencies for the two methods (*a*) and (*b*) are evaluated as follows:

|  | (*a*) | (*b*) |
|---|---|---|
| Power input | $P_{im} = V_m I_1$ | $V_m I_m$ |
| Power loss | $p_m = \frac{1}{2}(V_b I_1 + p_s)$ | $\frac{1}{2}(V_b I_g + V_1 I_1)$ |
| Efficiency | $\eta_m = 1 - (p_m / P_{im})$ | $1 - (p_m / P_{im})$ |

### Commutation

Methods are described in Sect. 5.8. The tolerance on the compole m.m.f. and flux density is inversely proportional to the operating speed, and in high-speed machines may be as little as $\frac{1}{2}\%$. If the compole current can be varied (e.g. by a booster or shunt diverter) the upper and lower limits of sparkless commutation can be observed, and the compole shims adjusted for normal series connection. The objective is so to adjust the compole flux (and possibly also the brush position) as to obtain commutation within the 'black band' over the working range of speed and load.

### Temperature-rise

The heating time-constant of a large machine may be several hours, and unless a back-to-back test can be arranged a 'heat run' is too expensive and wasteful. Small machines can be type-tested using a brake or a resistance-loaded generator. A method by which the ultimate temperature-rise can be inferred is described in Sect. 10.5.

## Other tests

The integrity of the *insulation* is checked by application of test voltages between windings, and between windings and frame. Test-voltage levels are specified for machines of various types, operating conditions and rated voltages.

For *rectifier-fed* machines, a type test is applied with rectified current followed by a test on d.c. supply to establish standards. A 50 Hz test at an agreed voltage level may also be performed to obtain the voltages across the field and armature windings.

Oscillographic records are sometimes required for the *response* of the machine to an interruption of rated current for a duration between 0.5 and 1.0 s. Blower motors for traction machines are tested for starting capability at the maximum and minimum line voltages that may occur in such service.

## 10.4 HEAT TRANSFER

Transfer to ambient of the heat originating in losses is a complex phenomenon, a combination of conduction, convection and radiation to the ambient air and surroundings which are, in the ultimate, taken to constitute a heat sink having the character of an 'infinite thermal busbar'. The ambient temperature is defined to be that derived from thermometers placed at a distance of 1–2 m from the machine and protected from direct radiation and air currents.

The table below sets out the symbols used in this section and typical values of some relevant physical quantities:

## Symbols

| | | | | *Subscripts*: | |
|---|---|---|---|---|---|
| $a$ | area | $t$ | time | | |
| $b$ | pressure ratio | $u$ | speed | $a$ | absorption |
| $c$ | cooling coeff. | $w$ | width | $c$ | conduction |
| $c_p$ | sp. heat capacity | $x$ | dimension | $d$ | duct |
| $G$ | mass | $\delta$ | density | $f$ | fluid |
| $J$ | current density | $\epsilon$ | radiation | $i$ | iron |
| $k$ | coefficient | | emissivity | $r$ | radiation |
| $l$ | length | $\theta$ | temperature, | $s$ | space |
| $p$ | power | | temperature-rise | $s$ | surface |
| $q$ | quantity | $\kappa$ | thermal | $v$ | convection |
| $S$ | surface area | | conductivity | | |
| $T$ | thermodynamic | $\lambda$ | surface emissivity | | |
| | temperature | $\tau$ | time-constant | | |

| Material | Specific heat $c_p$ J/(kg K) | Density $\delta$ kg/m³ | Thermal conductivity $\kappa$ W/(m K) | Thermal resistivity $1/\kappa$ (m K)/W | Expansion coefficient per K |
|---|---|---|---|---|---|
| Aluminium | 920 | 2700 | 200 | 0.0050 | $25.5 \times 10^{-6}$ † |
| Copper | 390 | 8900 | 380 | 0.0026 | $16.7 \times 10^{-6}$ † |
| Steel | 500 | 7800 | 150 | 0.0067 | $11.0 \times 10^{-6}$ † |
| Air* | 1000 | 1.3 | 0.025 | 40 | $3.66 \times 10^{-3}$ ‡ |
| Water | 4200 | 1000 | 0.63 | 1.6 | – |
| Mica | – | – | 0.33 | 3 | – |
| Micanite | – | – | 0.12 | 8 | – |
| Paper | – | 800 | 0.12 | 8 | – |

\* At s.t.p. † Linear ‡ Volume

### Conduction

The conductive flow of heat, from the region in which it originates to the surface at which it is transferred to the coolant medium, is important in determining hot-spot temperatures and the thermal gradients to which associated insulants are subjected. In an industrial machine a 50°C hot-spot temperature-rise above ambient might, for example, be made up of a rise of 11°C in the coolant, 25 at the heat-exchange surface, and 14°C internally from that surface to the hot-spot.

Thermal conduction is a function of the temperature-gradient $d\theta/dx$, the heat capacity of the path and its thermal conductivity (rate of energy flow along a path of unit length and unit area per kelvin of temperature difference). For steady-state one-dimensional heat transfer, the rate of flow through a path of length $x$ and area $a$ between thermal equipotential planes having a temperature-difference $\Delta\theta$ is $p_c = \kappa(a/x) \cdot \Delta\theta$. Heat flows fairly readily along thick conductors and along (but not across) the steel laminae of an iron core, but its flow through insulating materials demands a much higher temperature-gradient.

When conducted heat reaches a cooling surface, the temperature-drop $\Delta\theta$ is dependent upon the flow conditions (streamline or turbulent) of the coolant, the surface of the hot body, and the efficacy of the 'scrubbing' action of the fluid on the boundary surface. At the interface there may be interfering oxide films and a stagnant layer of coolant. If the hot body has a surface temperature $\theta_s$ and the bulk temperature of the adjacent coolant is $\theta_f$, the thermal transfer across the interface approximates to $p = k_c(\theta_s - \theta_f)$, where the coefficient $k_c$ is typically in the range $10 - 50$ W/(m² K) for air.

*Quantity of Coolant* Where heat is convected away from a surface by a fluid coolant in motion, a flow of 1 kg/s can absorb a power $\theta c_p$ for a temperature-rise $\theta$. For a power $p$ the required rate of flow in mass or volume per unit time is

$$q = (p/c_p\theta) \text{ [kg/s]} \quad \text{or} \quad q = (p/c_p\delta\theta) \text{ [m}^3/\text{s]} \tag{10.2}$$

The volume of a gas coolant depends on its thermodynamic temperature $T$ and its pressure. For air at the fraction $b$ of standard atmospheric pressure, the volume rate required is

$$q = 8 \times 10^{-4}(p/\theta)(T/273)(1/b) \text{ [m}^3/\text{s]} \tag{10.3}$$

## Radiation

The radiant energy interchange between two surfaces involves the 'view' that they have of each other, their surface emissivities and their respective thermodynamic temperatures, according to the Stefan-Boltzmann relation. A typical expression for radiated power is $p_r = 2.9\ \epsilon\theta^{1.17}$ [W/m$^2$], where the emissivity $\epsilon$ depends on the surface and its treatment (e.g. by painting), and varies from about 0.2 for steel sheet to 0.9 from a grey-painted surface. Radiation is a two-way process: machines that work outdoors and unshaded in a tropical climate may absorb radiant heat from the sun.

## Convection

Heat picked up from a hot surface by an ambient fluid coolant causes the fluid to expand, creating a thermal 'head' with consequent upward circulation. Convection is a complicated phenomenon related to the thermal power density; the dimensions, orientation and surface condition of the hot body; the temperature difference between the emitting surface and the bulk fluid; the thermal conductivity, specific heat capacity, density, viscosity and volume-expansion coefficient of the fluid; and the value of the gravitational constant. Practical estimates are based on the use of 'dimensionless parameters', with formulae set up empirically. Natural convection can be augmented by blowers and fans, which greatly increase the fluid flow that would otherwise be generated only by the change of fluid density and volume with temperature.

## 10.5 TEMPERATURE-RISE

All the losses listed in Sect. 10.1 generate heat, raising the temperature of the region of origin above that of its immediate environment in accordance with (i) the rate of local heat production, (ii) the rate of transfer and (iii) the thermal capacity. The temperature reached in service at a given point in a machine must not exceed that above which the part (in particular the insulation) is endangered. Thus temperature-rise determines rating, which becomes a compromise between cost and 'life'. Industrial machines are therefore rated on a basis of specified limits of temperature-rise above ambient. The thermal temperature/flow field of a machine, with its thermal conductance and capacity, could by analogy be represented by a conductance-capacitance electrical network, with current and p.d. as the analogues of heat-flow and temperature-difference. But the parameters are not readily formulated, and in any case the network would be formidable if the heat sources (slot and overhang conductors, field coils, teeth, core, poles, commutator, brushes, bearings etc.) were to be individually included. Empirical methods are generally relied upon for estimating temperature-rise, because to a reasonable approximation the rise is linearly related to the rate of heat production.

### Ideal temperature-rise/time relation

An ideal homogeneous body, internally heated and surface cooled, has a surface temperature-rise proportional directly to the rate of heat production and inversely to the surface emissivity. Using the same symbols as in Sect. 10.4, let the surface of a body of mass $G$ be at a temperature $\theta$ above ambient. In a time element $dt$ the heat energy $p \cdot dt$ is produced, a heat storage $Gc_p \cdot d\theta$ occurs as a result of a temperature-rise $d\theta$, and a heat quantity $S\theta \lambda \cdot dt$ is emitted from the surface to the coolant on the assumption that dissipation is proportional to the surface/coolant temperature difference. The energy balance is $Gc_p \cdot dt + S\theta \lambda \cdot dt = p \cdot dt$, which solves to the exponential relation

$$\theta = \theta_m [1 - \exp(-t/\tau)] \tag{10.4}$$

in which $\theta_m = p/S\lambda$ is the final steady temperature-rise reached when all internal heat is dissipated from the surface and there is no further thermal storage; and $\tau = Gc_p/S$ is the *heating time-constant*. Thermal time-constants range from a few seconds to several hours depending on the size of the machine, the cooling surface and the efficacy of heat transfer to the coolant; it is the time taken for the surface temperature to rise to $\theta_m$ if the initial rate of rise were maintained, i.e. if all the heat were stored in thermal capacity and none were dissipated. Fig. 10.6 shows a typical exponential temperature-rise/time curve.

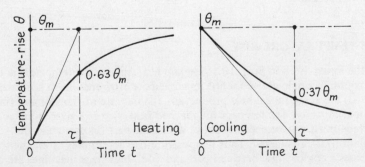

10.6 Exponential heating and cooling temperature-rise/time curves

When the cooling surface is at some temperature $\theta$ above ambient or coolant, its instantaneous rate of rise is

$$\mathrm{d}\theta/\mathrm{d}t = (p - S\lambda\theta)/Gc_p = (\theta_m - \theta)/\tau \tag{10.5}$$

At the start from cold ($\theta = 0$) of a heating cycle, the rate of rise depends on the heat-rate and the thermal capacity alone, so that $\mathrm{d}\theta/\mathrm{d}t = \theta_m/\tau = p/Gc_p$. Similar conditions hold approximately when there is a sudden increase in the loss rate, as may occur when a small motor is rapidly accelerated or a load peak is applied. If, for example, a rotor has a winding material of density $\delta$ and a resistivity $\rho$ at the initial temperature $\theta$, the specific $I^2R$ loss (power per unit mass) for a current density $J$ is $J^2\rho/\delta$. With all heat stored in the winding, the temperature-rise in a short time $\Delta t$ is

$$\Delta\theta = (J^2\rho/\delta c_p)\cdot\Delta t \tag{10.6}$$

where $\rho$ applies to the temperature $\theta + \frac{1}{2}\Delta\theta$ if $\Delta\theta$ is large.

When a hot body cools owing to a reduction or cessation of internal loss, the temperature-fall/time relation is

$$\theta = \theta_m \exp(-t/\tau) \tag{10.7}$$

from an initial $\theta_m$. For the ideal homogeneous body postulated, the cooling and heating time-constants are the same. In the practical case of a machine which cools after shunt-down, and so lacks the dissipation effects of rotation and fanning, the cooling time-constant is greater than that for the running condition.

EXAMPLE 10.3: A machine with a full-load loss of 7.5 kW is considered as a homogeneous mass of 950 kg, with a thermal capacity of 600 J/(kg K) and a surface area 8 m$^2$ of emissivity 11 W/(m$^2$ K). During a full-load heat-run its surface temperature-rise is observed to be 40°C: estimate the rise 1 h later. The heating time-constant is $\tau = 950 \times 600/8 \times 11 = 6480$ s = 1.80 h, and the final steady rise is $\theta_m = 7500/8 \times 11 = 85$°C. The rise after 1 hr is therefore $(85 - 40)\ [1 - \exp(-1/1.8)] + 40 = 59$°C.

EXAMPLE 10.4: An insulated copper field coil, with a gross cross-section of 61.3 cm$^2$ and a mean turn length of 0.80 m, dissipates 140 W continuously. The cooling surface is 0.116 m$^2$ and the surface emissivity is 34 W/m$^2$ per °C rise above ambient. The space factor is 0.545. Estimate the final steady temperature-rise of the coil surface and the time-constant.
The mass of the conductor material is $61.3 \times 10^{-4} \times 0.8 \times 0.545 \times 8900$ = 23.8 kg. The final steady temperature-rise is

$$\theta_m = p/S\lambda = 140/(0.116 \times 34) = 35.5\text{°C}$$

and the heating time-constant is

$$\tau = Gc_p/S\lambda = 23.8 \times 390/(0.116 \times 34) = 2350 \text{ s} = 39 \text{ min.}$$

EXAMPLE 10.5: A motor on constant load gave the following readings of surface temperature:

Time $t$           (h):      0    0.25  0.5   0.75  1.0    1.25  1.5   1.75
Temperature $\theta_s$ (°C):          42.3  45.0  47.4  49.5  51.4  53.0  54.4  55.6

The ambient air temperature was 20°C. Estimate the final steady temperature and temperature-rise, and the heating time-constant.
The exponential relation in eq. (10.4) applies to the temperature-rise $\theta$, not to the observed scale temperature $\theta_s$; the rises are given by $\theta = \theta_s - 42.3$. Applying eq. (10.4) to the adjusted figures ($t_1 = 0.75$ h, $\theta_1 = 7.2$) and ($t_2 = 1.5$ h, $\theta_2 = 12.1$):

$$7.2 = \theta_m \; [1 - \exp(-0.75/\tau)] = \theta_m(1 - k)$$
$$12.1 = \theta_m \; [1 - \exp(-1.5/\tau)] \; = \theta_m(1 - k^2)$$

Thus $\theta_2/\theta_1 = 1 + k = 1.68$, whence $k = \exp(-0.75/\tau) = 0.68$ and the heating time-constant is $\tau = 1.8$ h. Also $\theta_m = \theta_1/(1 - k) = 22.5$°C. This is a rise above the initial scale temperature of 42.3°C, giving a final *scale* temperature of 65°C and a *rise* above ambient of 45°C.


**Thermal cycles**

If a machine, operating in a load condition to which $\theta_{m1}$ and $\tau_1$ apply, has the load changed to a second condition with different values $\theta_{m2}$ and $\tau_2$, the temperature-rise/time curve comprises successive exponential segments. If the duty cycle is repeated indefinitely, the temperature-rise eventually becomes cyclic, the rise from $\theta_2$ to $\theta_1$ in time $t_1$ being equal to the fall from $\theta_1$ to $\theta_2$ in time $t_2$ : then

$$t_1 = -\tau_1 \ln \frac{\theta_{m1} - \theta_1}{\theta_{m1} - \theta_2} \quad \text{and} \quad t_2 = -\tau_2 \ln \frac{\theta_2 - \theta_{m2}}{\theta_1 - \theta_{m2}} \qquad (10.8)$$

from eq. (10.4). If $(\theta_1 - \theta_2)$ is small, the exponential rise and fall segments approximate to straight lines.

EXAMPLE 10.6: A motor has a heating time-constant $\tau_1 = 2.0$ h and a cooling time-constant (when stationary) $\tau_2 = 3.0$ h. When run on constant 1.0 p.u. load the final steady temperature-rise is 60°C, and the losses are proportional to $(\text{load})^2$. Find (i) the temperature-rise (starting from cold) of a duty of 2.0 h on 1.0 p.u. load, 1.0 h shut-down, and 0.5 h at 1.5 p.u. load; and (ii) the eventual temperature-rise limits attained on a repeated duty cycle of 1.5 p.u. load for 0.5 h followed by a 1.0 h shut-down.

(i) $\theta_1 = 60[1 - \exp(-2/2)] = 60 \times 0.63 = 38$°C; $\theta_{m2} = 0$, $\theta_2 = 38 \exp(-1/3)$ $= 38 \times 0.72 = 27$°C; $\theta_{m3} = 60 \times 1.5^2 = 135$°C, $\theta_3 = 135 - (135 - 27)$ $\exp(-0.5/2) = 135 - (108 \times 0.78) = 51$°C.

(ii) From eq. (9.8) with $\theta_{m1} = 135$ and $\theta_{m2} = 0$,
$0.5 = -2 \ln[(135 - \theta_1)/(135 - \theta_2)]$, whence $\theta_1 = 0.78 \, \theta_2 + 30$;
$1.0 = -3 \ln [\theta_2/\theta_1]$, giving $\theta_2/\theta_1 = 0.72$, from which the required limits are $\theta_1 = 60$°C and $\theta_2 = 49$°C.

## Practical temperature-rise/time relation

A rotating machine is not a homogeneous body. Small machine can be modelled in terms of the sum of two exponentials with different time-constants, one short (and roughly the same for both heating and cooling), the other longer (with the cooling time-constant greater than that for heating). For large machines matters are considerably more complex, for there are several regions in which heat losses are developed, each with a characteristic surface dissipating area, mass and thermal capacity, a loss related to the loading and a thermal emissivity affected by the operating conditions — more particularly, the speed. It is not valid to assign a simple exponential temperature-rise/time relation to each region, because the parts are not isolated: they interact. Further difficulties arise with machines on variable speed and load, or on intermittent duty, as in lathe-drive, crane and traction motors.

Suppose that the temperature-rise of one member, say the armature, of a machine has been determined for a base-load and base-speed condition. For a different load at base speed the 'potential temperature-rise' (p.t.-r.) is readily calculated. But the actual rise is higher or lower than the p.t.-r. according as the mean speed is lower or higher than the base speed. Tustin *et al* [31] suggest that, to a good approximation, the correction is a function only of the ratio (mean-speed/base-speed). Thus the armature temperature-rise on intermittent duty can be obtained from two graphs: (i) the p.t.-r. for various base-speed loads, and (ii) a correction factor.

The p.t.-r. is obtained from test. The machine is run, as indicated in Fig. 10.7, for a period $t_1$ of a few minutes duration at a specified load and

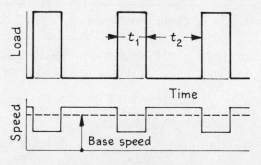

10.7 Temperature-rise test

speed, followed by a period $t_2$ during which the machine is on no load (and as far as possible with no loss) but at a speed such that its mean over the cycle $(t_1 + t_2)$ is the base speed. The mean loss is then $t_1/(t_1 + t_2)$ of that in the operating condition and the mean speed is base speed. If the average temperature-rise is $\theta_m$, then the p.t.-r. is $\theta_p = \theta_m (t_1 + t_2)/t_1$. Similarly the correction factor is obtained by keeping the operating condition constant during the 'on' period $t_1$, and giving the mean speed an appropriate range of values by control of the 'off' speed during $t_2$. The p.t.-r. has the practical advantage that the mean rise for the test cycle is obtained by averaging the p.t.-r.s weighted according to the proportion of the total time that each lasts in a required duty. The result is corrected for mean speed, if necessary, as a final step.

A number of standard duty-cycles has been internationally agreed in order to clarify the appropriate thermal ratings, as described in Sect. 10.7.


## 10.6 COOLING METHODS

Natural convection is relied upon for the cooling of miniature and small machines. In most industrial machines, convection is forced by the provision of rotor fans. For large machines the rotor core mass is too great for purely external surface dissipation to suffice, so that radial and axial air-ducts are formed in the building process with means for directing the coolant through the labyrinth. Heat transfer between a surface and the coolant air is based empirically on the speed of one relative to that of the other, multiplied by a coefficient based on experience and on confirmatory tests. The maximum temperature-rise of a given surface is $\theta_m = p/S\lambda = c(p/S)$, where $c = 1/\lambda$ is the cooling coefficient. The table gives representative ranges of $c$ in terms of a characteristic speed $u$ [m/s] and per unit area of the dissipating surface $S$ [m$^2$].

Cooling Coefficient

$u$ = armature peripheral speed; $u_c$ = commutator peripheral speed; $u_d$ = air speed in duct

| Part | Cooling coefficient | Notes |
|---|---|---|
| Cylindrical surface of rotor | $\dfrac{0.015 \text{ to } 0.035}{1 + 0.1u}$ | Smaller values for large open machine |
| Commutator | $\dfrac{0.015 \text{ to } 0.025}{1 + 0.1\,u_c}$ | Surface includes risers |
| Field coil | $\dfrac{0.06 \text{ to } 0.09}{1 + 0.1u}$ | Based on total coil surface |
| Ventilating ducts in cores | $\dfrac{0.05 \text{ to } 0.2}{u_d}$ | $u_d \simeq 0.1\,u$ for self-ventilation |

Natural convection dissipates typically about 7 W/m$^2$ per °C of temperature difference between surface and air, and radiation somewhat less. In a simple totally-enclosed machine the combined outer-surface dissipation is of the order of 12 W/m$^2$ per °C. With high-speed commutators ($u_c$ up to 45 m/s) the brush-arms baffle the air so that the relative speed of commutator surface and ambient air approaches $u_c$; but low-speed commutators may require forced axial cooling. In ducts, the velocity $u_d$ is taken as (air-flow rate/duct cross-section); even at the low speed of 5 m/s, the duct surface can transfer 25 W/m² per °C.

## Enclosure

The scheme of ventilation is closely related to the construction, in particular the kind of enclosure. Briefly, these are designated

| | |
|---|---|
| SV | Self-ventilated, by means if necessary of rotor-mounted fans. |
| FV | Forced-ventilated, by external fans or blowers. |
| TE | Totally-enclosed: only the carcase is exposed to ambient air. |
| TEFC | Totally-enclosed, fan-cooled; the external surface is swept with air by means of a shaft-mounted fan. |
| TESAC | Totally-enclosed, separate air cooled by external blower. |
| FLP | Flame-proof, with joints and glands capable of retaining and withstanding the effects of explosion of flammable gas. |

## Cooling-air circuit

Miniature machines rely on the fanning action of the rotor to circulate cooling air. For small machines it is common to fit a rotor fan to augment the flow of coolant. Problems of heat dissipation become more difficult at higher ratings. Consider two machines, similar except that the linear dimensions of one are $k$ times as great as those of the other. If flux and current densities are the same for each, the losses are proportional to volume, i.e. to $k^3$, but the cooling surfaces are proportional only to surface area, i.e. to $k^2$. The loss per unit dissipating area is thus proportional to $k$, and although a larger machine may have a higher efficiency, its cooling system must be more elaborate.

The paths devised for the passage of cooling air depend on the size of a machine and its enclosure. Typical arrangements are shown diagrammatically in Fig. 10.8.

*Axial* (*a*) A fan is mounted on the rotor shaft at the end remote from the commutator. It draws air over and through the commutator and end windings, and through the gaps between the stator poles, directed by the end-covers and guides. Dissipation from the rotor core may be augmented by axial channels.

*Axial/Radial* (*b*) Where it is necessary to subdivide the rotor core, air is drawn through radial ducts to pick up heat from the cylindrical surface. With long cores, starvation of the central radial ducts may be countered by the provision of two separate air circuits in parallel.

*Total Enclosure* (*c*) In (i) a shaft-mounted fan, external to the working parts, blows air over the carcase through a space between the main housing and a thin external cover plate. In (ii) an additional fan improves heat transfer to the carcase. The choice of frame material is usually either aluminium or steel:

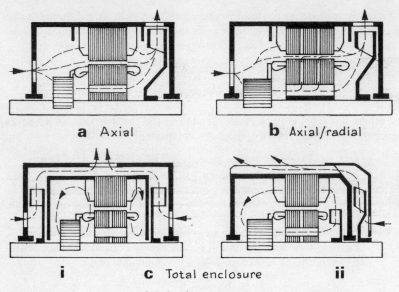

**a** Axial          **b** Axial/radial

**i**          **c** Total enclosure          **ii**

10.8 Cooling-air circuits

the former has the higher thermal conductivity, but the latter provides the greater thermal capacity.

An air-circuit comprises a flow generator feeding a series-parallel air-resistance network. The generator may be an external blower, a rotor fan, or (in small machines) simply the rotor itself. The design of a blower can be optimized, but that of a rotor fan is constrained by the rotor diameter and speed. Simplified, a fan of outer and inner blade diameters $D$ and $d$, running at speed $n$, develops a static head $h_0 = k(D^2 - d^2)n^2$. Applied to an air-circuit this converts to a velocity head $h_1$ which delivers a flow-rate $q_1$ with a characteristic like that in (*a*) of Fig. 10.9. The load characteristic of a given

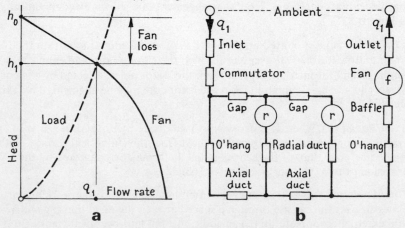

**a**          **b**

10.9 Fan characteristic and air circuit

flow path approximates to a square law, the intersection of the fan and load characteristics determining $h_1$ and $q_1$. The typical equivalent air-resistance network in $(b)$ for a machine with radial ventilation has its fan generator $f$ supplemented by the fanning action $r$ of the rotor. The circuit is divided into an arbitrary number of elements, for each of which an estimate of air resistance must be made. It is necessary to design the many channels to secure that the rotor is not starved of cooling air, and that the temperature-rise of the coolant between inlet and outlet is not excessive; but air-directing baffles can cause a marked loss of head.

Design of the air circuit does not complete the problem of cooling because it is necessary, by means of further thermal networks, to estimate the hot-spot temperatures reached at the heat-producing parts themselves.

EXAMPLE 10.7: Fig. 10.10 shows the dimensions (in mm) of a rotor slot. The rotor is cooled by air (assumed to be at ambient temperature) flowing

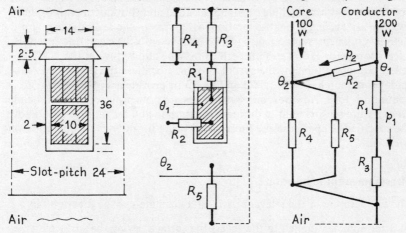

10.10 Thermal network: Example 10.7

over the inner and outer surfaces of the core. Relevant data are: Loss per metre of axial length; conductor $I^2R$, 200 W; core loss in slot pitch width, 100 W. Thermal conductivity: slot insulation and wedge, 0.2 W/(m K); conductor and core-plates, infinite. Thermal transfer at core/air surfaces: outer, 140 W/(m² K); inner, 100 W/(m² K). Obtain values for the thermal resistance per metre of axial length for the heat-flow paths (i) conductor to air through the wedge, (ii) conductor to core through the slot insulation, and (iii) core-surface to air at each boundary with the ambient coolant. Construct an equivalent circuit of thermal resistances and from it estimate the temperature-rises of the conductor and core.

The thermal resistances indicated in Fig. 10.10 have the values

$$R_1 = 0.0045/(0.2 \times 0.014) = 1.6 \quad R_2 = 0.002/(0.2 \times 0.094) = 0.1$$
$$R_3 = 1/(140 \times 0.014) = 0.5 \quad R_4 = 1/(140 \times 0.01) = 0.7$$
$$R_5 = 1/(100 \times 0.024) = 0.4$$

The thermal resistance network and its energy-rate supplies of 200 and 100 W are inserted. Ordinary electric network theory is used for the solution, yielding $p_1$ = 39 W and $p_2$ = 261 W. The temperature-rises are then: conductor, $\theta_c$ = 39 × 2.1 = 82°C; core, $\theta_i$ = 261 × 0.25 = 65°C.

## 10.7 THERMAL RATING

The thermal rating of an industrial machine is based on limits of temperature-rise above a specified ambient temperature [32]. If a machine runs continuously on a fixed load it is assigned a *continuous* rating such that the temperature-rises specified for its various parts are not exceeded. If the machine is loaded for a stated period (e.g. 0.5 h or 1 h), and is then shut down for such a time that all its part temperatures sink to ambient, it can be given a *short-time* rating. More usually, machines are subject to cyclic conditions that may include overload, no-load and shut-down periods. On such a *duty cycle,* the heat-generating losses vary in the several parts, and the thermal emissivities of cooling surfaces change with the operating speed. Brief excess-torque or excess-current periods may be allowed when the temperature-rise is below the specified limit. For these the level of current-limit protection has to be raised and does not provide adequate safeguard for continuous load. However, many machines are now provided with embedded temperature sensors built in during construction and set to, say, 10°C above the specified temperature-rise limit, so allowing for duty-cycle operation on overload.

### Measurement of temperature

The temperature of the relevant parts of a machine may be measured by:

T: Thermometer applied to the accessible surface, giving the surface temperature at the point of application
E: Embedded temperature detector (thermo-couple or resistance coil), giving the temperature at one internal point, usually at or near a hot-spot
R: Resistance, involving the measurement of the resistance of a winding at known temperatures, and estimating temperature by use of the resistance-temperature coefficient

These methods give results that do not refer to the same thing and furnish different bases for the estimation of hot-spot temperatures. Embedded sensors can be monitored during the operation of a machine, but thermo-metric and resistance measurements are made immediately after shut-down, and may be corrected for the temperature drop that occurs in the interval between shut-down and measurement.

## Temperature-rise limits

The international IEC limits of temperature-rise are based for industrial machines on an ambient temperature not exceeding 40°C, and for traction machines 25°C with the limits raised by 10°C for totally enclosed machines. The methods T and/or R are specified for temperature measurement.

Limits of Temperature-Rise, °C

| Part | | Insulation Class | | | | |
|------|---|---|---|---|---|---|
| | | A | E | B | F | H |
| **Industrial machines** | | | | | | |
| Field winding: | | | | | | |
| normal with insulated conductors | T | 50 | 65 | 70 | 85 | 105 |
| | R | 60 | 75 | 80 | 100 | 125 |
| low-resistance 1-layer insulated | TR | 60 | 75 | 80 | 100 | 125 |
| bare conductors | TR | 65 | 80 | 90 | 110 | 135 |
| Armature winding | T | 50 | 65 | 70 | 85 | 105 |
| | R | 60 | 75 | 80 | 100 | 125 |
| Commutator | T | 60 | 70 | 80 | 90 | 100 |
| Core: in contact with insulated winding | T | 60 | 75 | 80 | 100 | 125 |
| not in contact with insulated winding | T | non-injurious value | | | | |
| **Traction machines** | | | | | | |
| Field winding | R | 95 | 115 | 130 | 155 | 180 |
| Armature | R | 85 | 105 | 120 | 140 | 160 |
| Commutator | T | 90 | 90 | 90 | 90 | 90 |

## Thermal rating

The temperature-rise is interpreted in accordance with the duty that a machine is called upon to perform. Although this may not be predictable with any precision, it may be comparable with one of the internationally agreed standard duty cycles designated as follows:

| | | | |
|---|---|---|---|
| S1 | Continuous load | S5 | As S4 but with electric braking |
| S2 | Short-time load | S6 | Repeated cycle |
| S3 | Intermittent load | S7 | As S6 but with electric braking |
| S4 | As S3 but with starting | S8 | As S6 but with related load/ speed characteristic |

These are illustrated, with associated temperature-rise/time curves, in Fig. 10.11. The loading designations referred to in the graphs are

N  Normal load;    R  Standstill, de-energized;    D  Starting;
F  Braking;    V  No-load, with machine rotating.

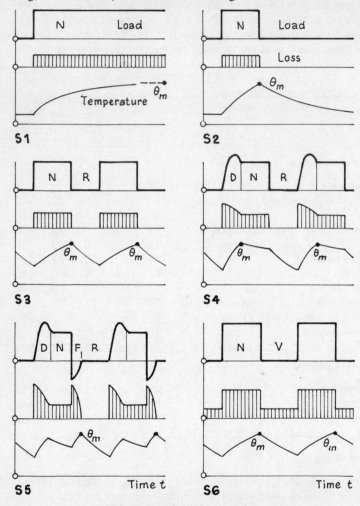

10.11 Standard duty cycles

*Cycle-duration Factor*  For the more elaborate operating conditions a cyclic-duration factor is defined from the length of time of each identifiable loading condition in Fig. 10.11:

S3    $N/(N + R)$                    S5    $(D + N + F)/(D + N + F + R)$
S4    $(D + N)/(D + N + R)$          S6    $N/(N + V)$

*Rating Classes*  These are abbreviated indications of the rating on which the design of the machine is based:

MCR: Maximum Continuous Rating, operating as in S1

STR: Short-time Rating. The load durations in S2 are usually 10, 30, 60 or 90 min, and the limits of temperature-rise may, by agreement, be increased by 10°C

ECR: Equivalent Continuous Rating, used for test purposes in the case of duties S1 and S2

DTR: Duty-type Rating, corresponding to S3–S8. The time for a complete cycle is taken as 10 min, with cycle-duration factors 0.15, 0.25, 0.40 or 0.60. The information required for the d.t.r. includes the actual cycle time, the cycle-duration factor and the kinetic energy stored in the motor and load inertia.

# 11 Control Machines

## 11.1 TYPES AND CHARACTERISTICS

As the subject of automatic control has an extensive literature, this chapter deals only with the characteristics of d.c. control machines and not with the analysis of control and servo systems in general.

Conventional permanent-magnet and current-excited machines are applied in open- and closed-loop control systems, together with such variants as split-field, cross-field and tachometer machines, all with commutators and brushgear. Fractional-kilowatt d.c. motors have higher efficiencies and smaller frame sizes than their a.c. rivals, and their performance is often superior. However, they have certain inherent disadvantages concerned with magnetic and commutation phenomena. Saturation in machines with field-current control results in a nonlinear relation between control current and working flux; hysteresis causes rising and falling flux/field-current relations to diverge, introducing an effect analogous to backlash in a gear-train with a consequent tendency to instability. The conventional commutator in a small machine has comparatively few sectors, giving rise to e.m.f. and torque ripple. Brushes are responsible for volt-drop, friction, wear and radio interference. Commutators and brushgear require maintenance, a particular problem where many small motors operate in a complex assembly. In aircraft, where reliability is a prime factor, the low pressure of the upper ambient air impairs commutation.

In some applications, brushless 'stepper' motors with electronic methods of commutation can be applied to circumvent the disadvantages of the conventional arrangement.

### Performance

Control and servo motors rarely operate in the steady state. Typically their duty involves high acceleration rates (e.g. 30 000 rad/s$^2$), rapid reversal, incremental positioning at several thousand steps per second, and 'plug' braking by reversal of polarity of the control voltage.

*Inertia* This limits the rate of change of speed. The inertia of conventional slotted rotors can be reduced by lengthening the core and reducing its diameter. Very low inertia can be achieved with the 'moving-coil' construction, which has the added advantages of reduced armature inductance and the elimination of 'cogging'; and occasionally the windings

278

are of aluminium. Electrical time-constants less than 1 ms and accelerations up to $10^6$ rad/s$^2$ are possible.

*Thermal Dissipation* The term 'rating' as used for industrial machines in Chapter 10 has little significance for control machines, the operating conditions of which are so different. The causes of loss are included in the list in Sect. 10.1, but they are proportionally larger than could be tolerated in an industrial machine and they have a different relative importance. Temperature-rise significantly increases the winding resistance, impairing regulation; and it affects the values of the e.m.f. and torque 'constants' by reason of the negative flux-temperature coefficient in permanent-magnet machines. The frequency response of the machine is altered because both the electrical and the mechanical time-constants are temperature-sensitive. Heat flow to ambient is through rotor/stator and stator/housing paths in series, the thermal time-constant of the former being short. It is possible for the rotor to reach excessive temperature on a rapid-acceleration duty without the fact being evident externally.

*Resonances* A rotor connected to an inertial load by a shaft transmits its torque by torsion. The system thus has stored kinetic and potential energy, and so is capable of torsional resonance. A periodic control signal having a harmonic close to the natural frequency may lead to abnormal response. Shaft 'whip' is also possible, and both phenomena necessitate shafts of adequate stiffness.

*Demagnetization* A permanent-magnet motor may suffer some loss of magnetic field strength as the result of armature reaction, as when high acceleration rates or plug braking impose high current levels.

*Mechanical Characteristics* To preserve operational accuracy, motors must be dynamically balanced and rigidly mounted. Sleeve bearings are quiet, but ball bearings are better able to give centrality and withstand axial thrust.

### Model

For conventional commutator machines, we adopt the model described in Sect. 7.2 with the system equations

$$\text{Field:} \quad v_f = (r_f + L_f \text{p}) i_f$$
$$\text{Armature:} \quad v_a = (r_a + L_a \text{p}) i_a + G\omega_r i_f \tag{11.1}$$
$$\text{Torque:} \quad M_e = G i_f i_a = -(M_l + M_a + M_f + M_d)$$

Where the field is derived from permanent magnets, $v_f$ does not apply and the motional e.m.f. term in $v_a$ becomes $k_e \omega_r$. Brushless machines are considered separately in Sect. 11.4.

## 11.2 PERMANENT-MAGNET MACHINES

Let the magnets produce a gap flux $\Phi$, constant and unaffected by armature reaction, and let the rotor be conventional with brushes in the q-axis. The rotational e.m.f. is $e_r = k_e\omega_r$ and the gross torque is $M_e = k_t i$ for a rotor current $i$, with $k_t = k_e$. In a machine with $2p$ poles and a rotor with $N$ turns in $2a$ parallel paths between brushes, $k_e = k_t = (p/a) N \Phi/\pi$. In the following a 2-pole machine is assumed and its response as a motor to an armature voltage $v$ is evaluated, using the subscripts

| | | | | | |
|---|---|---|---|---|---|
| $a$ | acceleration | $f$ | friction | $o$ | output |
| $d$ | damping | $i$ | input | $s$ | stalling |
| $e$ | electrical | $l$ | load | $t$ | torque |
| $em$ | electromechanical | $m$ | mechanical | $0$ | no-load |

*General Equations*  The armature has an inductance $L$ and a resistance $r$ (including brush volt-drop effect). The electromagnetic torque provides the load torque $M_l$, the torque $M_f$ lost in friction and windage, the eddy-current damping torque $M_d = k_d\omega_r$ proportional to speed, and the accelerating torque $M_a = Jp\omega_r$. The general equations, eq. (11.1), reduce to

Armature:   $v = (r + Lp) i + k_e\omega_r$

Torque:   $M_e = M_l + M_f + k_d\omega_r + Jp\omega_r = k_t i$

which are linked by the relation $i = M_e/k_t$.

### Steady-state characteristics

For a constant armature voltage $V = E_r + rI$ the voltage equation is $V = k_e\omega_r + (r/k_t) (M_l + M_f + k_d\omega_r)$, whence the constant speed is

$$\omega_r = \frac{V(k_t/r) - M_l - M_f}{k_e(k_t/r) + k_d} \tag{11.2}$$

A series of constant voltages gives a corresponding series of linear speed/torque or speed/current characteristics, Fig. 11.1(a). Their slope,

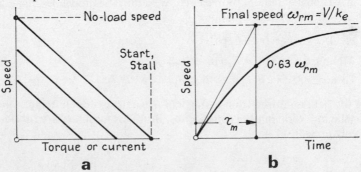

11.1 Permanent-magnet motor: characteristics

$1/[k_e(k_t/r) + k_d]$, defines the speed regulation. The *stalling torque* at $\omega_r = 0$ is

$$M_{es} = V(k_t/r) - M_f \cong V(k_t/r) \tag{11.3}$$

and the *no-load speed* for $M_l = 0$ is

$$\omega_{r0} = \frac{V(k_t/r) - M_f}{k_e(k_t/r) + k_d} \cong \frac{V}{k_e} \tag{11.4}$$

The approximations are for negligible friction and damping.

### Dynamic characteristics

We deal now with instantaneous voltage, current and speed. In view of the high acceleration rates demanded, a duty in which inertial torque is paramount, we may simplify the analysis by assuming the motor and its load to form a *purely inertial* system. Then $v = k_e\omega_r + (r + Lp) i$, and $M_e = k_t i = Jp\omega_r$, whence $i = (J/k_t)p\,\omega_r$. Substituting for $i$ in the voltage equation gives

$$v = k_e [p^2 L(J/k_e k_t) + pr(J/k_e k_t) + 1]\,\omega_r$$

and in terms of Laplace transforms, with $J/k_e k_t = g$ for simplicity, the transfer function relating speed to voltage is

$$\frac{\omega_r(s)}{V(s)} = \frac{1}{k_e(Lg\,s^2 + rg\,s + 1)} \tag{11.5}$$

The denominator has the roots $(r/2L)\,[-1 \pm \sqrt{(1 - 4L/r^2 g)}]$. The inductance is usually small enough to replace the square-root term by $(1 - 2L/r^2 g)$ from the binomial expansion. The poles of the denominator are then

$$-\frac{k_e k_t}{rJ} = -\frac{1}{\tau_{em}} \quad \text{and} \quad -\frac{r}{2L}\left[2 - \frac{2L}{r^2 g}\right] \cong -\frac{r}{L} = -\frac{1}{\tau_e}$$

where $\tau_{em} = rJ/k_e k_t$ and $\tau_e = L/r$ are respectively the *electromechanical* and the *electrical time-constants*. The approximations are valid if $\tau_{em} > 10\,\tau_e$. The speed/voltage t.f. is now

$$\frac{\omega_r(s)}{V(s)} = \frac{1}{k_e(1 + s\tau_{em})(1 + s\tau_e)} \tag{11.6}$$

If the armature inductance can be ignored, the electrical time-constant is zero. In such a case, the response to a step-function voltage $V$ applied to the armature at rest is an exponential rise, Fig. 11.1(*b*), to a final steady-state speed $V/k_e$.

The angle of rotor rotation is the time-integral of the speed. The t.f. relating angle $\theta$ to control voltage $v$ is eq. (11.6) multiplied by $(1/s)$.

The analysis above has been simplified by omission of the load, friction and damping torques. Their inclusion complicates the t.f. considerably, obscuring the essential presence of two inherent time-constants, namely $\tau_e$ depending on the armature inductance, and $\tau_{em}$ a function of inertia and electrical parameters in combination.

EXAMPLE 11.1: A p.m. motor and direct-connected load have the following parameters. Armature, resistance $r = 1.2\ \Omega$, inductance negligible. Torque constant $k_t = 45$ mN-m/A. Damping coefficient $k_d = 0.28$ mN-m per rad/s. Inertia, motor $J_m = 4.6 \times 10^{-6}$ kg-m$^2$, load $J_l = 1.4 \times 10^{-6}$ kg-m$^2$. Constant torques, load $M_l = 80$ mN-m, friction $M_f = 40$ mN-m. (*a*) Evaluate (i) the electromechanical time-constant, (ii) the no-load speed for an armature voltage of 25 V, and (iii) the voltage for which the load torque stalls the motor. (*b*) Estimate the armature voltage, loss and efficiency for a constant speed of 3000 r/min ($\omega_r = 314$ rad/s). (*c*) The system moves in 0.50 rad steps at 50 steps/s as shown in Fig. 11.2(*a*) to give a mean speed of 25 rad/s, the

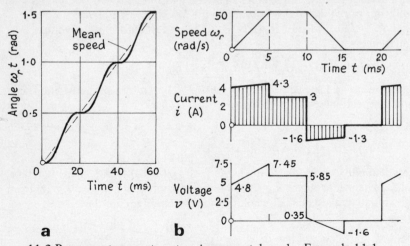

**a**　　　　　**b**

11.2 Permanent-magnet motor: incremental mode: Example 11.1

speed/time profile being assumed to be trapezoidal: find the applied-voltage profile, and estimate the loss and efficiency.

(*a*)(i) *Time-constant* The total inertia is $J = (4.6 + 1.4)10^{-6} = 6.0 \times 10^{-6}$ kg-m$^2$, whence $\tau_{em} = rJ/k_t k_e = 3.6$ ms.

(*a*)(ii) *No-load Speed* The motor has to provide the friction and damping torques $(M_f + k_d\omega_0) = k_t i_0$. Hence $e_r = k_e\omega_0 = V - r i_0 = V - r$ $[(M_f + k_d\omega_0)/k_t]$, which gives $\omega_0 = 456$ rad/s (4350 r/min).

(*a*)(iii) *Stalling Torque* At zero speed the gross torque is the sum of the load and friction torques, i.e. 120 mN-m, requiring a current $i_a = 120/45 = 2.67$ A and therefore an armature voltage $V = 2.67 \times 1.2 = 3.2$ V.

(*b*) *Slewing* From eq. (11.1) with constant speed $\omega_r$ the current is

$$I = [M_l + M_f + k_d \omega_r]/k_t = [120 + 0.28 \times 314]/45 = 4.6 \text{ A}$$

The $I^2 R$ loss is $p_c = 1.2 \times 4.6^2 = 25$ W. The mechanical loss is $p_f + p_d = 10^{-3} [40 + 0.28 \times 314] \, 314 = 40$ W. In the slewing (constant-speed) mode the mechanical loss predominates. With $k_e = 45$ mV per rad/s, the rotational e.m.f. is $e_r = k_e \omega_r = 14.1$ V, the volt-drop in resistance is $4.6 \times 1.2 = 5.5$ V, and the control voltage applied to the armature is $14.1 + 5.5 = 19.6$ V. The output power is $p_o = M_l \omega_r = 25.1$ W, the input power is $p_i = 19.6 \times 4.6 = 90$ W, whence the efficiency is $25.1/90 = 0.28$ p.u.

(*c*) *Stepping* The initial acceleration is $50/(5 \times 10^{-3}) = 10\,000$ rad/s$^2$, which demands an inertial torque $(4.6 + 1.4)10^{-6} \times 10\,000 = 60$ mN-m. From eq. (11.1)

$$i = [80 + 40 + 0.28 \, \omega_r + 60]/45 = 4.0 + 0.0062 \, \omega_r$$

The initial current is 4.0 A, rising in the acceleration period to 4.3 A. The r.m.s. current is about 4.2 A and the conduction loss is $4.2^2 \times 1.2 \times 5 = 106$ mJ. During the constant-speed period the inertial torque vanishes and the current is now 3.0 A giving a conduction loss of 54 mJ. During the retardation period the torques $M_l$, $M_f$ and $M_d$ all assist the reduction of speed: thus

$$i = [-80 - 40 - 0.28 \, \omega_r + 60]/45 = -1.33 - 0.0062 \, \omega_r$$

which changes from $-1.63$ A to $-1.33$ A and results in a conduction loss of about 14 mJ. The mechanical loss is the time summation of the term $(M_f + k_d \omega_r)\omega_r$ over the three 5 ms periods, giving 22 mJ. For 50 steps/s the mean rates of energy dissipation are the loss powers

$$p_c = 50(106 + 54 + 14)10^{-3} \quad = 8.7 \text{ W}$$
$$p_f + p_d = 50 \times 22 \times 10^{-3} \quad = 1.1 \text{ W}$$

Thus in the stepping mode the $I^2 R$ loss resulting from the large accelerating current predominates. The output, taken to be the product of the load torque $M_l = 80$ mN-m at the mean speed of 25 rad/s is $P_o = 80 \times 25 \times 10^{-3} = 2.0$ W. The input $P_i = 2.0 + 8.7 + 1.1 = 11.8$ W, whence the efficiency is 0.17 p.u. Fig. 11.2(*b*) shows the current profile, and the voltage profile constructed from the relation $v = k_e \omega_r + ir$. A simpler input voltage profile, such as 6.0 V for 10 ms and zero for the succeeding 10 ms in a step, would give a substantially similar result.

## Practical machines

Details of two typical p.m. motors are listed in the table below. Machine A has a moving-coil rotor enclosing a fixed central 2-pole magnet, and a commutator with 7 silver-alloy sectors, of diameter 2 mm, and a pair of 3-leaf gold brushes to reduce friction and volt-drop. It is continuously rated on constant voltage to drive small portable tape recorders, models and medical instruments; and intermittently rated for variable voltage to drive integrators, counters and miniature servomechanisms. Its maximum efficiency approaches 0.85 p.u. Machine B has a conventional slotted rotor and operates at voltages between 18 and 27 V. Its test results on nominal 22.5 V are shown in Fig. 11.3.

| Machine: | | A | B |
|---|---|---|---|
| Housing diameter/length | (mm): | 20/25 | 35/68 |
| Mass | (kg): | 0.035 | 0.29 |
| Nominal voltage | (V): | 3.0 | 22.5 |
| Nominal current | (A): | 0.5 | 0.75 |
| Nominal output | (W): | 1.0 | 15 |
| Armature resistance | ($\Omega$): | 2.0 | 6.4 |
| Armature inductance | (mH): | 0.14 | 9.0 |
| Torque constant | (mN-m/A): | 2.7 | 21.4 |
| No-load speed | (rad/s): | 1150 | 1050 |
| | (r/min): | 11 000 | 10 000 |
| Stalling torque | (mN-m): | 4.0 | 67.5 |
| Moment of inertia | ($10^{-6}$ kg-m$^2$): | 0.14 | 1.32 |
| Maximum acceleration | (krad/s$^2$): | 29 | 51 |
| Mechanical time-constant | (ms): | 39 | 18.5 |
| Electrical time-constant | (ms): | 0.07 | 1.4 |

*Operational Limitations*  The significant characteristics in operation are those of no-load speed, maximum continuous and maximum intermittent torques, commutation, and heating. The general effect of these in delimiting the speed/torque operating range is shown in Fig. 11.4.

Machines normally form part of a motor-tachogenerator-amplifier system and are connected to loads which, in view of the very high inherent acceleration capability of the motor, have appreciable inertial effect. The initial acceleration rate of an unloaded motor of inertia $J$ at a current that gives the peak (stall) torque $M_s$ is $(d\omega_r/dt) = \alpha = M_s/J$, but when connected to a load of inertia $J_l$ is naturally reduced. It is usual to employ a motor for which $J$ is considerably greater than $J_l$ (e.g. $J = 4J_l$) in order to avoid sacrificing acceleration rate unduly. The product $M_s\alpha = M_s^2/J$ is a *power-rate* parameter (in W/s), used sometimes as a guide in the selection of a motor to drive a given load inertia.

In high-acceleration machines, the distribution of the gap flux density may be markedly deformed by armature reaction, so that each point on the magnet surface has a different recoil line. Some units are provided with tapered airgaps to mitigate demagnetization and to improve commutation. Rapid stopping duty, such as a stop within 1000 revolutions from a speed of 10 000 r/min, demands stabilized magnets.

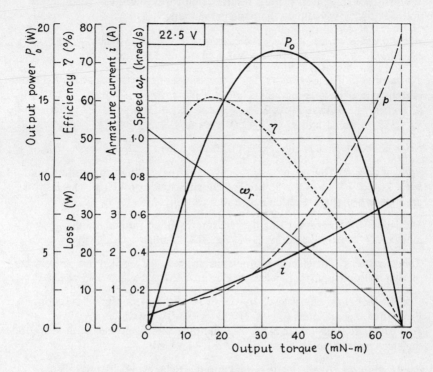

11.3 Permanent-magnet motor B: test characteristics at 22.5 V

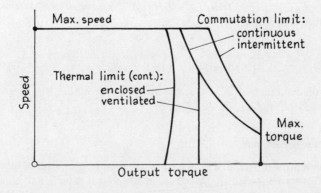

11.4 Permanent-magnet motor: operating limits

EXAMPLE 11.2: Given the test results of motor B, Fig. 11.3, for a supply voltage of 22.5 V, evaluate the basic parameters.

*Armature Resistance*   At stall, the current is $I = 3.5$ A, whence $r = 22.5/3.5 = 6.4\ \Omega$, an effective value that includes brush-drop.

*Torque and E.M.F. Constants*   These are assessed from the range in which the current/output-torque relation is linear. Taking the $(M_e, I)$ pairs (20 mN-m, 1.10 A) and (50, 2.5) gives $\Delta M_e/\Delta I = 30/1.4 = 21.4$; hence $k_t = 21.4$ mN-m/A and $k_e = 21.4$ mV per rad/s.

*No-load Speed*   This is 1050 rad/s (10 000 r/min).

*Stalling Torque*   The current is $I = 3.5$ A, and the stalling torque should be $k_t I = 21.4 \times 10^{-3} \times 3.5 = 75$ mN-m. The measured torque (67.5) is less on account of armature reaction and static friction.

*Maximum Acceleration*   Using the measured stalling torque and the known inertia, the initial acceleration is $M_s/J = 51\ 000$ rad/s$^2$.

*Time-constants*   The electrical time-constant is $\tau_e = L/r = 1.4$ ms. The electromechanical time-constant, calculated from $Jr/k_t k_e$, is $\tau_{em} = 18.4$ ms.

*Viscous Damping*   This is given by the slope of the torque/speed relation. From the $(M_o, \omega_r)$ pairs (0, 1050) and (37, 500), the value of $k_d$ is $37/550 = 0.067$ mN-m per rad/s. The damping coefficient can be used to find the *mechanical* time-constant, $\tau_m = J/k_d = 19.7$ ms.

*Power and Loss*   At the load for maximum efficiency $M_o = 16.0$ mN-m, $\omega_r = 805$ rad/s and $I = 0.90$ A. Then the input is $P_i = 22.5 \times 0.90 = 20.2$ W, the output is $16.0 \times 805 = 12.9$ W, the efficiency is 0.64 p.u., and the loss is $20.2 - 12.9 = 7.3$ W. The $I^2 R$ loss accounts for $0.90^2 \times 6.4 = 5.2$ W, the remaining 2.1 W being the load (stray) loss.

*Transfer Function*   For the motor unloaded, using eq. (11.6), $g = J/k_e k_t = 2.9 \times 10^{-3}$. Then $Lg = 0.4 \times 10^{-6}$ and $rg = 18.6 \times 10^{-3}$. The transfer function is therefore

$$\frac{\omega_r(s)}{V(s)} = \frac{1000}{21.4[0.4 \times 10^{-6} s^2 + 18.6 \times 10^{-3} s + 1]} \cong \frac{46.7}{[1 + 0.019s]}$$

For the steady state and 22.5 V at the armature terminals, this gives the no-load speed, 1050 rad/s (10 000 r/min).

## Tachometer generators

A tachogenerator is a small d.c. machine designed to give a terminal output voltage $V$ proportional to the speed $\omega_r$ at which it is driven, and to act as an analogue speed indicator, velocity-feedback device or signal integrator. In some applications the accuracy is of prime importance, but in servo damping duty the fidelity is secondary to freedom from 'noise'. Normally a p.m. field system is employed for its constant flux, compact construction, reliability and high output – typically of the order of 10 mV per rad/s (10 V per 1000 r/min). Moving-coil armature windings suit conditions requiring low inertia, and the low inherent inductance benefits commutation. Conventional rotors have slots skewed to reduce ripple. Where long periods of low-speed running might impair the contact surface of a copper commutator, silver is substituted. Tachogenerators associated with control motors may be built into the motor housing, and on occasion both motor and tachogenerator windings can be carried on the one rotor core. If, however, the generator has a moving-coil winding it is necessary to adopt separate units with a stiff coupling shaft.

*Characteristics* For constant p.m. flux, the *ideal* linear relation $V = k_e \omega_r - ri$ of (*a*) in Fig. 11.5 is obtained. With a load of resistance $R$, as

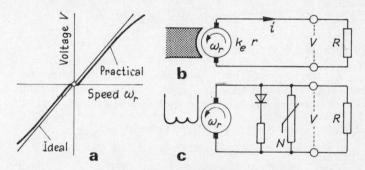

11.5 Tachometer generators

in (*b*), the output voltage is $V = k_e \omega_r [R/(r + R)]$, showing that the load affects the voltage/speed gradient and that $R$ should be large, typically $5 - 10 \text{ k}\Omega$.

The $V/\omega_r$ characteristic of a *practical* machine is not precisely linear. At high speeds the commutation time is short, raising the reactance voltage and reducing the output current in comparison with the ideal. Further, the brush volt-drop (which is minimized by a suitable choice of brush material) introduces a zero error, and at high speeds may be aggravated by aerodynamic 'lift'. The table gives test results for a precision tachogenerator having a nominal voltage gradient of 3.027 V per 1000 r/min (28.91 mV per rad/s):

| Speed | | (r/min): | 1000 | 2000 | 3000 | 4000 | 5000 | 10 000 |
|---|---|---|---|---|---|---|---|---|
| Voltage: | ideal | (V): | 3.027 | 6.055 | 9.082 | 12.11 | 15.14 | 30.27 |
| | measured (V): | | 3.05 | 6.20 | 8.99 | 12.02 | 15.15 | 29.11 |

Voltage *ripple* results from the finite number of armature coils and commutator sectors, vibration, magnetic and mechanical asymmetry, current pulsation in coils undergoing commutation, and sparking. The high-frequency components may cause radio interference and call for a resistance-capacitance filter (e.g. 10 kΩ, 1 $\mu$F). Rise of *temperature* increases the armature resistance and reduces the p.m. flux. Precision tachogenerators with stable p.m. materials and a compensating alloy insert in the magnetic circuit can achieve a change of output voltage with temperature as low as 0.01% per °C. With cheaper ceramic magnets and a thermistor temperature-compensation network the variation is of the order of 0.05% per °C. Greater temperature coefficients may be tolerated in tachogenerators for position-servo damping.

*Separate Excitation*  This may be used in systems operating with unstabilized power supplies because voltage variations tend to produce self-compensating changes of field current. If magnetic-circuit hysteresis causes the output voltage to differ with direction of rotation, the effect may be mitigated by means of a shunted diode, Fig. 11.5(c); and a temperature-sensitive element N may be included to compensate for change of armature resistance with temperature.

## 11.3 CURRENT-EXCITED MACHINES

Permanent-magnet machines can be built economically in ratings up to several kilowatts, but above about 1 kW the current-excited machines become competitive. Assuming linearity, the general equations of behaviour are

Field: $\qquad v_f = r_f i_f + L_f(\mathrm{d}i_f/\mathrm{d}t) = (r_f + L_f \mathrm{p})\, i_f$

Armature: $\quad v_a = r_a i_a + L_a(\mathrm{d}i_a/\mathrm{d}t) + e_r = (r_a + L_a \mathrm{p})\, i_a + G\omega_r i_f \qquad (11.7)$

Torque: $\qquad M_e = M_l + M_f + k_d\omega_r + Jp\omega_r = Gi_f i_a$

As the torque is proportional to the product of $i_f$ and $i_a$, either current can be the control variable.

### Field control

With the armature supplied at *constant voltage,* control of the field current gives the speed/torque relations of Fig. 11.6(a). But the field current for low

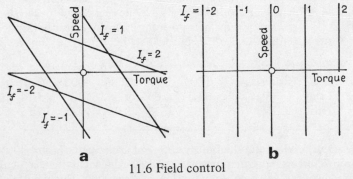

a                                                   b

11.6 Field control

speeds may be excessive, and change of the direction of rotation involves reversing the field current through zero with a transitional loss of control. If, instead, the armature is fed with a *constant current* $I_a$, the torque is $M_e = (Gi_f)I_a = K_a i_f$, which is independent of speed and yields the speed/torque relations in (b). These suit position-control applications, especially if viscous friction gives adequate damping. For the armature current to be constant, its terminal voltage must be $v_a = r_a I_a + G\omega_r i_f$. For small machines a constant armature supply voltage $V_a$ feeds the armature through a high resistance $R$ to reduce the effect of rotational e.m.f. For larger machines an electronically controlled armature supply is used to avoid undue waste. As the field power is much less than the armature output, there is a good power-gain capability. The response to field control is, however, slowed by field inductance.

*Split-field Machine* A machine with constant armature current has two identical but opposing field windings with different field currents. In the quiescent state, the field currents are equal and nominally develop no gap flux. A control signal unbalances them to develop a torque, the direction of which depends upon which field current is the larger, making for simple reversibility. With multiturn windings the quiescent currents may be below 100 mA, but the field inductances are high.

*Velodyne* This is a split-field machine, Fig. 11.7, for speed stabilization and for giving a total armature angular rotation proportional to the time-integral

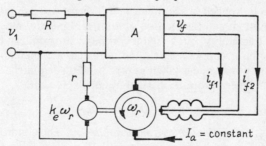

11.7 Velodyne

of a varying voltage signal $v_1$. The fields are fed from an amplifier of high gain $A$ having as input the difference between $v_1$ and a speed-voltage $k_e\omega_r$ from a tachogenerator on the main shaft. Assuming the mechanical system to be purely inertial, the amplifier input voltage is $(v_1 r - k_e\omega_r R)/(R + r)$, its output is $A$ times as much, and the torque equation is

$$M_e = \frac{K_f A(v_1 r - k_e\omega_r R)}{(r_f + L_f p)(R + r)} = Jp\omega_r \tag{11.8}$$

For a steady signal voltage $V_1$ the torque vanishes so that $V_1 R = k_e\omega_r R$, whence $\omega_r = V_1(r/k_e R)$. The total angular revolution of the armature in time $t$ is thus proportional to the product $V_1 t$.

## Armature control

With a fixed field current $I_f$ and a controlled armature voltage $v_a$, the steady-state torque is $M_e = (GI_f)i_a = K_f i_a$, where $i_a = (v_a - K_f \omega_r)/r_a$. The speed/torque relation for various armature voltages is naturally identical with that for the p.m. motor, Fig. 11.1(a). The linearity gives effective control and reasonable efficiency, so that the method is widely used for control motors of large rating. But as the whole armature power has to be controlled, its supply requires a thyristor or equivalent equipment.

*Series Motors*   These are uncommon, although their inherently large accelerating torque may occasionally be useful. Some aircraft servo systems employ series motors as actuator elements. To reverse the direction of rotation it is necessary to reverse the field connections by, for example, a split field winding with a contactor relay. For a machine of total resistance $r$ and inductance $L$, the equations of behaviour on a linear assumption are

$$v = G \omega_r i + (r + Lp) i \quad \text{and} \quad M_e = Gi^2$$

but in a practical machine the magnetic-circuit saturation and hysteresis effects make $G$ a complicated function of $i$, so that a linear analysis is valid only over a limited range of current.

## Dynamic characteristics

Eq. (11.7) can be applied in accordance with the motor parameters, the operating conditions and the method of control. Linearity must be assumed for analytical tractability, but a digital-computer program can produce numerical results with more realistic and nonlinear parameters. To simplify analysis, the output torque is taken as represented adequately by inertial and viscous-friction (speed-proportional) components, so that

$$M_e = Gi_f i_a = k_d \omega_r + Jp \omega_r = (k_d + Jp) \omega_r$$

*Field-voltage Control, Constant Armature Current*   The conditions are $M_e = (GI_a)i_f = K_a i_f$, with $i_f = v_f/(r_f + L_f p)$. Thus $K_a \cdot v_f/(r_f + L_f p)$ $= (k_d + Jp) \omega_r$ and the speed/field-voltage t.f. is

$$\frac{\omega_r(s)}{V_f(s)} = \frac{1}{(r_f k_d/K_a)(1 + s\tau_m)(1 + s\tau_e)} \tag{11.9}$$

where $\tau_m = J/k_d$ is the *mechanical time-constant* describing the effect of inertia and speed-proportional damping; and $\tau_e = L_f/r_f$ is the electrical time-constant of the field winding. As there is no equivalent electrical damping, all inherent damping is provided by $k_d$, augmented in a practical machine by bearing (coulomb) friction. The field time-constant may be large, delaying response, and it will vary with the saturation level.

*Velodyne*   As has already been shown, this can give a speed-time integral. If the input voltage is variable, the speed/voltage t.f. is

$$\frac{\omega_r(s)}{V_1(s)} = \frac{K_f AR}{JL_f(R+r)} \cdot \frac{1}{s^2 + (r_f/L_f)s + K_f k_e AR/JL_f(R+r)}$$

This is oscillatory if the last term in the denominator is greater than $(r_f/L_f)^2/4$. For accurate voltage-time integration in the presence of coulomb-friction or load torque, the gain $A$ must be large. Further, the tachogenerator time-constant, ignored in the expression above, must be short, a condition that requires $(R+r)$ to be high. In practice the amplifier is subject to gain-limitation during rapid changes of input voltage, slowing the response.

*Armature-voltage Control, Constant Field* The torque is now given by $M_e = (GI_f)i_a = K_f i_a$, with $i_a = (v_a - K_f\omega_r)/(r_a + L_a p)$. The t.f. relating speed to armature voltage takes the form

$$\frac{\omega_r(s)}{V_a(s)} = \frac{K_f}{JL_a s^2 + (Jr_a + k_d L_a)s + (k_d r_a + K_f^2)} \tag{11.10}$$

The t.f. for the output position angle, $\theta(s)/V_a(s)$, is eq. (11.10) multiplied by $(1/s)$. These t.f.s are for a machine in which only viscous friction, inertia and armature inductance have been taken as the demands on the electromagnetic torque. If viscous friction can be ignored, eq. (11.10) becomes

$$\frac{\omega_r(s)}{V_a(s)} = \frac{K_f}{JL_a s^2 + Jr_a s + K_f^2} \tag{11.11}$$

If, instead, the armature inductance is negligible, then

$$\frac{\omega_r(s)}{V_a(s)} = \frac{K_f}{(k_d r_a + K_f^2)\,[1 + \{Jr_a/(k_d r_a + K_f^2)\}\,s]} \tag{11.12}$$

with a single composite time-constant $\tau_a = Jr_a/(k_d r_a + K_f^2)$, which simplifies still further to $\tau_a = Jr_a/K_f^2$ if viscous friction can be ignored. As $K_f$ in the current-excited machine corresponds to $k_t = k_e$ in the permanent-magnet machine, the equations above have their counterpart in the expressions derived in Sect. 11.2.

EXAMPLE 11.3: A motor has a constant separate excitation giving $K_f = 0.40$ V per rad/s, and is controlled by the applied armature voltage. The armature parameters are $r_a = 0.30\ \Omega$ and $L_a = 0.15$ H. The load and motor have a combined inertia $J = 0.25$ kg-m$^2$ and a viscous-friction damping coefficient $k_d = 0.030$ N-m per rad/s. (i) Obtain the transfer function relating output angular speed $\omega_r$ to applied armature voltage $v_a$; (ii) Calculate the response in the steady state to a sinusoidal applied voltage of 25 V peak and angular frequency $\omega$ over the range zero to 5 rad/s; (iii) find the response to a step-function armature voltage of 10 V applied at $t = 0$ with the machine at rest.

(i) *Transfer Function* Inserting values in eq. (11.10) gives

$$\frac{\omega_r(s)}{V_a(s)} = \frac{10.7}{s^2 + 2.12s + 4.51}$$

(ii) *Response to Sine Voltage* Putting $s = j\omega$ and $v_a = 25$ V, the peak sinusoidal swing of the motor speed becomes

$$\omega_r(j\omega) = 267/[(4.51 - \omega^2) + 2.12\, j\omega]$$

At $\omega = 0$, the d.c. condition, the steady-state constant speed is $267/4.51 \cong 60$ rad/s (565 r/min). For low applied frequencies the peak swing of speed is augmented by resonance, but above 2.5 rad/s (0.4 Hz) it rapidly falls, the speed excursions lagging the applied voltage. The lag is 90° for $\omega = 2.12$ rad/s (0.34 Hz). At high frequencies the rotor speed oscillations become very small and the lag approaches 180°. Speed/frequency and angle/frequency graphs are given in Fig. 11.8($a$), together with their combination in a polar locus diagram.

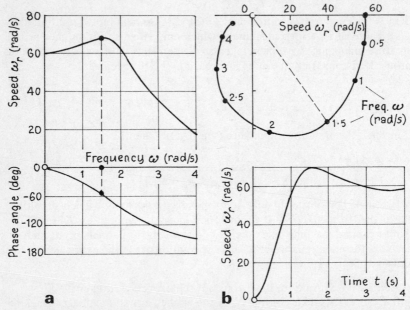

11.8 Separately excited motor: responses to sine and step armature voltages

(iii) *Response to Step Voltage* When the step $V_a$ is applied to the armature at rest, the parameters $L_a$ and $J$ in the denominator of eq. (11.10) initially predominate: the current begins to rise at rate $(di/dt) = V_a/L_a$, whence $i_a = (V_a/L_a)t$; and the torque is $K_f i_a = J(d\omega_r/dt)$ giving $\omega_r = \frac{1}{2}(K_f V_a/JL_a)t^2$. With the appropriate numerical values, $\omega_r = 135\, t^2$, determining the start of the speed/time curve in Fig. 11.8($b$). The final steady-state speed is obtained from the last denominator term, i.e. $K_f V_a/(k_d r_a + K_f^2) \cong 60$ rad/s. As the speed and current rise from zero, viscous-friction and armature-resistance effects appear and the rotor develops a rotational e.m.f.; all these slow the rates of rise. Overshoot results in a damped oscillation. The speed for $V_a(s) = 25/s$ is given by

$$\omega_r(s) = \frac{10.7 \times 25}{s \ [(s + 1.06 + j \ 1.83) \ (s + 1.06 - j \ 1.83)]}$$

which, by standard inverse-transform methods, yields

$$\omega_r(t) = 60 \ [1 - \exp(-1.06 \ t) \{ \cos 1.83 \ t + 0.57 \sin 1.83 \ t \}]$$

The final speed is first crossed in 1.05 sec, the oscillation frequency is 1.83 rad/s (0.2 Hz) and the settling time is about $3\frac{1}{2}$ sec.

EXAMPLE 11.4: A motor with a constant armature current has a field circuit with $r_f$ = 30 $\Omega$ and $L_f$ = 15 H. The torque is $(Gi_a) i_f = K_a i_f$ where $K_a$ = 61.2 N-m/A. The viscous-friction coefficient is $k_d$ = 0.025 N-m per rad/s and the inertia is $J$ = 0.16 kg-m$^2$. The motor is to drive a load of inertia $J_l$ = 62.5 kg-m$^2$ and speed-proportional torque $M_l$ = 40 N-m per rad/s. Derive the transfer function relating load-shaft speed to the field voltage $V_f$ with the motor (i) direct-coupled, (ii) connected to the load through matched gearing to give maximum acceleration.
Transforms for the field current and motor torque are

$$I_f(s) = V_f(s)/(15s + 30) \quad \text{and} \quad M_e(s) = 61.2 \ I_f(s)$$

(i) The system inertia and load torque are

$$J + J_l = 62.5 + 0.16 = 62.7, k_d + M_l = 0.025 + 40 \cong 40$$

Then $61.2 \ I_f(s) = V_f(s) \ [61.2/(15s + 30)] = \omega_r(s) \ [62.7s + 40]$,

$$\frac{\omega_r(s)}{V_f(s)} = \frac{0.065}{(s + 2) \ (s + 0.64)} \tag{i}$$

With direct coupling the motor and load shaft speeds are the same.

(ii) The gear-ratio has to make the motor inertia referred to the load shaft equivalent to the load inertia, i.e. $n = \sqrt{(J_l/J)} \cong 20$. The effective motor mechanical parameters are then $J' = n^2 J = 64$ and $k_d' = n^2 k_d = 10$. The effective motor torque is $M_e' = nM_e = 1220 \ i_f$. For the system referred to the load shaft

$$J' + J_l = 126 \ \text{kg-m}^2 \quad \text{and} \quad k_d' + M_l = 50 \ \text{N-m per rad/s}$$

whence

$$\frac{\omega_r'(s)}{V_f(s)} = \frac{0.65}{(s + 2) \ (s + 0.4)} \tag{ii}$$

Comparison of eq. (i) and (ii) shows that the mechanical time-constant is slightly reduced and that the acceleration rate is raised ten-fold.

EXAMPLE 11.5: In Example 11.4 the field inductance is taken as constant. But for higher field currents the magnetic circuit saturates. Let a step voltage $V = 15$ V be applied to a field winding of resistance $r = 10 \, \Omega$ and a field-current/flux-linkage relation given by Fig. 11.9($a$) and the following table:

| Current $i$ (A): | 0 | 0.1 | 0.2 | 0.3 | 0.4 | 0.6 | 0.8 | 1.0 | 1.2 | 1.5 |
|---|---|---|---|---|---|---|---|---|---|---|
| Linkage $\Psi$ (mWb-t): | 5 | 15 | 43 | 71 | 87 | 96 | 100 | 102 | 104 | 106 |

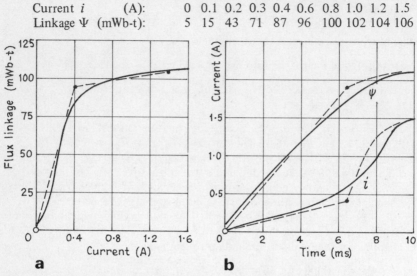

a          b

11.9 Field build-up: Example 11.5

Derive the linkage/time and current/time relations using (i) a step-by-step and (ii) a linearization method.

(i) The finite-difference equation $\Delta\Psi/\Delta t = V - ri$ has to be solved in steps. At $t = 0, i = 0$ and $\Psi$ has the remanence value 5 mWb-t; $\Delta\Psi/\Delta t = V = 15$. Then at $t = 1$ ms the linkage is $15 + 5 = 20$ mWb, which requires the current $i_1 = 0.12$ A. For the next step $\Delta\Psi/\Delta t = 15 - (10 \times 0.12) = 13.8$, giving $\Psi_2 = 20 + 13.8 = 33.8$, for which $i_2 = 0.17$ A. Proceeding in this way gives the full-line linkage/time and current/time curves in Fig. 11.9($b$).

(ii) The dotted lines in ($a$) give a possible linearization of the $\Psi/i$ curve. The two lines range from 0/0 to 95/0.4, and from 95/0.4 to 105/1.4, corresponding to field inductances 95/0.4 = 240 mH and 10/1.0 = 10 mH. The current is then obtained from

$$i = 1.5 \, [1 - \exp(-0.042 \, t)] \text{ up to } 0.4 \text{ A} \tag{i}$$

$$i = 1.1 \, [1 - \exp(-1.0 \, t)] + 0.4 \text{ thereafter} \tag{ii}$$

with $t$ in millisec. Using the exponential series for (i) gives $i = 1.5 \times 0.042 \, t = 0.063 \, t$ approximately. The resulting current/time relation, with the linkage/time curve derived from it, are shown in ($b$).

**Practical machines**

The linear analyses on which eqs. (11.10) and (11.11) are based have an accuracy limited by magnetic saturation, structural asymmetry (which introduced deleterious interwinding couplings), eddy-currents (which oppose rapid changes) and commutation effects.

Example 11.5 shows that magnetic nonlinearity can be reduced by working with low saturation densities. For 'constant field' operation, however, it may be advantageous to run at high saturations to render the field insensitive to armature reaction and to small changes in the field current. Eddy-currents are reduced by the complete lamination of the magnetic circuit and the use of low-loss steel. Commutation introduces a measure of unpredictability.

Where practicable, motors have compoles and compensating windings, providing a short-time capability of several times full-load current without sparking. Machines are normally field-controlled for high-speed and armature-controlled for low-speed ranges. The inductive field time-lag may be reduced by field-strengthening. Compared with industrial machines, the number of slots and commutator sectors is large to improve commutation, the diameter/length ratio low to reduce inertia, and oversize shafts employed to withstand the high torque stresses. Forced ventilation may be necessary. The advantage of matching the motor and load inertia values by gearing will usually be exploited.

Cross-field machines (Chapter 7) may be employed in the middle range of ratings. Small machines, especially those with p.m. fields, can rarely accommodate compoles, and the small rotor diameters make necessary the use of a limited number of commutator sectors with the result that the torque is sensitive to rotor position. A significant operating feature of small machines is the source impedance, which can sometimes be considerable.

Miniature machines can be controlled by transistors. The closed-loop system, Fig. 11.10, is a typical example of speed control. The reference voltage $V_r$ drives a current through the tachogenerator rotor and transistor

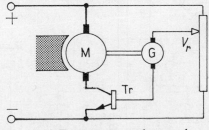

11.10 Transistor speed control

base causing the transistor to conduct and allowing an armature current to flow through the motor. The tachogenerator e.m.f. $e_r$ opposes $V_r$, affecting the base/emitter voltage and therefore the motor armature current. With a d.c. supply the control is continuous.

## 11.4 BRUSHLESS MACHINES

Brushless polarized motors (Sect. 4.8) have electronic commutation of an armature winding carried on the stator, commonly a 4-pole arrangement with the individual pole windings connected to a 'star' point. A permanent-magnet rotor provides the working flux. The stator field is pulsed to change the magnetic axis in steps in conformity with signals from a rotor position sensor such as a Hall-effect device. The torque arises from the misalignment of the stator and rotor magnetic axes, and the sensors provide for a roughly constant torque angle.

The torque/speed characteristic resembles that of a p.m. machine with a conventional commutator. The starting torque, however, is limited by the method of commutation. The maximum continuous torque is determined by thermal dissipation: it is of the order of one-half of the starting torque. Although the sensors ensure that the rotor does not fall out of step, a sudden starting demand may make it necessary to feed stator input pulses from an intermediate counter store to withhold a step pulse until the previous one has been completed.

The design of the electronic commutation system has a strong influence on performance characteristics. The conventional system employs analogue processing of data from the rotor position sensors, but some progress in digital processing has been reported by Hanitsch and Meyna [33]. During start-up the motor behaves like a stepper motor (Sect. 11.5). When running at full speed it has the characteristics of a normal shunt motor. The stator windings are switched not by Hall generators or photoconductive cells but by the stator induced e.m.f. applied to logic circuits. Rotor position sensing takes place when the e.m.f. passes through zero.

Another system has a voltage controlled oscillator in the switching circuit, Fig. 11.11. Again the armature e.m.f. is used for the sensing process, its

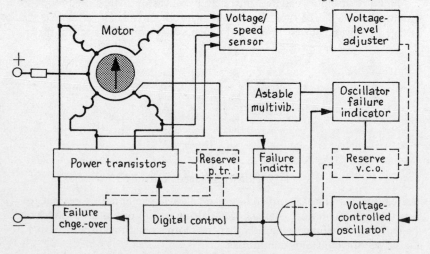

11.11 Brushless motor control

instantaneous value being passed through diodes to oscillators. These produce a pulse train of voltage-controlled frequency. The main components of the digital control circuit are the shift registers and the main logic circuit.

The complicated and varying waveform of the stator currents makes close assessment of the torque very difficult. The control circuits give rise to intermittent currents which contain both odd and even harmonics, which in turn develop a series of airgap radial force waves that produce acoustic noise.

Sophisticated control systems raise the problem of reliability. While greater reliability can be achieved by standby elements, a penalty is paid in power, cost, mass and volume if redundancy is implemented by standard components. The use of microcircuits may offer advantage. A number of standby and redundancy components are included in Fig. 11.11.

## 11.5 STEPPING MOTORS

Fig. 4.9 shows a brushless d.c. motor with a rotor field, a stator armature, and commutation by means of a rotor position sensor and switched stator windings. If the position sensor is omitted the rotor will, for any pair of excited stator coils, move into a minimum reluctance position for which its m.m.f. axis aligns with that of the stator, This principle is applied in the stepping motor, a rotary machine that moves in discrete angular steps in response to the application of d.c. pulses to its stator 'phase' windings. The shaft can be made to stop in any one of a number of positions of angular separation typically $0.9°, 7\frac{1}{2}°\ldots 180°$ with precision, because each is a stable alignment of m.m.f. axes. A single pulse advances the shaft position by one angular step; a train of pulses produces a rotation in rapid steps, each with a non-cumulative error of, say, ± 3%. For example, with a nominal step angle of $1.8°$, one pulse will move the rotor position by $1.8 ± 0.05°$; a train of 1000 pulses will give a rotation angle of $1800 ± 0.05°$ equivalent to 5 complete revolutions. With such an accuracy, a stepper motor when properly matched to its pulse driver and load may not require closed-loop control, for positional precision is determined by the number of pulses in the sequence.

A single pulse develops a *detent* torque which turns the rotor by one angular step, and it comes to rest after a short damped oscillation of frequency and duration dependent upon the system inertia, elasticity and damping. A properly timed pulse sequence sustains a *slewing* rotation, the rotor passing through a corresponding number of angular steps and developing an effective mean torque and speed (or acceleration), all with positional integrity. With torques in the range 0.5 $\mu$N-m to 100 N-m, outputs from a few milliwatts to several kilowatts, and pulse rates of 1200 or more per second, the stepper motor can be applied to a wide field of open or closed-loop position control in wrist-watch drives, numerically controlled machine-tool operation, colour registration in printing, pen positioning on chart plotters, and similar data-system applications.

## Basic forms

One realization of the stepping principle utilizes a *permanent-magnet* rotor (p.m.), another relies on the *variable reluctance* (v.r.) of the airgap. Both are embodied in the *hybrid inductor* (h.i.) machine. The motor may be supplemented by ratchet or nutating-gear attachments or by hydraulic devices.

*Permanent-magnet*  Fig. 11.12. The prototype p.m. motor (*a*) has a 2-pole rotor magnet within a 4-pole stator with windings (called 'phases') marked

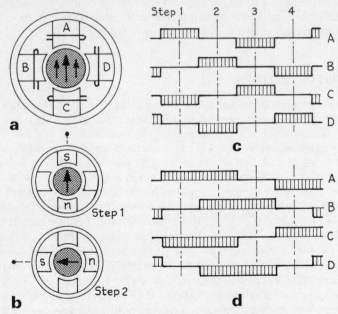

11.12 Permanent-magnet stepper motor

ABCD. Excitation applied to A and C develops the magnetic polarities indicated for Step 1 in (*b*) and the rotor sets itself vertically. If now A and C are switched out, and B and D excited as in Step 2, an alignment torque is developed on the rotor to turn its axis to the horizontal by a 90° step. With B and D off, and A and C re-energized in reverse, the rotor turns a further 90°, and so on. The direction of rotation is counterclockwise, but can be reversed by changing the current directions suitably. The stator currents (ignoring build-up delay) with uniform pulse frequency and equal on/off periods are shown to a time-base in (*c*). The stepping angle can be made 45° by energizing successive phases simultaneously as in (*d*).

*Variable-reluctance*  Fig. 11.13. A 6-pole prototype (called 'three-phase' because opposite poles are normally excited together) has its rotor structurally polarized but unexcited. The rotor saliency must differ from that of the stator, the difference determining the step angle. On Step 1, with AD

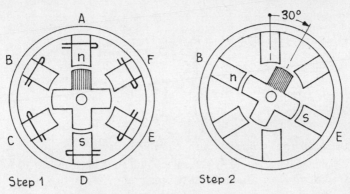

11.13 Variable-reluctance stepper motor

energized, two opposite rotor poles are aligned. On Step 2, AD is switched out and BE energized: the rotor moves to a minimum-reluctance position with its other two poles in alignment. The step angle is the difference between the change (60°) in the stator m.m.f. axis and the angle (90°) between successive rotor poles: i.e. $60° - 90° = -30°$, the rotor moving in a direction opposite to that of the stator m.m.f. switching.

*Hybrid* There are several arrangements. The h.i. motor has an axially magnetized p.m. providing homopolar excitation between the rotor (which is in two differently oriented parts) and a heteropolar excited stator. Some hybrids are built for large step angles, with an integral gearing system capable of output steps down to 0.18°.

### Drive

Stepping motors are essentially digital, with one angular step for each input pulse supplied by power transistors switched by digital controllers or computers. Drive circuit design strongly influences overall system performance. The three main drive elements are the *controller* which generates the pulses, the *translator* which directs the pulses to the phase windings, and the *power supplies* which feed the actual phase currents in accordance with logic level signals from the translator. A phase winding can be represented by a resistance $r$ and an inductance $L$ in series, having a time-constant $L/r$ typically of the order 10 ms. If power is supplied at constant voltage, then ignoring rotational effects the current reaches 95% of its maximum in the time of three time-constants (e.g. 30 ms), imposing severe limitations on stepping rate. Methods of achieving faster current rise are:

(i) A high-voltage supply and a series forcing resistance $R$ to reduce the time-constant to $L/(R + r)$. The circuit, Fig. 11.14, includes a parallel diode to quench the phase current at the end of a step.

(ii) A chopped supply waveform having a high initial voltage for fast build-up and subsequent pulse modulation for maintenance.

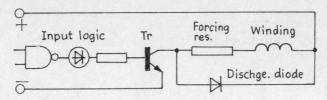

11.14 Stator drive circuit

(iii) A two-level drive voltage with the same effect as (ii).

(iv) A variable-voltage supply (e.g. from a phase-controlled thyristor) to produce constant step currents at all stepping rates.

A motor can be started from rest at a limited rate (e.g. 400 steps/s). For fast slewing it is therefore necessary to give the drive a 'ramping' characteristic, to provide an automatic increase in stepping rate as the speed rises. If rapid braking and overshoot limitation are called for, the drive system may have special pulse sequences to reverse the motor torque at timed intervals. For the highest precision the control loop may be closed by a shaft-position sensor with feedback to adjust the pulse timing to the optimum.

### Response

The response of a stepper motor to a single pulse is illustrated at (*a*) in Fig. 11.15. A detent alignment torque is developed which moves the rotor

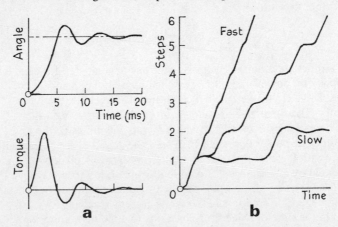

11.15 Stepper motor response

towards the angular position demanded. Inertial overshoot reverses the torque and generates an oscillation about the final position. Damping is essential to avoid prolonged oscillation at a frequency typically several hundred hertz. In (*b*) the response to slow, intermediate and fast stepping rates is graphed: at low rates the motion is a succession of the step responses in (*a*), while at high rates the motor runs at a nearly constant slewing speed. The maximum achievable slew speed is limited by the load torque, inertia and the rate at which successive pulses can be established and suppressed without undue reduction of the stator fluxes on which the torque depends.

*Torque* A stepper is required to start and stop at various stepping rates and with various conditions of load, friction and inertia; and there may be fast-slewing and settling-time limits to fulfil. The two idealized torque/pulse-rate curves in Fig. 11.16 (*a*) are (i) for *starting* (pull-in) and (ii) for *slewing*

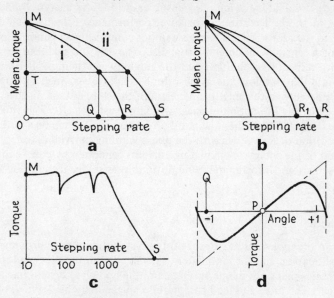

11.16 Stepper motor characteristics

(pull-out) for a system with low inertia. For starting, the load plus friction torque must lie within curve (i): a torque demand OT cannot be started at a rate exceeding OQ. Curve (ii) shows that the load torque to pull out a motor when slewing is greater for a given stepping rate than would prevent it from starting. At low step rates the curves approach the point M of maximum torque. The curves in (*b*) indicate the restrictive effects of inertia. The low-inertia curve MR is the same as in (*a*), but with increasing inertia the step rate for starting becomes more limited. Curve $MR_1$ is for a load inertia equal to that of the stepper motor alone.

The maximum stepping rate of a motor, load and drive system may lie between a few hundred and as much as 30 000 per sec. At high rates several factors become significant. Natural mass-stiffness resonances and drive impedances may appear, phase-winding inductance may resonate with stray capacitance, and the time to establish each working flux may become an appreciable fraction of the time per step. These affect the torque/step-rate characteristic, Fig. 11.16(*c*), to impose 'dips' in the available torque. In some applications these are troublesome; in others it may be possible to accelerate the stepping motor through them, particularly if adequate damping is provided. Lawrenson and Kingham [34] discuss the mechanisms of resonance-induced maloperation and analyse the damping factors necessary to suppress them.

*Damping*   Several methods have been used to shorten the duration of rotor oscillation, which undamped may last as much as 1 sec. *Electronic* damping involves the drive circuits, which furnish a sequence of braking pulses opposing the oscillatory swing. Lawrenson *et al* [34, 35] analyse *friction* and *resistance* methods. The former, for lower-performance systems, involves coupling to the motor shaft a viscous damper comprising a cylindrical shell containing a centrally mounted inertial disc. The disc and shell can rotate freely with respect to each other, but the gap between them is filled with a viscous fluid that develops energy-absorbing drag torques. Such a damper adds considerably to the length and inertia of the stepper unit. In the resistance method the optimum phase-winding resistance $(R + r)$ is found: in most cases it is higher than the natural resistance $r$ and can be augmented by the addition of $R$ in series with the phase winding, providing for dissipation of the energy in circulating currents generated by e.m.f.s induced by the oscillation. Unfortunately, the optimum resistance depends on the stepping rate.

## Performance

Analysis of the dynamics of stepper-motor systems involves approximation and simplification, and is so intimately associated with the drive and load characteristics that only the performance of the complete system has significance. The drive and load parameters can generally be defined, but the motor parameters are more difficult to assess and most analyses rely on the results of static and dynamic tests. Operational aspects of importance are the pull-in and pull-out characteristics. Loads may be assumed to be mainly inertial or viscous, or constant and unidirectional; in the latter a problem is the tendency of the rotor to move backward when the stator is not energized, and to augment the electromagnetic reverse torque on overshoot. Stepper dynamics has been described by Hammad and Mathur [36] in terms of computed phase-plane diagrams.

Ellis [37] deals with a p.m. stepping motor with a drive either from a high-impedance source giving a constant current, or from a low-impedance source giving a constant voltage. In each case the rotor torque is taken as $M_e = k_t i \cos \theta$, where $\theta$ is the rotor electrical position angle. The system equation can then be written

$$J\mathrm{p}^2\theta + k_d\mathrm{p}\theta + k_t i \cos \theta + M_f = 0$$

where $k_d$ is the viscous damping coefficient, and $M_f$, a function of time $t$, takes account of static and coulomb friction, and any effects due to unbalance in a gravitational field. For constant current $i = I$, while for constant voltage

$V$ the current is obtained from $V = ri + L(\mathrm{d}i/\mathrm{d}t) - k_e\omega_r \sin\theta$, where $\omega_r$ is the instantaneous speed. Assuming $r$ and $L$ to be constants, the step response is obtained. Slewing conditions require a computer program to solve iteratively the succession of steps, each of which has initial conditions depending on the previous steps.

Analysis of the v.r. motor requires that the phase inductance be expressed as a function of both rotor position and current, because saturation effect is prominent. Pickup and Tipping [38] employ a polynomial in the two variables, and alternatively a Fourier series. They ignore hysteresis but include eddy-currents (which delay the growth and decay of winding current) and damping. Prediction of performance requires two tests to establish the equivalent circuit: (i) measurement of the switch-on flux-linkage and winding-current transients with the rotor in the minimum-reluctance position and the magnetic circuit unsaturated, and (ii) measurement of the flux-linkage/rotor-angle relation for a series of steady input currents.

Techniques for the prediction of performance from test data are given by Lawrenson:

1. *Steady-state Torque/speed Characteristic*  From test curve (*a*) in Fig. 11.17 for the pullout-torque/speed relation for a motor with its drive circuit, the

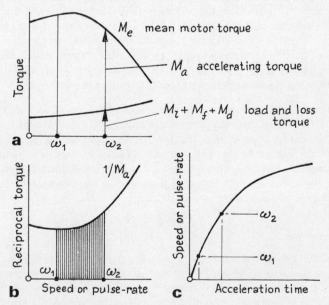

11.17 Stepper motor performance from torque/speed characteristic

acceleration and retardation capability can be assessed for a specified load. The load and damping torques, $M_l$ and $M_d$, are deducted from the pullout torque $M_e$ to give a curve of net accelerating torque $M_a$ for any speed. Then for a total system inertia $J$,

$$M_a = J\,(\mathrm{d}\omega_r/\mathrm{d}t) \quad \text{whence} \quad t = J \int_{\omega_1}^{\omega_2} (1/M_a)\cdot\mathrm{d}\omega_r$$

for a change of speed from $\omega_1$ to $\omega_2$. The term $(1/M_a)$ is drawn in (b), the shaded area being the required integral. From the results for a series of speed-intervals, the minimum time for a given rise (or fall) in speed can be obtained, as in (c). The integration can be programmed for computation. The method is optimistic because the test pullout-torque/speed characteristic is assumed to apply to operation on load, requiring optimum rate of increase ('ramping') of the stepping; and the effect of torque dips and resonances is ignored.

2. *Steady-state Torque/displacement Characteristic*   This is used to predict the shaft angle at any instant following the application of one step or a train of steps, and so to find the starting rate. A typical characteristic for one excited phase is shown in Fig. 11.16 (d). The rotor tends to settle at the position P of zero torque; stepping the drive to the next phase shifts the torque/displacement curve by one step and puts the rotor into a high-torque position, from which it approaches a new zero Q. (For rapid stepping it may be necessary to allow for phase current build-up). To predict the instantaneous rotor position it is necessary to know the timing of the torque shift, and the rotor position and speed. The equation of motion can then be solved. If it is adequate to take the torque/displacement curve as linear, the equation of motion becomes a simple second-order system characterized by a natural frequency $\omega_n = \sqrt{(k_t/J)}$ and a damping ratio $c = \frac{1}{2}k_d/\sqrt{(k_t J)}$, where $k_d$ is the viscous-friction coefficient (friction torque per unit speed).

3. *Complete Model*   A stepper system is described by sets of differential equations, (i) electrical for the drive and windings, (ii) for the load torque and (iii) mechanical for torque, load, friction, inertia, acceleration, speed and displacement. The resistances and inductances must be expressed as functions of displacement, the voltage waveforms defined, and the initial conditions specified. Solutions are found by digital routines. Saturation can be taken into account at the expense of computer time.

### Stepper systems

Details of stepper design and performance are given in Ref. [39]. The stepping motor is in active development: for example, a pole interpolation technique provides for rotor movement from pole to pole in a series of 'ministeps'. A 200-step motor with 64 ministeps per pole has 12 800 effective rotor positions per revolution, each of less than $0.03°$, the excitation being forced by capacitor discharge. Another method of increasing starting and stopping rates is to use optimally timed pulses of 10 ms or less, applied to each phase winding only when it can produce positive torque on the rotor.

## 11.6 TESTING

The parameters of interest are *Mechanical* (eccentricity, play, static and viscous friction, inertia, time-constants), *Electrical* (resistance, inductance, torque and e.m.f. constants, time-constant, no-load speed, voltage, current, losses, ripple, regulation) and *Thermal* (thermal resistance, time-constant, temperature-rise).

### Mechanical

End-play, and radial deflection of the shaft, are checked by simple mechanical tests; shaft eccentricity by sensitive feeler indicator. Static friction is obtained by fixing a spring-loaded torque-meter to the shaft and turning it until the shaft begins to move, for a number of shaft positions and for both directions of breakaway. One method of determining rotor inertia is to suspend it by a steel wire and observing the periodic time of rotary oscillation. The electromechanical time-constant is $\tau_{em} = Jr_a/k_e k_t$. If the electrical time-constant is short, $\tau_{em}$ can be obtained from the time for the rotor to attain 0.63 of final speed following the application of a step-voltage, or from the 'break-frequency' derived from a test with a variable-frequency applied voltage.

### Electrical

For control motors and tachometer generators, the resistance $r_a$ over the armature brushes is measured with the machine driven slowly — for example, 0.5 r/s — in order to accommodate variations due to slotting. The armature is supplied with a current $I$ at a voltage $V$: then $r_a = V/I$. The resistance may be specified at a standard temperature such as 25°C. An inductance bridge may be used to measure the armature inductance $L_a$, the average being taken for a number of rotor orientations. The electrical time-constant is then $\tau_e = L_a/r_a$: it can be checked by oscilloscope from the current response to a step voltage input. For machines with windings in rotor slots, the operational impedance is $z_a(s) = r_a + sL_a$, but with moving-coil types the impedance is better represented by assuming the inductance to be parallelled by a resistance $R$, giving $z_a(s) = r_a + sL_a[R/(R + sL_a)]$.

The torque and e.m.f. constants $k_t$ and $k_e$ (which are identical in SI units) can be obtained by driving the machine as a generator and observing the armature-voltage/speed relation, giving $k_e = E_r/\omega_r$. Alternatively, the torque/current relation at constant speed is found by use of a sensitive dynamometer: the result is a straight line of slope $k_t = \Delta M_e/\Delta i$.

The input power and speed on no load are readily measured. By varying the applied voltage, the rotational-loss/speed relation can be obtained. The minimum starting current is measured by slowly increasing the armature applied voltage until the current $I_0$ is just sufficient to initiate a very slow rotation. The static friction torque is then given by $k_t I_0$.

Demagnetization in a p.m. machine is based on the reduction of the steady-state voltage required to drive the machine on no load at a specified speed, before and after the application of a step voltage, for which the impulsive starting current may cause a small reduction of the p.m. flux. Torque ripple is assessed by a rotary torque meter of high sensitivity and very low inertia, with a pickup winding to give a signal to an oscilloscope.

The regulation is the slope of the speed/torque graph for rated applied voltage and with the loading raised from zero to, say, 1.5 times full-load condition. It is expressed as $\Delta\omega_r/\Delta M_e$, which should give the same result as $r_a/k_t k_e$. The efficiency for a given operating condition is obtained by the aid of a sensitive dynamometer and a speed counter.

The frequency-response test requires an amplifier with a frequency range typically between 5 Hz and 10 kHz. A reference response is found for the lowest frequency, after which the response can be measured over the range of frequencies of interest. The transfer function of the machine in a drive system is a function of the total inertia, the torque and e.m.f. constants, the load and loss torques, the amplifier characteristics and the operating conditions.

By developing the analysis in Sect. 11.2, Pasek [39] has devised a comparative test for the parameters of p.m. (or separately excited) armature-controlled motors, useful for quality control in motor production. The method can be automated, and requires only a single oscillogram of the armature-current/time response to a step voltage $V$ applied to the motor at rest. Fig. 11.18($a$) shows the essential test rig. The storage oscilloscope

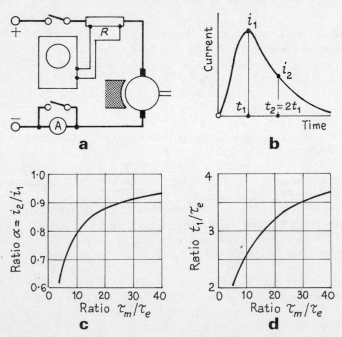

11.18 Comparative testing

indicates current by means of the volt drop across a resistor $R$, which must be small enough not to mask the natural behaviour of the machine. The ammeter, normally short-circuited, may be used in calibrating the oscillogram and checking the rotor resistance $r$. A typical current/time response is shown in (*b*): peak current $i_1$ occurs at time $t_1$, and at time $t_2 = 2t_1$ the current has fallen to $i_2$, the current ratio being $\alpha = i_2/i_1$. The test is completed by observing the final steady-state speed $\omega_1$. Two standard curves, (*c*) and (*d*), serve to evaluate the inertia and time-constants: (*c*) gives the ratio $\tau_{em}/\tau_e$ of the electromechanical and electrical time-constants, and (*d*) the ratio $t_1/\tau_e$. Then from the values $V$, $\omega_1$, $\tau_{em}$ and $\tau_e$, and a current defined by $I_s = i_1^2/i_2 = i_1/\alpha$, the remaining parameters are obtained from

$$r = V/I_s \quad k_e = V/\omega_1 = k_t \quad J = \tau_{em}k_ek_t/r$$

As a check, $I_s$ (equivalent to the current with the rotor at rest) can be measured directly by means of the ammeter. An advantage of the technique is that the parameters relate to dynamic conditions. The effective resistance and inductance of the armature will differ somewhat from those obtained from static tests and bridge measurements, but are more relevant to practical operating conditions.

The Pasek method assumes the load to be purely inertial, but has been extended by Lord and Hwang [39] to include viscous damping and dead-zone effects.

### Thermal

In manufacture, the temperatures in a sample rotor are obtained from detectors built in at expected hot-spot positions. Thermal resistances and time-constants can then be determined from a temperature-rise/time graph. The heat path comprises the rotor/housing and housing/ambient thermal resistances in series, each resistance being parallelled by a thermal capacity.

### Stepping motors

The testing requirements are severe because the essential behaviour is dynamic. Rotor positioning is of basic importance, its measurement demanding high resolution and a frequency response up to five times maximum stepping frequency in order to detect positional harmonics. An optical position indication (e.g. by an encoder) may be employed to avoid loading the motor and to achieve good resolution by delivering output in the form of one pulse per 0.0001 rev. In essence the position indicator measures also the time between successive encoder pulses by means of a crystal-controlled clock. From a record of the pulse/time relation, the position/time, velocity/time and acceleration/time curves can be computed.

Static-torque/displacement characteristics can be obtained by applying a strain-gauge load cell to measure the force required to produce a series of precision settings of the shaft. The test rig must be designed to deal with the falling as well as the rising region of the torque/displacement curve. The most common stepping characteristic is that relating mean pull-out torque to stepping frequency, and it must be possible to locate the various resonance 'dips'. The test gear must be capable of producing a stable and accurate stepping frequency, to apply a wide range of accurate and speed-independent loading torques, and to impose minimal additional inertia. A magnetic-powder brake satisfies such demands, but may not have the peak torque requirement for the upper stepping ratings.

# 12 Special Machines

## 12.1 EDDY-CURRENT DEVICES

If a metal plate is moved laterally through the magnetic field of a pair of flanking poles, a motional e.m.f. is generated in it, causing an eddy-current to flow. Interaction between the current and the magnetic field in this elementary homopolar device develops a force opposing the motion.

### Eddy-current brake

Consider the system in Fig. 12.1($a$). A thin non-magnetic but conducting disc, of thickness $d$ and conductivity $\sigma$, is mounted on a central spindle and

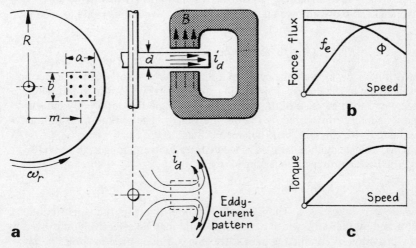

12.1 Eddy-current disc brake

rotated at angular speed $\omega_r$. A magnetic flux of density $B$ in a small area $ab$ at a mean radius $m$ will generate a radially directed e.m.f. $e_r = Bam\omega_r$ in accordance with eq. (1.7). The e.m.f. drives a radial current through the path of conductance $g = \sigma bd/a$, the circuit being completed in the less restricted (but somewhat ill-defined) remainder of the disc external to the pole region. The eddy-current is $i_d = e_r g$, and in the pole region produces a tangential force $f_e = Bai_d$ and a torque

309

$$M_e = f_e m = Ba(Bam\omega_r)(\sigma bd/a) m = [B^2 m^2 \sigma abd]\,\omega_r \qquad (12.1)$$

opposing the rotation. If the working flux is provided by a permanent magnet, then $M_e$ is directly proportional to speed. Such a system is employed in integrating energy-meters to provide a braking torque proportional to speed, the driving torque generated by alternating voltage and current elements being proportional to the mean power. As a result the number of disc revolutions in a given time is proportional to the energy to be recorded. In this application the pole area is restricted to increase $B$, and the disc made of high-conductivity metal (aluminium or copper). The braking torque is adjusted by varying the mean radius $m$, or by modifying the flux distribution with a magnetic deflector plate.

Eq. (12.1) is oversimplified. It disregards the resistance of the external current path, the disc inductance, the skin effect, and the 'armature reaction'. Both skin effect and reaction rise with speed. At high speeds and currents there may be a significant fall in effective pole flux to such an extent that a further rise in speed actually reduces the torque. Further, as a disc element moves rapidly into the polar region it is 'shock-excited', and a wave of eddy-current diffuses inwards from the surface with attenuation and phase-lag. Schieber [40] deals with thin discs at low speeds for which reaction and skin effects can be ignored; for the disc (a) in Fig. 12.1 (but with the rectangle ab replaced by a circular pole of effective radius $r$ taking account of the flux fringing) he obtains

$$M_e = B^2 m^2 \sigma(\tfrac{1}{2}\pi r^2)\, d\omega_r(1 - k_1^2)$$

which has a clear resemblance to eq. (12.1). The position of the magnet pole is included in the term $k_1 = (r/R)/[1 - (m/R)^2]$. The torque is naturally increased with increase of $m$, but at the same time the outer eddy-current path becomes constricted by the disc periphery. The optimum position is obtained with the ratio $m/R$ between 0.8 and 0.9.

For a single cylindrical pole placed centrally over a long strip of width $w$, the braking force at linear speed $u$ is

$$f_e = B^2 \sigma(\tfrac{1}{2}\pi r^2)\, du(1 - k_2^2)$$

where $k_2 = (r/w)\,\pi/\sqrt{6}$. The factor $(1 - k_2^2)$ results from the constriction imposed by the finite width of the strip, and may be effectively improved by the provision of conducting bars at the edges. For an array of poles of *like* polarity, set a pole-pitch apart along the centre-line of the strip, there is an optimum pole radius $r$ for which the force per pole is a maximum. With a *heteropolar* array, the force reduces (as might be expected) as the pole-pitch is increased.

Venkataratnam *et al* [41] treat the armature-reaction effect to derive the typical flux/speed and force-speed relations in Fig. 12.1(b). Singh [42] analyses the skin effect in thick discs with reference to the eddy-current brake dynamometer in which the disc must be mechanically strong, using the 'transmission-line' equation in terms of the depth of penetration $\sqrt{(1/\mu\sigma\omega)}$, where $\omega$ is the product of the speed and the number of poles.

Typical torque/speed relations in (c) indicate the loss of braking force at high speeds.

Advanced analyses apply the Maxwell equations to the system in two (or occasionally three) dimensions, but to be tractable there must still be simplifying assumptions. Carpenter [43] suggests a finite-element network model to represent the conducting member in which the eddy-currents are generated.

### Eddy-current coupling

The basic elements, Fig. 12.2, comprise a *loss drum* concentric with a multipolar d.c. excited *field* system providing a homo- or heteropolar gap-field pattern. If the drum is driven at an input speed $\omega_1$, eddy-currents induced in it interact with the gap flux to produce a torque in the direction of rotation. The field member then rotates at an output speed $\omega_2$ which is less than $\omega_1$ by the *slip* speed $\omega_s$, for without a speed-difference there would be no eddy-currents and no torque. The coupling is a torque transmitter in that, neglecting windage and friction, the input and output torques are identical. For a transmitted torque $M_e$, the input power is $M_e\omega_1$, the output power is $M_e\omega_2$ and the slip power, which is dissipated in the loss-drum resistance, is $M_e(\omega_1 - \omega_2) = M_e\omega_s = M_e s\omega_1$. The per-unit slip is defined as $s = \omega_s/\omega_1$, and the power efficiency of the coupling is $(1 - s)$ per-unit. The action gives the torque/output-speed characteristic of Fig. 12.2(c), which has some resemblance to that of a polyphase induction motor. For a field excitation current $I$ producing a gap flux $\Phi$ per pole, the torque is proportional to $\Phi^2$ and therefore roughly to $I^2$. As the eddy current in the loss drum is inversely proportional to the effective drum resistance, the torque characteristic will vary with the drum temperature.

Torque transmission will take place if the field system drives the drum, but it is usual to make the drum the input member so that its outer position and higher speed facilitate its cooling. The drum serves both to complete the magnetic circuit and to provide the eddy-current path, so that it should have both high permeability and low resistivity. Wrought iron is a common choice, sometimes with brazed copper end-rings to raise the effective conductance. In large couplings the field system comprises laminated salient poles (a), particularly if rapid field build-up is a requirement. Smaller sizes have imbricated poles (b) excited by a single field coil, stationary or fed through slip-rings. Ratings range from 1 kW up to 1 MW or more, based usually on a slip between 0.04 and 0.08 p.u. and a torque of about 0.5 of the maximum. Cooling by air or water is essential, for temperatures exceeding about 300°C endanger the insulation of neighbouring field windings and cause loss of torque by excessive drum resistivity.

The operation of the coupling is controlled by adjustment of the field excitation, which is normally provided from an a.c. main supply through a controlled thyristor rectifier. For closed-loop control of speed, the typical system in Fig. 12.2(d) may be used. A stabilized d.c. supply provides a reference voltage for comparison with the speed-proportional voltage from a

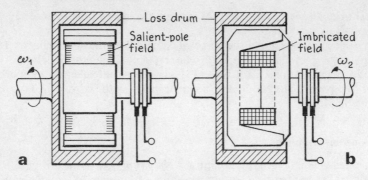

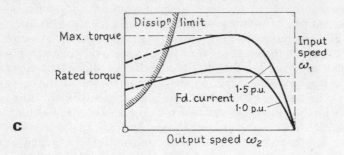

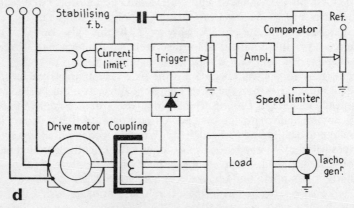

12.2 Eddy-current coupling

tachometer generator, the error voltage being amplified to set the field excitation by controlled triggering of the thyristors. Such a system can give a wide speed range with a regulation of about $\frac{1}{2}$%. Bloxham and Wright [44] describe coupling applications to torque, position, load-sharing and process controls.

Saturation, reaction, skin-effect and the imprecision of eddy-current path patterns make analysis difficult. Lawrence and Ralph [45] develop a general three-dimensional Maxwellian solution for cylindrical couplings. Davies *et al*

[46] have considered optimum pole configurations and have obtained normalized torque/slip evaluations based on the assumption that eddy-currents enter the drum end-rings at right-angles and that the magnetization characteristic of the drum material can be represented by $B = aH^x$.

### Dynamometer brake

An eddy-current coupling becomes a dynamometer brake if one of its members is fixed. If the 'fixed' member is mounted on bearings and its movement is restrained by a spring-balance or other equivalent device, the brake torque can be measured directly. A dynamometer brake can be applied to the load-testing of small rotating machines, of rating up to the limit of thermal dissipation of the brake, which converts the whole of its input into $I^2 R$ loss.

## 12.2 CLUTCHES AND BRAKES

Conventional electromagnetic clutches and brakes comprise essentially two members: (1) a ferromagnetic disc, and (2) a field member carrying an electromagnet system which can be d.c. energized. The field member is usually mounted on the drive shaft with the disc on the driven shaft in a clutch. In a brake, one member is fixed. The two members are held apart with a very small gap by spring loading, which is overcome by exciting the field member for torque transfer or braking through friction linings.

In computers, tape transport control can be effected by such clutches. In machine-tools, brakes in boring machines stop the spindles at the end of a cut to avoid helical grooving of the workpiece during the return stroke of the table. Clutch drives that obviate 'snatch' on the work are employed in textile looms, railway truck tippers, paper reeling and slitting machines and sheet-steel flatteners. High rates of acceleration and braking of work-table positioning in automatic machine-tools can be achieved.

Operation is direct and rapid. Industrial units are made with diameters from 25 mm to 0.5 m or more, developing static torques from 0.25 N-m up to 3 kN-m, to give smooth starting, stopping, tensioning, positioning, indexing and 'inching' with very simple control methods. Other applications are to hoists, cranes and winches. Excitation is usually derived from an a.c. mains supply through a diode bridge rectifier with nonlinear resistor protection against overvoltage. The de-energizing time depends on the method of connection, the shortest time being obtained by open-circuiting the magnet winding.

## 12.3 LINEAR MACHINES

Linear forms of d.c. machine for motor and generator modes require both 'field' and 'armature' members to be conductively fed (except where permanent magnets can be employed), and in this respect are inferior to a.c. devices working on the induction principle. However, several useful linear d.c. machines have been devised.

### Liquid-metal pump

The electromagnetic pump utilizes the electrical conductivity of a liquid metal to establish forces that are transferred directly to the liquid itself. Applications are to the pumping of the liquid-metal coolant in nuclear-energy plants, in which sodium, sodium-potassium and lithium are the metals chiefly concerned. Industrial use is normally restricted to low-melting-point metals, as in die-casting, or the pumping of mercury in chlorine plants. The wide differences in physical properties — conductivity, viscosity and corrosive effect — of the liquids to be pumped will influence practical design, a major part of which is concerned with forming or fabricating the channel. Its wall thickness must be chosen to give a proper balance between mechanical strength, durability and pump efficiency; it may be as much as 2–3 mm in a large pump. Channels may be formed by flattening part of a length of normal round pipework, or fabricated from sheet. High-conductivity copper electrodes may be cast or brazed to the end-walls.

The basic dimensions of a d.c. conduction pump are given in Fig. 2.2, and the distribution of the operating flux and current densities $B$ and $J$ in Fig. 3.17. The performance of a practical pump diverges from the ideal, eq. (2.3), by (i) shunting of part of the channel supply current through the tube walls, (ii) distortion of the flux and current distributions, and (iii) end effects.

*Shunting*  If the liquid metal has a resistivity $\rho$, and is contained in a channel of wall resistivity $\rho_c$ and thickness $d$, then the channel-wall and liquid-metal path resistances between electrodes are respectively $r_c = \rho_c \, l/2dw$ and $r = \rho l/bw$, and their ratio is $r_c/r = (\rho_c/\rho)(b/2d)$. This determines the proportion $I$ of the supply current $I_2$ that flows usefully through the liquid. However, this assumes that the current components flow directly between electrodes of width $w$ and are uniformly distributed, which is not in fact the case.

*Distortion*  Fig. 3.17 shows that where $B$ is high, $J$ is low, leading to a reduction of hydraulic pressure and an increase in $I^2R$ loss. With some liquids and high speeds it is advantageous to introduce *compensation* by returning the current from one electrode through conducting sheets placed between the channel wall and the pole face on each side.

*End Effects*  Some of the supply current diverts through both channel and walls into regions beyond the influence of the pole flux, where the only restriction to flow is the ohmic resistance because the motional e.m.f. is zero. It may be advantageous to shape the pole-faces so as to develop a fringing field and make more effective use of the stray current. For this to be effective it is necessary for the motional e.m.f. in the active part of the pump to be small.

Pole excitation is by means of a series winding of a few turns around the pole shanks, utilizing the large channel-supply current on which the pump action depends. The equivalent circuit of the arrangement comprises a resistance $r_f$ for the field winding in series with a parallel combination of three branches, (i) the liquid-channel resistance $r$ and an e.m.f. $e_r$ due to motion of the liquid in the field flux; (ii) the channel-wall resistance $r_c$; and (iii) a resistance $r_e$ to represent the path of the diverted current. Only the product $e_r I$ of the motional e.m.f. and the current in the liquid is useful.

D.C. pumps are versatile, and can cover pumping speeds from 1 mm$^3$/s to 1 m$^3$/s, with efficiencies from very low values up to about 0.6 p.u. in the largest sizes, for which the operating current might be 100 kA at 2–3 V. A detailed analysis of the design technology is given by Blake [47].

## Magnetoplasmadynamic generator

The liquid-metal pump operates in the motor mode: the principle can be inverted to the generator mode. A conducting fluid is caused to flow in a magnetic field, and a motional e.m.f. appears across the electrodes. The 'prime mover' need not be mechanical, and in the magnetoplasmadynamic (m.p.d.) generator, a thermally expanded moving gas provides the 'drive'.

A cold gas is non-conducting. A gas at very low pressure but at a temperature of $10 \times 10^6$ K has an electrical conductivity that can be higher than that of metals at normal temperature, as the result of thermal ionization. But the conductivity is too low at the limiting temperature (e.g. 3000 K) obtainable with the presently available thermal energy sources, unless it is substantially raised (e.g. to a few score siemens per metre) by injecting a small concentration of an alkali metal such as caesium. The working gas becomes a *plasma* consisting of equal numbers of free electrons and ions, together with a much larger number of uncharged atoms and molecules. When such a fluid sweeps past a transverse magnetic field at a speed $u$ an electric field strength $E = Bu$ is generated. The speed may be of the order of 1 km/s.

In one realization of m.p.d. generation, air enriched with oxygen is compressed and preheated to 1200 K, then mixed with natural gas in a combustion chamber where it reaches a temperature of 3000 K to become a plasma. A nozzle, akin to a rocket motor, accelerates the flow through the m.p.d. channel to about 1000 m/s. The e.m.f. between the channel electrodes reaches some tens of kilovolts. Thereafter the expensive 'seeding' material, caseium or potassium carbonate, is removed, and the gas, still very hot, is used to raise steam for a conventional condensing steam turbogenerator. The d.c. power from the m.p.d. stage is inverted to a.c. form.

*M.P.D. Channel*   The configuration considered to be the most suitable for the extraction of electrical energy from a plasma flow is a duct of rectangular section, with electrodes on opposite walls and a magnetic field directed transverse to the flow across the other two walls, Fig. 12.3. There are alternative methods of arranging the electrodes. In (*a*) they span the full

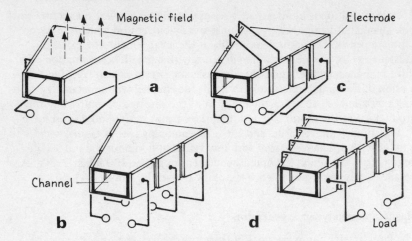

12.3 Magnetoplasmadynamic channels

length of the channel and are connected together through a single external load. Performance is limited by the Hall effect which gives rise to axially directed currents, particularly with high magnetic flux densities and low gas pressures. In (*b*) the Hall effect is suppressed by segmenting the electrodes, but electrode pairs must feed separate loads, and as there may have to be upwards of 50 pairs in a practical m.p.d. generator, the complication is a disadvantage. The modification (*c*) introduces an electrode pitch and inter-connection matched to the Hall effect for a particular flux density and gas pressure. In (*d*) the electrodes are segmented and short-circuited in pairs, with the output taken from the inlet and outlet pairs. The object here is to exploit the Hall effect by extracting power developed by *axially* directed e.m.f. and current. All four arrangements are substantially equivalent for a given geometry, magnetic field, plasma conductivity and fluid-flow characteristics.

The magnet design may require to produce flux densities of 5 T or more over an active length of several metres. Superconducting magnets are considered to be essential for large-scale m.p.d. generation because of the great mass of a conventional magnet system.

**Thrustor**

The tubular short-stroke thrustor of Fig. 3.16(*a*), devised by Green and Paul [8], has a performance that cannot be derived from eq. (2.7) because the stator and runner windings are coupled and do not conform to purely d-axis and q-axis systems. We therefore return to eq. (1.9), which gives e.m.f.s in terms of the rate of change of flux linkage $\Psi$. Then for a given runner position relative to the stator and for given stator and runner currents $i_1$ and $i_2$, we have

Stator: $v_1 = r_1 i_1 + p\Psi_1$     where     $\Psi_1 = \Psi_{11} + \Psi_{12} = L_{11} i_1 + L_{12} i_2$

Runner: $v_2 = r_2 i_2 + p\Psi_2$     where     $\Psi_2 = \Psi_{22} + \Psi_{21} = L_{22} i_2 + L_{21} i_1$

with $L_{12} = L_{21}$ only if saturation can be neglected. The linkages change both with current and with runner position, as indicated in eq. (1.10) of Sect. 1.3. Writing p for $\partial/\partial t$ and q for $\partial/\partial x$, then $e_1 = p\Psi_1 + u \cdot q\Psi_1$, giving

$$e_1 = (L_{11} p \, i_1 + L_{12} p \, i_2) + u(i_1 q \, L_{11} + i_2 q \, L_{12})$$

and similarly for $e_2$. The instantaneous power input to the system (apart from $I^2 R$ loss) is $p = e_1 i_1 + e_2 i_2$, i.e.

$$p = [(L_{11} i_1 + L_{21} i_2) \, p \, i_1 + (L_{22} i_2 + L_{12} i_1) \, p \, i_2]$$
$$+ u [i_1^2 \cdot q \, L_{11} + i_2^2 \cdot q \, L_{22} + 2 i_1 i_2 \cdot q \, L_{12}] = p_p + p_r$$

This is the sum of $p_p$, the pulsational power concerned with the change of linkage (and therefore of field energy) with current, and $p_r$ resulting from a change of the relative position of runner and stator (and therefore with the product of thrust and motion). The magnetic energy stored in the field is

$$w_f = \tfrac{1}{2}\Psi_1 i_1 + \tfrac{1}{2}\Psi_2 i_2 = \tfrac{1}{2}L_{11} i_1^2 + \tfrac{1}{2}L_{22} i_2^2 + L_{12} i_1 i_2$$

on the assumption that $L_{12} = L_{21}$. A change of linkage requires the power $(dw_f/dt) = p_f$ where

$$p_f = [(L_{11} i_1 + L_{21} i_2) \, p \, i_1 + (L_{22} i_2 + L_{12} i_1) \, p \, i_2]$$
$$+ \tfrac{1}{2} u \, [i_1^2 \cdot q \, L_{11} + i_2^2 \cdot q \, L_{22} + 2 i_1 i_2 \cdot q \, L_{12}] = p_p + \tfrac{1}{2} p_r$$

which is less than the system input power by $\tfrac{1}{2} p_r$. This difference is therefore the converted power, whence the translational force is

$$f_e = \tfrac{1}{2} i_1^2 \cdot q \, L_{11} + \tfrac{1}{2} i_2^2 \cdot q \, L_{22} + i_1 i_2 \cdot q \, L_{12} \tag{12.2}$$

For the thrustor of Fig. 3.16(a) it is necessary to make a series of intricate flux plots (and to include the effects of saturation) in order to evaluate the position-varying inductances. The static thrust/position relation can then be derived, but for the dynamic performance it is essential to write a computer program, which must include the mechanical effects of mass, friction and (where appropriate) spring compliance.

The device can be modified in several ways. Instead of a uniform armature winding, it can be graded to alter the thrust/position relation to suit the application; the field and armature currents can be adjusted for optimum efficacy; and the 2-pole structure can be extended to three or more poles.

### Traverser

Long-stroke traversers like that in Fig. 3.16(b) can be analysed more directly, as the configuration of the magnetic circuit is independent of position. But there will still be the effects of saturation in the stator core and of demagnetization of the runner magnets by armature reaction.

For analytical simplicity, assume linearity and constant gap flux. The armature voltage equation is $v_2 = (r_2 + L_2 p) i_2 + e_r$ for a winding with an effective resistance $r_2$ and self inductance $L_2$. From eq. (2.9) modified for a tubular machine, the motional e.m.f. is $e_r = B_1 \pi D N_2 u$ based on the mean gap-flux density. If, as in Fig. 3.16($b$), each pole arc subtends an angle $\beta$ and the gap density under the poles is $B_g$, then $B_1 = B_g(\beta/\pi)$. Thus $e_r = k_e u$ where $k_e = B_1 \pi D N_2$. The instantaneous armature current and gross torque are consequently given by

$$i_2 = (v_2 - k_e u)/(r_2 + L_2 p) \quad \text{and} f_e = k_e i_2$$

and $f_e$ provides for the forces $f_l$ demanded by the load, $f_f$ for static or Coulomb friction, $f_a = mpu$ for accelerating the mass $m$ of the moving system and $f_d = k_d u$ for viscous friction and damping. (In some cases there may also be gravitational and spring-compliance forces.)

*Static Condition*    At rest the gross force is clearly proportional to $i_2 = v_2/r_2$, and the net force developed is $f_e - f_f$.

*Steady-state Condition*    For constant speed $u$, voltage $V_2$ and current $I_2$ the relevant equations are

$$I_2 = (V_2 - k_e u)/r_2 \quad f_e = k_e I_2 = f_l + f_f + k_d u$$
$$u = [(k_e V_2/r_2) - (f_l + f_f)]/[k_d + (k_e^2/r_2)]$$

*Transient Condition*    Expressed in terms of Laplace transforms, the equations become

$$I_2(s) = [V_2(s) - k_e U(s)]/[r_2 + L_2 s]$$
$$f_e(s) = k_e I_2(s) = f_l(s) + (k_d + ms) U(s) + f_f$$

from which the transient speed can be obtained for a given applied armature voltage and load, in a manner strictly comparable with the treatment of rotary motors in Sects. 7.3–5 for separate excitation.

**Reluctance motor**

The linear propulsive system shown in Fig. 12.4($a$) comprises a number of coils wound on C-shaped cores associated with a toothed 'rack' of mild steel. The pitch of the C-cores differs from that of the rack teeth, and the coils are energized in sequence to produce forces of attraction that cause relative motion in one or other direction. A variety of performance characteristics can be produced in accordance with the switching sequence: motion may be made synchronous or oscillating, or with feedback control be a fairly precise excursion over a required distance. Static thrust up to 8 N per cm$^2$ of pole area can be achieved, and a typical 1 kW unit can develop a thrust of 125 N at 5 m/s with a loss (mainly in the rack teeth) of 0.3 kW.

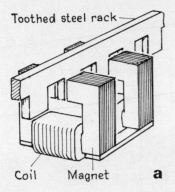
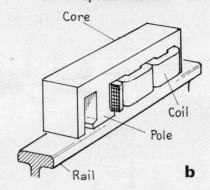

12.4 Linear reluctance motor and brake

**Brake**

Boldea and Babescu [48] have analysed the d.c. linear brake shown in
Fig. 12.4(*b*), applicable to railway vehicles and using a primary magnet
system with the rail as a secondary. The device is an eddy-current brake with
some analytical difficulties. The closed eddy paths in the rail-head saturate
the iron at the higher vehicle speeds, affecting both the magnitude and the
distribution of the rail currents. As a result, the brake force for a given
inductor excitation initially rises in proportion to the speed, but levels off at
higher speeds owing to skin effect and saturation.

## 12.4 HOMOPOLAR MACHINES

The first homopolar machine was the Faraday disc (1831). Other forms, such
as that in Fig. 1.8, are possible, the requirement being the motion of a
conducting member in the flux of a stationary magnet pole to develop a
motional e.m.f. which normally is only a few volts. Homopolar machines
have found application in control and instrumentation, and in large ratings
for generating very high currents at low voltage for feeding electrochemical
production cells and d.c. liquid-metal pumps. Current collection from the
moving member presents considerable technological difficulty for large
currents, with the result that most homopolar machines employ 'liquid
brushes' to minimize brush-wear and to limit the collector volt-drop.

**Generator**

A 10 kW, 10–16 kA, 1.0–0.625 V experimental generator for d.c. pump
supply has been described by Watt [49]. It employs a vertical-axis nickel-
plated copper rotor of diameter 0.32 m, driven by an induction motor at
76 rad/s (725 r/min) in the radial field of an electromagnet, Fig. 12.5.
Concave collector rings, with a rhodium finish to resist wear and secure an
almost resistance-free contact, are formed at top and bottom of the cup-
shaped rotor. When the machine is running the collector rings are kept filled

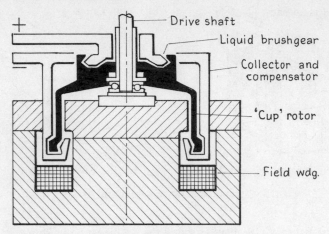

12.5 Homopolar generator

by a small flow of mercury which also provides cooling. The space
surrounding the rotor and brushes is purged with nitrogen to prevent
oxidation of the mercury. The efficiency of the generator including field
excitation is about 0.90 p.u., or 0.77 p.u. overall from terminal d.c. output
to a.c. input.

A recirculated eutectic alloy of sodium and potassium, which is a liquid
at normal ambient temperatures and has low viscosity and contact drop, is
used in the 'acyclic' generator, Fig. 3.18 (G.E., U.S.A.). The rotor is a solid
steel conductor, driven by a motor or turbine. The separately excited field
provides two homopolar fluxes to double the output voltage. The flux is
limited by the rotor diameter, as it must pass axially through the rotor cross-
section. The current rating depends wholly on the resistivity of the rotor steel
and the thermal dissipation capability. A typical machine comprises a 60 MW
steam turbine driving six generators on a common shaft, each rated at 150
kA, 67 V to provide a total output of 150 kA at 400 V for the electrolytic
production of aluminium. All field windings are fed in parallel from a single
exciter, with load or fault control by regulation or interruption of the field
current. The efficiency of the homopolar generator set approaches 0.98 p.u.

**Motor**

A notable achievement in 1969 was the construction of a 400 V homopolar
segmented-disc motor to develop 2250 kW at 21 rad/s (200 r/min), using a
*superconducting* field system. Later applications have been made in ship
propulsion with improved cryostat design and the use of metal-plated carbon-
fibre brushes operating at current densities up to 100 A/cm$^2$, together with
segmented rotors capable of series connection to give outputs at 2 kV. Marine
machinery requires high low-speed outputs, speed control and reversibility:
these features can be obtained with superconducting-field generator-motor
units.

## 12.5 HETEROPOLAR MACHINES

### Transportation auxiliaries

Road-transport vehicles, trains and aircraft require basic standby batteries that require continuous charging. A characteristic feature is the wide range of operating speed, and in the case of railway coaches also the reversal of direction. The modern method is by a.c.-generator/rectifier equipments, which compared with their d.c. equivalents have a better power/mass ratio (100 compared with 50 W/kg), a higher maximum speed, a robust commutatorless rotor, a simpler control because the a.c. generator is self-regulating and the rectifier prevents reverse current when the vehicle is idling or at rest. However, d.c. systems are still common in cars and small aircraft. Traditional d.c. charging generators are the three-brush machine (Sayers, 1896) and the cross-field generator (Rosenberg, 1905).

*Three-brush Generator,* Fig. 12.6(*a*). This uses the effect of armature reaction to accommodate wide speed variations. The shunt field winding, instead of

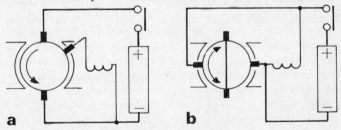

**a**          **b**

12.6  Three-brush and Rosenberg generators

being connected across the q-axis brushes, is fed from one main brush and a third brush, to span the 'leading' region of the pole-face. Provided that the voltage of the battery does not vary unduly, a rise in speed tends to increase the armature e.m.f. and current, but augmented armature reaction shifts the flux from the leading pole-tip and so weakens the field excitation. A further reduction is contributed by reaction of the current in the coils short-circuited by the third brush. At low speeds the armature e.m.f. is not high enough for battery charging, and the generator is open-circuited by a reverse-current cutout.

*Rosenberg Generator,* Fig. 12.6(*b*). This device is an application of armature excitation, Sect. 4.11, Fig. 4.16, and was formerly used for train-lighting because (i) it develops an e.m.f. the polarity of which is independent of the direction of drive, and (ii) above the speed at which normal operation begins, the output current is substantially independent of speed. The main field winding is shunt excited, and the normal q-axis brushes are short-circuited. The q-axis m.m.f. $F_q$ therefore produces a q-axis flux in which the armature rotates, developing a rotational e.m.f. across the d-axis brushes to which the load is connected. The load current imposes an m.m.f. $F_d$ in opposition to the main-field m.m.f. $F_f$. The action stabilizes for a given speed when the resultant d-axis m.m.f. $F_f - F_d$ is sufficient to generate a q-axis current and d-axis brush e.m.f. for the required battery charging current. If the direction of rotation is reversed, so also is the q-axis short-circuit current: the double reversal maintains the polarity of the output voltage.

*Starter Motor* Though intermittent, the duty of a starter motor is severe. The machine is a series motor with its details optimized for mass production. The field winding may be produced from one continuous length of aluminium strip, avoiding the inter-connecting links that conventional individual pole windings would require. Face-type moulded commutators have been devised, capable of carrying the several hundred amperes demanded for vehicle starters; they save copper and production time, are more reliable, have higher bursting speeds and give lower rates of brush wear. The brushgear consists of four wedge-shaped brushes in a single insulating moulding for accurate positioning and brush pressure. It is the practice to earth one end of the series field winding to limit its voltage to earth.

## Universal motor

The universal motor is a series commutator machine suitable for use on d.c. or 1-ph a.c. supply, and confined to fractional-kilowatt ratings because of commutation difficulties, particularly on a.c. The motor is applicable to suction cleaners, hair-dryers, portable drills and kitchen machines. It is most satisfactory at speeds between 3000 and 20 000 r/min and for intermittent duty so that carbon-brush and bearing wear is not too rapid. The design is dominated by the a.c. requirements, and as neither compoles nor compensating windings are fitted, good commutation rests heavily on the brush grade and the commutator design. The number of turns per coil must be limited, which implies a large number of sectors; low slot permeance is achieved by wide and shallow slots; but the advantage so gained in commutation has to be balanced against the increased gap reluctance, the fluctuation of which can be mitigated by skewing the rotor slots. Brush bounce resulting from commutator eccentricity is countered by high brush pressure, and the brush grade must be capable of wearing down the 'micas' at the same rate as the sectors. Brush positions are fixed to the frame with no allowance for adjustment, calling for careful positioning of the commutator bars relative to the armature slots. The magnetic circuit has to be fully laminated, and should not be unduly saturated as otherwise the number of field turns is excessive.

EXAMPLE 12.1: A universal motor has the linear parameters $L_f = 0.13$ H, $L_a = 0.07$H, $(r_f + r_a) = 20$ $\Omega$, $G = 1.0$ V/A per rad/s. Plot its speed/current and torque/current characteristics when connected (i) to a 240 V d.c. supply, (ii) to a 240 V 50 Hz supply. In each case the loss torque is $(0.1 I^2 + 0.5/I)$ N-m.

(i) *D.C.* With $I = 2.0$ A we have $E_r = 240-40 = 200$ V, $\omega_r = 200/(1.0 \times 2.0)$ $= 100$ rad/s (or 950 r/min), $E_r I = 400$ W, gross torque $M_e = 4.0$ N-m, net torque $M = 4.0 - 0.6 = 3.4$ N-m, efficiency $3.4 \times 100/240 \times 2.0 = 0.71$ p.u. The characteristic is plotted in Fig. 12.7.

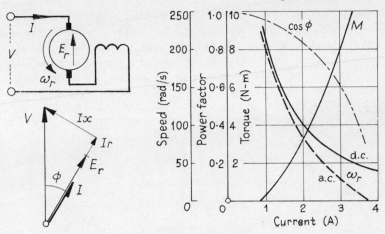

12.7 Universal series motor

(ii) *A.C.* The inductive reactance is $x = \omega L = 2\pi \times 50 \times 0.20 = 63\ \Omega$. The calculation is based on the phasor diagram, shown to scale for $I = 2.0$ A r.m.s. in Fig. 12.7. Here $Ix = 126$ V, giving $(E_r + Ir) = \sqrt{(240^2 - 126^2)} = 204$ V, whence $E_r = 164$ V and $\omega_r = 82$ rad/s (or 780 r/min); $E_r I = 328$ W, $M_e = 4.0$ N-m, $M = 3.4$ N-m; efficiency $3.4 \times 82/240 \times 2.0 = 0.58$ p.u. The power factor is $204/240 = 0.85$ lagging.

For a given current, the d.c. and a.c. torques are the same (but in the a.c. case it pulsates at 100 Hz). The volt-drop in reactance lowers the speed on a.c., especially at higher load current.

## D.C. induction motor

In an inverted (rotor-fed) *polyphase* induction motor, with $p$ pole-pairs and a supply of frequency $f_1$, the rotor windings develop a travelling wave of gap flux density moving with respect to the rotor at a 'synchronous' angular speed $\omega_1 = 2\pi f_1/p$ in, say, a clockwise (c.w.) direction. When the rotor is at rest, the travelling wave induces in the short-circuited stator phase windings e.m.f.s and currents of frequency $f_1$, which interact with the gap flux to develop on the rotor a counterclockwise (c.c.w.) torque. Stable conditions for a given load torque are reached when the c.c.w. rotor speed is $\omega_r$, the gap flux density wave moves c.w. at $\omega_1 - \omega_r = \omega_1(1 - s)$ with respect to the stator, and the induced stator currents have the frequency $s\omega_1$, where $s$ is the slip. If the c.c.w. rotor speed $\omega_r$ increased to become equal to the c.w. synchronous speed $\omega_1$, the gap-flux distribution would be stationary with respect to the stator windings. The slip would be zero, no stator current would be induced, and the torque on the rotor would vanish.

The so-called *d.c. induction motor* with its rotor at rest resembles the stationary polyphase machine if a direct current is fed to a commutator armature through brushgear that is rotated bodily clockwise at a 'synchronous' angular speed $\omega_1$. The rotor m.m.f. acts along the brush axis and therefore rotates at $\omega_1$, inducing e.m.f.s of that frequency in a set of short-circuited stator 'phase' windings and developing a counterclockwise torque on the rotor. But here the similarity ends, for regardless of rotor speed, the travelling wave of gap density remains at speed $\omega_1$. In the stator the frequency of the currents is always $\omega_2$; in the rotor the current in an armature coil is reversed as it passes a brush, and the rectangular coil-current wave has the frequency $(\omega_1 + \omega_r)$. The characteristics in Fig. 12.8 relate (*a*) rotor speed $\omega_r$ to output power $P_o$ for a rotor voltage $V$ and brushgear

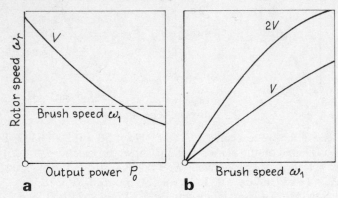

12.8 D.C. induction motor characteristics

speed $\omega_1$; and (*b*) no-load rotor speed to brush speed for rotor voltages $V$ and $2V$.

## 12.6 SUPERCONDUCTIVITY IN D.C. MACHINES

Certain metals and alloys exhibit *superconductivity;* they conduct a current without measurable resistance and $I^2 R$ loss when cooled to very low temperatures (less than 20 K, or $-253°$C). Engineering application became possible with the development of alloys that could preserve superconduction in the presence of strong magnetic fields, enabling coils to be constructed to give flux densities up to about 15 T for the expenditure of a few kilowatts to keep the winding refrigerated at operating temperature. Such flux densities in a copper coil at normal ambient temperature would require many megawatts.

One possible superconductor is a niobium-tin alloy, but it is brittle. The present most useful material is a niobium-titanium alloy, mechanically strong especially if used in the form of fine filaments. A conductor may, for example, be constructed with 60 filaments of Nb-Ti, 0.025 mm diameter, encapsulated in an epoxy resin and enclosed in a copper matrix 0.5 mm in diameter, working with a filament current density of 120 A/mm². Alternatively, some thousands of filaments of diameter 0.5 $\mu$m may be used.

For the present, superconductivity is restricted to the field system of the d.c. machine, and a homopolar structure adopted for which a single field coil is sufficient, simplifying the refrigerating arrangements. Up to 5 MW a disc armature is employed, but above this rating a cylindrical rotor should be feasible. Superconduction for heteropolar machines, using a pair of field windings, may well be a future development.

High working flux densities enable the size and mass of a machine to be greatly reduced, the cost saving being balanced against that of the field refrigerator. The field winding must be braced against self-disruptive electro-magnetic forces. There will be considerable flux leakage, for ferromagnetic materials at high flux densities are little better than air. Problems of super-conducting field-system design, discussed by Maddock and James [50], concern the stability and cooling of the superconductor, and its protection against the temperature-rise and voltage breakdown consequent upon a refrigeration failure.

EXAMPLE 12.2: The superconducting field coil of a 2250 kW 420 V homopolar motor, Fig. 12.9, has $N_f$ = 3500 turns carrying $I_f$ = 600 A at a field voltage of 4.2 V (due mainly to interconnections). The double-disc copper rotor, of outer diameter 2.0 m and inner diameter 0.4 m, runs at 200 r/min ($\omega_r$ ≈ 21 rad/s). Each disc comprises 20 series-connected sectors,

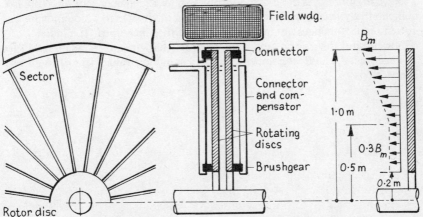

12.9 Homopolar motor with superconducting field system

and the discs are in parallel. Fig. 12.9 shows the approximate distribution of the axial field flux density over a disc radius. Evaluate (*a*) the maximum flux density $B_m$ required, (*b*) the total useful flux, and the field-circuit inductance, time-constant and stored energy (assuming that the useful flux is 0.7 of the total), and (*c*) the full-load rotor current and torque.

(*a*) The rotational e.m.f. per sector is 420/20 = 21 V. At a radius $x$, the axial flux density is 0.3 $B_m$ for 0.2 $<x<$ 0.5 m, and $B = B_m$ (1.4$x$ − 0.4) for 0.5 $<x<$ 1.0 m. Applying eq. (2.4), the rotational e.m.f. per sector is

$$e_r = B_m \omega_r \left[ 0.3 \int_{0.2}^{0.5} x \cdot dx + \int_{0.5}^{1.0} (1.4x - 0.4)x \cdot dx \right]$$

The integration gives

$$e_r = B_m \omega_r \, [0.032 + 0.263] = 21 \text{ V}$$

whence $B_m = 3.47$ T, nearly five times as great as in a conventional machine in spite of the magnetic circuit being entirely of air.

(*b*) The useful flux is the integral of $B \cdot 2\pi x \cdot dx$ over the active radius, i.e. $\Phi_m = 2\pi B_m [0.032 + 0.265] = 6.3$ Wb, and the total flux is $\Phi = 6.3/0.7 = 9.0$ Wb. The field inductance is $L_f = N_f \Phi_f / I_f = 52.5$ H. The effective resistance is $r_f = 4.2/600 = 7$ m$\Omega$ and the time-constant is $\tau_f = L_f/r_f = 7500$ s (about 2 h). The field energy is $w_f = \frac{1}{2} L_f I_f^2 = 9.45$ MJ.

(*c*) The rotor current is $2250/420 = 5.36$ kA, or 2.68 kA per disc. The full-load torque is

$$M_e = E_r I_a / \omega_r = 107 \text{ kN}$$

The mean peripheral current density at the disc edges is 426 A/m (outer) and 2150 A/m (inner). As the e.m.f. contribution by the inner region of a disc is comparatively small, some increase in the inner diameter to accommodate current collectors does not involve undue sacrifice in output. The radial sector currents develop an 'armature reaction' that gives to the main field flux an axial 'twist', an effect compensated by a suitable arrangement of the brush connections.

# 13 Construction and Design

## 13.1 CONSTRUCTIONAL VARIETY

The d.c. machine is constructed in many forms and for innumerable purposes, from the 3 mm stepper taking a few microamperes at 1.5 V in a quartz-crystal watch to the powerful mill motor and ship-propulsion plant. Constructional variety is found mainly in the micro-, miniature and control-motor range, but even in the small-motor field, Ref. [51], there has been considerable development and application of such special forms as the disc motor. Both small and large industrial machines have generally the conventional heteropolar cylindrical-rotor structure, although some unconventional homopolar machines have been devised. The following sections can give some details only of a few typical methods of construction.

## 13.2 MINIATURE MOTORS

Primary-battery driven motors of one or two watts output are made with the minimum of production-run parts and have wide tolerances. Fig. 13.1 gives an 'exploded' view (full size) of a typical motor, 23 mm in diameter and 30 mm in overall length. The frame is formed by two plastic mouldings in a thin steel tube. The moulding at the commutator end carries a pair of metal-carbon brushes on spring bronze strips which also serve as terminals. The other moulding carries a pair of ferrite magnets arranged as in Fig. 3.4($a$), and it interlocks with the commutator-end moulding by axial extensions which locate the brushes with respect to the magnets. The bearings are simple central holes in the mouldings, carrying the shaft of a 3-pole polarized armature winding of the type in Fig. 4.8, and a 3-sector commutator. The parts are assembled and fixed with adhesive, and a small pinion is fitted on the drive end to step down the high speed.

A 9 V motor for a portable tape recorder, Fig. 13.2, has a 5-pole armature with a coil embracing each pair of adjacent poles. The magnet is an annulus with a magnetization like that in Fig. 3.4($e$). The brushes are of spring wire. The bearings contain oiled washers for lubrication. The motor casing comprises a pair of concentric cylinders with rubber-sheeting between them to minimize vibration and pickup. The tape spindle is mounted on the shaft of a steel inertia wheel of diameter 70 mm, the periphery of which is rubber-belt driven from a 6 mm diameter pulley on the motor shaft.

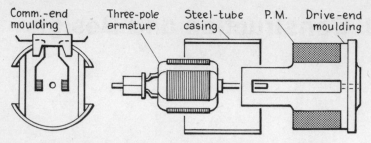

Comm.-end moulding | Three-pole armature | Steel-tube casing | P. M. | Drive-end moulding

13.1 Miniature 3 V permanent-magnet motor (full size)

Miniature p.m. machines in ratings up to 10 W have normally 3-pole rotors and ferrite magnets. For powers up to 100 W the armature is more conventional, with 11 or 13 slots. Typical applications are to battery-fed record players, clocks and windscreen wipers.

Precision machines have a more sophisticated design, with rare-earth magnets, face commutators, lever-arm graphite brushes, and sleeve or ball bearings. A typical motor with a 4 W output may take little more than 0.5 A at 12 V. Applications are extensive in office-machine drives and medical equipment. Two examples of the latter are the dental drill and the drug-injection device. The motorized dental drilling handpiece cannot compete with the air-turbine drill (which runs at speeds approaching 300 000 r/min) but is well established for operation at medium drilling speeds, stabilized by a simple feedback control to counter the inherent fall of speed with rise in torque. There are restrictions on the motor dimensions and mass to achieve a handpiece of convenient size and weight. In the drug-injection equipment, for continuous long-term use, a syringe driver in a small pack is carried by the patient while he leads a normal life. The pack contains a 12 V low-inertia moving-coil motor, pulsed at pre-set intervals from a battery-fed thick-film control circuit and driving the syringe through a 500/1 reduction gear.

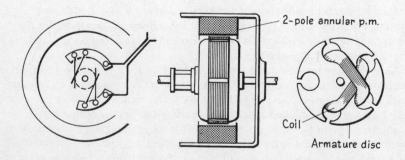

2-pole annular p.m.

Coil

Armature disc

13.2. Tape-recorder 9 V permanent-magnet motor (full size)

## Pole sensitivity

Permanent-magnet motors for record players and tape decks must have minimized pole-sensitivity, as it is desirable that the torque be independent of the turntable or tape position. Pole sensitivity is also undesirable for battery-operated clocks, which must be capable of starting even with a partially discharged battery.

Pole sensitivity arises by reason of the rotor slotting (particularly with the 3-pole rotor, Fig. 4.8), and also as a result of the spatial distribution of the p.m. flux in the airgap.

Consider the 2-pole annular magnet arrangements at (e) and (f) in Fig. 3.4, used in conjunction with a 3-slot polarized rotor. Both sources of pole sensitivity are present. Analysis of the *radial* magnetization form (e) shows that the flux distribution in terms of space harmonics will contain a fundamental (of half-period equal to the pole-pitch) and a succession of odd-order harmonics. The 3rd harmonic has little effect, and the torque is developed by the fundamental with the 5th and 7th space harmonics. All three introduce pole sensitivity.

The deliberate introduction in the magnetization process of a 3rd space harmonic into the gap-flux distribution can, however, be advantageous, because its effect is to produce a flattened space waveform such that the total pole flux can be increased without the remanent flux density $B_r$ being anywhere exceeded. A square-wave distribution (which has a strong 3rd harmonic) will therefore improve the performance of a motor by as much as 0.4 p.u. provided that the concomitant increase in pole sensitivity due to the 5th and 7th harmonics can be tolerated.

With the *transverse* magnetization pattern (f), the radial component of the gap flux density approximates to the sinusoidal. This almost eliminates pole sensitivity attributable to the gap-flux distribution, but for a given p.m. material has a smaller total flux than pattern (e) and consequently less torque.

Motors rated above about 10 W will be provided with multi-slot rotors. Here the pole sensitivity is a minor consideration compared with torque and efficiency, so that square-wave radial distribution is preferred as it gives the greatest working flux.

Annular magnets with radial magnetization are not easy to produce: contraction of the p.m. material accompanies the sintering process and is greater in the preferred flux direction than in directions perpendicular thereto, leading to strong internal stresses. An advantage of the alternative transverse magnetization is that it can be applied after the motor has been assembled. Either pattern, if subsequently dismantled, may suffer irreversible deterioration of the pole flux.

To take advantage of the p.m. capability, the radial airgap should be as small as is feasible, implying close dimensional tolerances.

## 13.3 CONTROL MACHINES

These comprise servo motors (conventional or brushless), tachometer generators and stepping motors. Start-stop duty, operating speed, reversal, acceleration, positional accuracy and damping have significant influence on design. Transistor and integrated-circuit developments have fostered a range of micromachines for instrumentation, regulation and control, typically with motors of 15 mm length and diameter, and powers of the order of 0.5 W. Brushless machines provide high-torque low-speed space and servo requirements with high overload capability, where a low level of rotor loss is wanted. Servo tachogenerators for rate feedback in industrial speed and position control have a low ripple content and a linearity within 0.5%. Stepping motors can upgrade mechanical systems by replacing cams, complex linkages, ratchet-and-pawl and similar mechanisms to give greater precision and production rate, sometimes without the need for closed-loop control.

### Servo motors

The mechanical features of fast-response servo motors can be classified in terms of the rotor structure: (i) cylindrical, (ii) disc and (iii) moving-coil. All have permanent-magnet fields, because current-excited field systems involve undue time-constants, slow response and greater loss, making them uncompetitive with p.m. fields and armature-current control.

*Cylindrical*  Most small motors have a 2-pole structure with maximum speeds of 500 rad/s and upward. Medium-sized machines have 4- or 6-pole fields, lap windings and speeds of about 300 rad/s. Slotted armatures are most common, but a few designs call for conductors cemented to the rotor surface to reduce inductance. As reversal by the sudden reversal of the drive voltage $V_a$ imposes nearly $2V_a$ across the armature, the e.m.f. constant $k_e$ must be limited to avoid excessive voltage between commutator sectors. Brushes are as for industrial motors except that silver-content material may be adopted to reduce brush volt-drop; but this makes the brush hard and leads to more rapid commutator wear. The brush pressure is important, for if it is reduced to avoid wear, the commutator surface may be eroded by sparking. In precision applications with direct coupling between motor shaft and load, the essential concentricity is achieved by use of precision ball bearings and a shaft diameter as large as the rotor can accommodate.

*Disc*  Most disc armatures have wave windings and use a single pair of brushes, except in the case of the punched-sheet winding in order to limit the current density in the armature conductors, as the brushes bear on them directly without a separate commutator. Bearings must be arranged to give axial accuracy to the rotor position in view of the very restricted gap clearances, and the shaft diameter must be at least 12 mm.

*Moving-coil*  The cup-shaped armature is skein-wound with copper or aluminium wire. The axial length is between 0.3 and 0.8 of the diameter, and for mechanical strength and rigidity the winding is encapsulated. The moving-coil construction, with its very low inertia and inductance, can be used in control systems having a bandwidth as high as 200 Hz, so that the compliance of the shaft becomes an important parameter. A typical arrangement is shown in Fig. 13.3.

The table gives some details of typical cylindrical (C), disc (D) and moving-coil (M) machines.

Parameters of Typical Servo Motors

| Size | Small | | | Medium | | | Large | | |
|---|---|---|---|---|---|---|---|---|---|
| Type | C | D | M | C | D | M | C | D | M |
| Motor mass (kg) | 1.0 | 2.5 | 1.6 | 5.5 | 13 | 8.5 | 48 | 30 | 43 |
| Inertia $10^{-6}$(kg-m$^2$) | 2.7 | 3.9 | 0.5 | 140 | 120 | 0.4 | 4400 | 360 | 190 |
| Peak torque* (N-m) | 1.1 | 1.4 | 1.6 | 14 | 16 | 5.6 | 135 | 49 | 42 |
| Acceleration†(krad/s$^2$) | 210 | 180 | 1700 | 50 | 70 | 660 | 15 | 67 | 110 |

*5 times continuous torque   † Equal load and motor inertia

**Tachometer generators**

Most precision tachogenerators have a 2-pole structure in order to obtain the

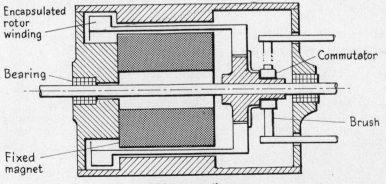

13.3 Moving-coil motor

greatest number of slots per pole. The slots are skewed to prevent cogging and to reduce ripple, and for the same reasons the poles may be chamfered. The rotor core steel is of high quality to limit hysteresis and eddy-current effects. Multiturn coils are required for adequate output voltage and they are connected to the commutator by high-resistance risers to limit coil short-circuit currents during commutation. Radial gap-lengths are of the order of 0.4–0.8 mm. The ratio slot-opening/gap-length may be about 0.7. Anti-friction bearings and low brush pressures keep down the mechanical loss.

The brushes are carefully set in the q-axis position because the rotor may spin in either direction. Occasionally a generator may have compoles, improving linearity. For example, a 2 W generator with a ± 2% nonlinearity over a ± 9000 r/min range was improved by the fitting of compoles to ± ½ % variation.

Tachogenerators are often directly associated with servo motors. A typical combination is shown in section in Fig. 13.4. The rotors and commutators are

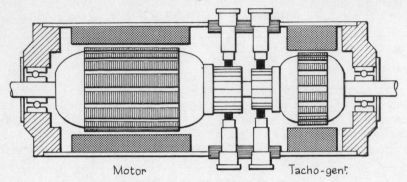

Motor                    Tacho-gen⸳

13.4 Servomotor/tachogenerator unit

mounted on a common shaft but are magnetically separate, although it is possible to accommodate both the motor and the generator windings in the same slots of a single rotor. Where a tachogenerator is to be attached to an existing motor, the two are connected by means of a flexible coupling to allow for slight deviations from exact alignment.

## Brushless and stepping motors

Drives for office machines, data-processing equipments, printers, tape machines and recorders may be better served by brushless motors, in respect of life, noise and radio interference, than by motors with rotating commutators and brushgear. The brushless motor requires position-sensing devices and electronically controlled supplies to the stator, the rotor being a permanent-magnet system. Essentially, a stepper motor is the same as a brushless motor without the position sensors.

Practical realization of the prototypes in Figs. 11.12 and 11.13 can have very considerable structural variety and a wide range of step-angles and ratings.

*Permanent-magnet Machines*  In general these are applied to duties requiring fairly low pulse-rates and 90° or 45° steps. With metal magnets it is not readily possible to obtain more than four poles. Pulse rates up to 350 per sec, torques up to 0.5 N-m and errors within ± 10% can be achieved, the inertia ranging from $1 \times 10^{-6}$ to $10 \times 10^{-6}$ kg–m². Inexpensive units with multipolar ceramic rotor magnets can give step angles down to $7\frac{1}{2}°$ but with lower accuracy and greater inertia.

*Variable-reluctance Machines* Simple v.r. motors have medium step angles between 5° and 15°, stepping rates up to 1000 per sec and a very small inertia because the rotor carries no magnets and can be of small diameter. Running torques up to 40 N-m can be obtained with a position error of ± 3%, in a frame of not more than 80 mm overall diameter. Multi-stack v.r. steppers have three or more separate stators operating on rotors mounted on a common shaft, Fig. 13.5. Either the sets of rotor or of stator teeth are axially out of alignment, so that very small resultant step-angles are obtained with a good slewing capability. The frame diameter is large (typically between 75 and 125 mm) in order to limit the overall shaft length on which flexure and eccentricity depend, but as a result the rotor inertia is inevitably high, in the range $2 \times 10^{-6}$ to $200 \times 10^{-6}$ kg–m$^2$.

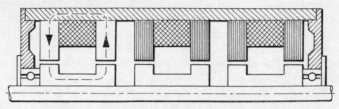

13.5 Three-stack variable-reluctance stepper motor

*Hybrid Machines* The performance generally achieves a high-torque, high-accuracy, high-inertia and small stepping-angle range (e.g., 150 to 8000 mN-m, $3 \times 10^{-6}$ to $4000 \times 10^{-6}$ kg–m$^2$, ± 3% error, 0.5 to 15° step). Typical constructional features are shown in Fig. 13.6. That in (*a*) is similar

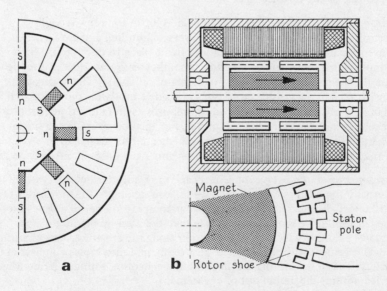

13.6 Hybrid stepper motors

to a p.m. machine but the rotor comprises multipole salient magnets. In (*b*) the rotor has an axial p.m. and a pair of cup-shaped axially-toothed steel 'shoes', the latter being mutually offset by one-half of a tooth-pitch. The stator poles are also dentated. The numbers of stator and rotor teeth are selected to suit the stepping angle required. The left- and right-hand rotor cups align themselves successively with the stator teeth by unlike-pole attraction and like-pole repulsion, the angle advancing by one-quarter of a tooth-pitch per step.

## 13.4 SMALL-POWER MACHINES

Direct-current motors of ratings between 0.1 and 1 kW, operated from secondary batteries, are manufactured in very large numbers for such applications as automobile-engine radiator fans and starter motors. Universal motors, operated normally on a.c. mains supplies, may on occasion also be operated on batteries or through rectifiers. Most small-power machines have a conventional cylindrical-rotor structure but in this field there has been some development in the disc configuration. An application of battery-fed traction motors is to the electric car or lorry, the commercial possibilities of which are likely to increase considerably in the future.

### Starter motors

The starter comprises a low-voltage series motor and a means of transmitting the torque to the flywheel gear ring.

*Motor* A large starting torque has to be developed for a few seconds intermittently, and the current may be several hundred amperes. The conventional cylindrical commutator is giving place to disc types, which economize in copper and production time; they are moulded and give higher bursting speeds, lower brush-wear and greater reliability. The brushgear consists of four wedge-shaped brushes in a single insulating moulding that accurately positions the brushes and minimizes earth faults. Constant brush pressure is maintained by helical springs. For the pole excitation a single winding of figure-8 form, made from a single continuous length of aluminium strip, can be used to facilitate production and avoid pole-coil interconnection.

*Transmission* For most petrol-driven cars, the traditional inertia system (an internally threaded pinion on a screwed sleeve splined to the motor shaft) is satisfactory. For diesel engines and higher-compression petrol engines the 'pre-engaged' drive is necessary. A solenoid mounted on the motor yoke actuates a linkage which slides the pinion along the armature-shaft extension into engagement with the flywheel gear-teeth, and then closes contacts in the battery feed to the motor. A roller clutch mechanism, acting as a free-wheel device, returns the pinion out of engagement.

**Disc motors**

Where rotary power has to be developed electrically in a restricted axial space, the disc construction offers advantages. It has been developed for small pumps, freezer compressor drives, low-inertia domestic machines such as sewing-machine drives, car radiator fans and electric-vehicle propulsion. Radiator fans give the advantage of controlled operation based on coolant temperature and a lower power demand than the usual belt-driven system; and in other applications a considerable saving in bulk and mass is made possible by the large diameter/length ratio.

There are three basic magnetic-circuit arrangements: (i) two magnet rings, one on either side of the armature, (ii) as (i) but with one magnet ring replaced by a fixed steel disc, (iii) a single magnet ring with a steel (or iron-dust) ring backing the armature. In (i) there is a negligible axial force on the armature and in (ii) it is very small, but in (iii) the pull requires a thrust bearing. A compensating advantage of (iii) is that its effective axial airgap is minimized, the other forms requiring two series gaps.

In small low-inertia armatures the windings are punched from 0.2 mm copper or aluminium sheet, and then placed on either side of an epoxy-resin impregnated glass fabric disc and the resin partly cured under pressure. The centre hole is drilled, the plate edge removed, and the end connectors welded at the inner and outer diameters to form lap or wave windings. Up to three discs may be assembled in series. The resin is finally given a completion cure. Silver-graphite brushes bear directly on the inner end-connectors, and for small machines a wave winding is chosen to simplify the construction and to employ a single pair of brushes. A typical small motor is contained within an aluminium tube 100 mm diameter with steel end-plates that form part of the magnetic circuit, eight magnets on each side of the armature, an effective airgap of 2 mm, an armature resistance of 1 $\Omega$ and an inductance of 100 $\mu$H. The machine could be used for tape recorders, computer spool drives and industrial instrumentation. At high speeds the thin, flat armature conductors generate some eddy-current loss, and the drive must be such as to limit the current on account of the current density at the inner radius of the conductors, on which the brushes bear, there being no separate commutator.

Larger disc motors have wire-wound armatures with face or barrel commutators. The length of the brushgear makes it advisable to arrange the winding end-connectors so as to enable the brushes to be fixed in the radial inter-magnet spaces. If a backplate for flux return is carried on the rotor it is subject to core loss; this may be mitigated by the use of compacted iron dust in an epoxy resin binder, but the increased reluctance reduces the available flux per pole. Some improvement is gained if the windings can occupy radial slots moulded into the face of the backplate.

Two disc motor arrangements are given in Fig. 13.7. The radiator fan (*a*) has to be short axially to fit the restricted space and to be fixed close to the cooling surface. The armature carries an iron dust flux-return disc. The magnetic pull is countered by the thrust of the fan when rotating, but the two forces must not be balanced, as otherwise the armature might oscillate axially. For simplicity and cheapness the fan may be moulded and the

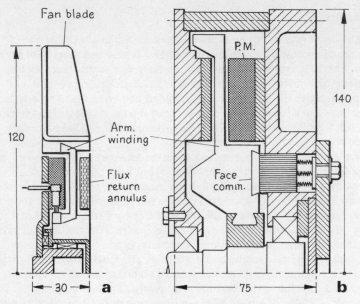

Fan blade

P.M.

140

120

Arm. winding

Flux return annulus

Face comm.

30   **a**

75   **b**

13.7 Disc motors (one-half size)

armature encapsulated in one operation. The machine illustrated has a shunt characteristic, developing a torque of 0.33 N-m at 2700 r/min when taking 12 A from a 12 V battery supply, producing an output power of 94 W at an efficiency of 0.65 p.u. Machine (*b*) is for application to the road wheels of a light electric vehicle, the motor driving one wheel directly without gearing. Outputs from 250 W to 3 kW are required. Such a motor for 900 W at 12 V and 2500 r/min has a 2-layer winding of 60 pre-formed 6-turn coils, lap-connected to a 60-sector face commutator. Twelve sector magnets bonded to a steel plate have each an area 1600 mm², thickness 17 mm, flux 0.37 mWb, and an angular pole-arc/pole-pitch ratio of 0.75. The flux-return disc is mounted on the stator to give a total gap length of 5 mm. The mechanical details are conventional, with an extensive use of aluminium alloy.

### Universal motors

The basic design requirements are discussed in Sect. 12.5. The 220/250 V fractional-kilowatt machine in Fig. 13.8 has only 8 commutator sectors and slots, a fully laminated magnetic circuit, and helical-spring loaded 3 mm square carbon brushes that span less than one-half of a commutator sector pitch.

### 13.5 CLUTCHES, COUPLINGS AND BRAKES

A typical design, Fig. 13.9, is applicable to all three functions. It has a stationary exciting winding to avoid the need for slip-rings, fed at a nominal

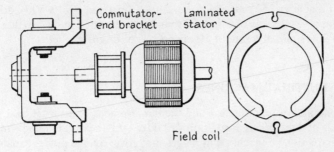

13.8 Universal motor (one-half size)

voltage (e.g. 6, 12 or 48 V) derived from an a.c. mains supply through rectifiers, with a capacitor-resistor combination across the coil terminals to dissipate the stored magnetic energy on switch-off. The winding is thermally insulated from the driving faces (the temperature of which may exceed 200°C) so that Class B insulation is adequate.

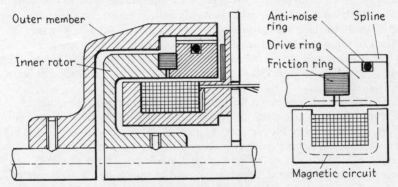

13.9 Electromagnetic clutch

Inherent self-adjustment over the wearing life of the driving faces is provided. The inner rotor is fitted to one shaft. The aluminium outer member, fitted to the other shaft, is provided with spline teeth. The drive ring, similarly splined, fits into the outer member and can slide axially towards the inner rotor. The groove on the drive ring is fitted with a high-temperature-resistant annular ring which has a small interference with the spline teeth and serves to reduce vibration and noise when the device runs disengaged. Both the drive ring and the inner rotor have an annular projecting pole ring. For torque to be transmitted, the drive ring is magnetically attracted towards the inner rotor and moves axially along the splines. Torque is then transferred through the pole rings and an intermediate friction ring of automobile clutch material. The device becomes a brake if one of the transmitting members is anchored. Typical ratings are

| Exciting-coil power (W): | 1 | 6 | 10 | 25 | 65 | 300 |
|---|---|---|---|---|---|---|
| Maximum torque (N-m): | 6 | 10 | 12 | 15 | 24 | 36 |

Units of diameter 250 mm can, for example, start and stop a 5 kW drive at 1500 r/min at intervals of 5 s. A 75 mm diameter unit can deal with a 100 W drive at 200 r/min at a rate of 10 operations of on-off per sec.

## 13.6. INDUSTRIAL MACHINES

The economics of manufacture and the development of international standards of dimensions, insulation and duty-cycle have generated the design and construction of machines in ranges that cover a wide power capability and have the optimum number of common features. Availability of a.c. mains supplies at 50 or 60 Hz is so well established that nine out of ten modern d.c. motors are built for operation through 1-ph or 3-ph rectifiers, normally fully-controlled. As d.c. motors are generally items in control systems for industrial processes, the normal design is the shunt or compound machine with a very wide speed range obtained by armature voltage variation, augmented by field weakening. The field and armature operating voltages are determined by the voltage of the a.c. supply and the type of rectifier connection. A small shunt motor fed from a 240 V 1-ph 50 Hz supply and to have a top speed of 1500 r/min would be designed for an armature voltage of about 170 V maximum, while a larger motor fed through fully controlled bridge rectifiers from a 415 V 50 Hz supply would have a rotor voltage up to about 480 V and a field voltage in the range 160–330 V obtained from a 1-ph uncontrolled or semi-controlled rectifier.

Rectifier supply imposes a fluctuating terminal voltage and harmonic currents. Although armature and supply-network inductance has a smoothing effect, a substantial ripple remains. As a result, (i) additional core and $I^2R$ losses occur, and (ii) the brush current and armature-reaction m.m.f. fluctuate and the inductive e.m.f. of commutation is increased. The effects in (i) can be accommodated by improved ventilation, and, possibly, by the inclusion of a series inductor; those in (ii) demand effective functioning of the compole flux, which must be capable of rapid change in synchronism with the variation of the armature current. Thus the compole core must be laminated and saturation nonlinearity avoided.

Certain fields of application demand specialized design and construction to satisfy particular operational conditions, characteristics and mechanical details, and to work in abnormal environments. Electric-vehicle motors are required for low battery voltages, special enclosure, ventilation, mounting and dimensions. Marine and naval machines have to withstand rolling and pitching and (for submarines) diving; they may be required to run when submerged, and be mechanically shockproof to transient accelerations up to 300 $g$. In rolling-mill drives the operating conditions impose very high mechanical and electrical stresses, demanding oversize shafts and strongly braced windings. Traction motors and auxiliaries are subject to persistent fluctuation and frequent interruption of contact-line voltage.

## Magnetic circuits

Normally the magnetic circuit is laminated throughout. For machines with ratings up to a few score kilowatts, the armature and yokes can be stamped from single sheets of core steel; for larger dimensions it is necessary to use segmented stampings. Pole punchings are separately assembled and bolted on. The stator core is clamped between collars and held by welded straps, and enclosed in a square, circular or octagonal frame. The rotor core, mounted on the shaft or on a 'spider', has slots skewed to reduce noise and low-speed torque fluctuations.

## Windings

Modern low-voltage windings for small production-run machines are made on high-speed winders and transferred mechanically into pre-insulated armature slots. Polyvinyl-acetal (p.v.a.) material is usual for Classes E and B and can meet the demand for higher outputs from a given frame. Sometimes the wire is double-coated, the outer being abrasion-resistant, for the temperature limit is not the only consideration: surface friction of the enamel and the springiness of the wire make high-speed winding more difficult, and friction may be reduced by use of a micro-crystalline wax. Pinholes always occur in enamel wire-coverings and their number may be substantial. In this respect a wire with a pinhole count of less than 1 per m length would generally be considered as 'good'. The thermal life of enamelled wire can sometimes be affected by incompatibility with adjacent connectors and tie-cords if these are of halogen-bearing material such as p.v.c.

The armature conductors of larger machines are of rectangular section copper, frequently with Class F insulation even if the machine rating is based on Class B temperature limits, because of the facility provided thereby for overload. Where, as in traction, the end-windings are subject to attack by cast-iron brake dust, the windings may be sealed with varnish or glass tape, and the overhang interstices filled to provide a smooth surface.

Standard lap and wave windings are commonly employed, but where conditions demand it a duplex winding may be necessary. Slot conductor insulation is typically based on glass-fibre material, surface treated with polyester or silicone varnish to reduce abrasion resistance during forming. High-temperature polyamide strips are preferred for inter-layer insulation. Glass-backed mica wrap around the slot portion of the coil-sides is applied to a thickness appropriate to the working voltage to earth.

Main and auxiliary field coils may be insulated from earth by silicone-varnish-bonded single-layer mica splittings on a glass backing, wound half-lap for 750 V, a further layer for 1500 V. Alternatively the enamel-covered conductors may be encapsulated in an epoxy resin moulding and fixed to the pole in the same operation. Suitable contouring of the windings can make a better utilization of the space available. Fig. 13.10 shows a field pole of this type, together with a typical slot for a traction auxiliary machine.

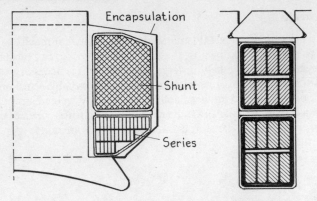

13.10 Main-pole and armature-slot details

## Commutator and brushgear

The problems associated with commutators are electrical, mechanical and
thermal. The mechanical stressing of the commutator sectors and V-rings is
dependent on the state of cure of the insulation between them and its ability
to withstand thermal cycling. The typical commutator assembly in Fig. 13.11
has fully cured epoxide resin-glass V-ring insulation moulded at high pressure.

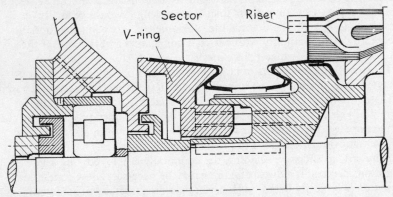

13.11 Commutator and bearing assembly

Micanite inter-sector insulation with a shellac bond is almost invariably
employed. The sectors are extruded copper containing about 0.05% silver, a
metal that has displaced hard-drawn copper as it can withstand the flood-
soldering of the armature coil-ends to the risers at 300°C. After assembly the
commutator is turned with a diamond tool following a rotational 'seasoning',
and the inter-sector insulation undercut to a depth of about 1 mm.

For small machines it is more usual to make moulded commutators, which
can contain centrifugal forces more readily and have more stable insulation.

The choice of brush grade rests considerably on experience. In general a
constant-pressure brush loading seems to yield a lower rate of brush wear, but

in the development of a motor range it is desirable to test several brush grades for wear-rate, commutator skin formation, bar marking or burning, grooving and copper drag. In totally-enclosed machines, silicone vapours may be slowly released when the machine warms up, and as they inhibit the formation of the commutator skin it is considered advisable to exclude silicone products from the winding insulation.

Two to five brushes per arm are provided in small machines, up to eight for 750 kW and as many as 16 for the largest ratings. Sometimes the brush-rocker can be rotated to facilitate brush maintenance; or the brushgear assembly can be swung out, with a simple mechanism to re-position the brush ring to a preset location.

## Cooling, ventilation and enclosure

Small machines are readily cooled by straightforward circulation of the ambient air. For larger machines a separate blower located on the frame drives air into the machine at the commutator end and out at the drive end, augmented where necessary by the provision of fans on the motor shaft; the fan may also be used as a balancing disc. Axial ventilation is often preferred to radial because there are relatively large gaps between pole windings, enabling these to be uprated and the pole length shortened. In the rotor, heat flows more readily along the laminations than across them, making radial ducts less effective per unit surface, and their omission gives a greater active rotor core length and a simpler mechanical construction.

Some machines have to work in a corrosive atmosphere. As good commutation relies on the formation and maintenance of an oxide film or skin on the copper, total enclosure of the brushgear or of the whole machine may be necessary. Total enclosure has the merit that the windings are well protected, but the cooling is restricted: while there is little difference in the dimensions of open and enclosed machines in ratings up to about 15 kW, it becomes increasingly difficult to obtain enough cooling surface for large ratings.

The choice of enclosure is naturally related to the environment, but there are other factors to be considered. The completely open machine is rarely justified in industry because of the danger of some foreign object falling into it. Where the ambient air is laden with explosive gases, the machine must either be flameproof or located outside in a safe area. The most common enclosure is the drip-proof, which can give adequate protection with minimum interference with cooling. Another common enclosure is the screen-protected form. Pipe-ventilated motors are suitable for conditions such as those encountered in sawmills. Machines to be operated in the open air are usually both totally enclosed and weatherproof; the alternative of providing water-tight enclosing boxes is not so effective and may render maintenance more difficult.

The nomenclature for machine enclosure is embodied in the international IP and IC Codes (BS 4999), indicating respectively the degree of protection and the type of cooling employed.

## Frame

Fig 13.12 shows typical forms, (*a*) for low-power and (*b*) for medium-power ratings. The frame contains the field system, provides bearings for the rotor, and accommodates feet or baseplate, lifting lugs, cable entry, air inlet and outlet ports, and exterior-mounted air-blowers. A tacho-generator is often

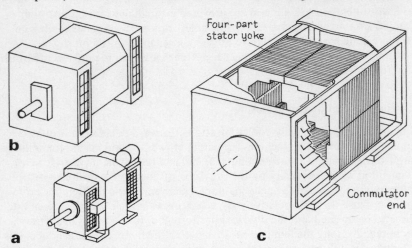

13.12 Typical frame and carcase forms

provided at the non-drive end. The frame in (*c*) is based on a welded structure of rectangular beams, closed at each end by steel-plate or cast-iron bearing end-shields. Sheet-steel panels give ready access to the brush-holders and commutator, and convenient mounting of the blower, air ports and fittings is facilitated. Ball or roller bearings are commonly used, the axial thrust being taken by a fixed bearing at the non-drive end. The arrangement is adaptable to most kinds of enclosure.

## Selection

The choice of a motor for an industrial drive requires a proper evaluation of the duty demanded. An overestimate of the rating may be justified in the interests of reliability but is disadvantageous for a duty that involves frequent starting. Among the conditions to be determined are (i) the supply available and its starting-current limitation, (ii) the ambient temperature, (iii) restraints on mounting or on physical size, (iv) operational hazards (such as explosive dust, corrosive chemicals, oil or water vapour, gaseous atmospheres, carbon dust), (v) speed and speed-range, (vi) power and duty-cycle, (vii) number of starts and stops per hour, and load inertia, (viii) acceleration and retardation rates, and stopping accuracy, (ix) severity of mechanical conditions that impose radial or axial bearing loads. With this information a manufacturer can advise on the most suitable motor types and options.

Although a wide range of enclosures is available, the tendency is to favour total enclosure where feasible, as improved cooling-fan design and heat-exchange techniques have enlarged the capabilities of totally-enclosed machines.

Where additional drive elements (gears, speed-change devices, clutches, brakes etc.) are required, selection is more complex. In general it is less costly and operationally more satisfactory to specify a complete 'package' rather than to assemble unmatched drive elements, with consequent additional cost, space demand and maintenance.

Frequent starting and stopping is a growing requirement, and fail-safe stopping is mandatory for certain high-risk drives under the factory regulations. Heating is more severe during starting, reversing or overload, and a proper assessment of cooling requires an analysis of the load, which may sometimes conform to one of the standard duty cycles set out in Sect. 10.7.

A high rated speed reduces the frame size for a given power, but in some cases a low-speed motor can reduce the overall drive ratio needed in a gear-box to achieve very low speeds; the per-unit starting current is usually lower, and starting and stopping rates can be higher because of the reduction in kinetic energy storage, permitting more starts per hour.

## 13.7 DESIGN

A d.c. machine must be a 'machine' in the mechanical as well as in the electrical sense. It must have a tough protective frame with bearings to with-stand axial and radial thrust, a shaft stiff enough to limit deflection, twisting and eccentricity, a rotor able to accommodate rotational and clamping stresses and vibration, a well-clamped stack of stator laminations, and a rigid and truly circular commutator.

Fig. 13.13 shows diagrammatically a section through a conventional

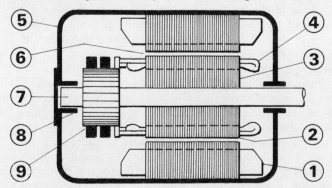

13.13 Design areas
    (1) Field windings. (2) Armature slots. (3) Armature core.
    (4) Armature winding. (5) Frame/carcase. (6) Airgap.
    (7) Shaft. (8) Bearings. (9) Commutator and brushgear.

industrial machine, with its principal constructional regions numbered. The electrical design areas are the magnetic circuit, the field and armature windings, the commutator and brushes, the insulation and the cooling system. There is an economic conflict for space between constructional, magnetic, conducting and insulating materials, and the ventilating channels for the coolant. Design limitations arise from the inherent material properties, such as magnetic saturation and the temperature-rise constraints on the insulation; and the commutator and brushgear present special problems of current collection, brush-wear and sparking. Many design parameters (such as the flux and current densities, the speed and the pole number) are interrelated, but the analytical correlations between them are often difficult to formulate.

A saleable machine must be optimized in relation to some specified objective. This is not necessarily 'minimum first cost': it may be a value assessment of many criteria that identify the 'best' design for a given duty. A single criterion, such as minimum mass or lowest manufacturing cost, might be subject to reasonable definition, but more often the weighted combination of many conflicting requirements has to be considered. Such formulations are inherently less amenable to quantization than are analyses of technical performance. The problems are aggravated by the practice, now normal, of designing machines in ranges, covering a wide span of ratings but with a number of standard features in common. Optimum design must then be related to the complete range rather than to a single machine.

Digital computation has not, so far, achieved completely automated design. However, *computer-aided design,* based on analytical and empirical formulations relating the variables in certain regions, has been comprehensively employed. Sub-system programs specifying performance and design parameters are run, and subsequently matched to evolve the complete design. Some design parameters are continuously variable, but others (such as standard frame dimensions, slot sizes, numbers of turns) can have only discrete values. There will also be absolute limits of dimensions or temperature-rise or mechanical stress that must be observed, and the sub-system programs must take account of them. After assembling the sub-systems, the overall design and performance is investigated for conformity with the master specification.

The computer accelerates the design process enormously. It makes possible 'trial' designs, and enables sophisticated calculations to be made without intolerable tedium and excessive time. It can store a great deal of technical data, and can facilitate the expression of complicated interrelations by evolving them from empirical data, so making possible the development of new and more comprehensive and sophisticated design procedures.

We do not deal here specifically with computer design methods. In the following examples a step-by-step evaluation is adopted, it being understood that a great deal of the longhand calculation is replaced in practice by computer print-outs.

## 13.8 DESIGN: PERMANENT-MAGNET MACHINES

Small p.m. motors are supplied from primary or secondary batteries, from an a.c. network through rectifiers, or from logic circuits, and it is necessary to take account of the internal resistance of the source. Friction loss is relatively large, especially at high speed, and the system efficiency tends to be low (e.g., 0.3 p.u. for a 1 W motor, 0.8 p.u. for a 50 W machine). The mechanical drive may be direct but is more often through step-down gearing or belt because, within limits imposed by commutation and brush wear, a high-speed motor achieves a better utilization of the structural material. The torque efficiency of a gear train is typically 0.6–0.7.

The necessary electromagnetic conversion power $P$ is estimated from the useful load power $P_l$ by $P = P_l/\eta_e\eta_m$, where the motor and drive efficiencies $\eta_e$ and $\eta_m$ have to be initially assumed. It would not be unusual in a small machine to find that the gross electrical power was three times the useful load power. There are also in general some additional requirements to be met in respect of starting torque, battery drain, ambient temperature and temperature-rise.

Let the motor be required to develop a conversion power $P$ at a speed $\omega_1$ when supplied from a source of internal e.m.f. $E$, and let $R$ be the combined resistance of the source, armature, brushes and ballast resistor (if used). Then the armature current is $I = (E - E_r)/R$, where the rotational e.m.f. $E_r = (\Phi N)\omega_1/\pi$ for a 2-pole machine with a useful flux $\Phi$ per pole and an armature winding of $N$ turns, eq. (2.6). The conversion power is $P = E_r I$ and the corresponding electromagnetic torque is

$$M_e = (\Phi N)(1/\pi R)[E - (\Phi N)(\omega_1/\pi)] \tag{13.1}$$

For the speed $\omega_1$ the conversion power is a maximum when $E_r = \frac{1}{2}E$ and $I = \frac{1}{2}E/R$. The electrical efficiency is then 0.5 p.u. and the torque is $M_1 = E^2/4R\omega_1$, whence

$$(\Phi N) = \frac{1}{2}E\pi/\omega_1$$

determines the flux-turn product required.

On *no load* the speed approaches $\omega_0 = 2\omega_1$. At *zero speed* (start or stall) the torque from eq. (13.1) is $M_{es} = (\Phi N)E/\pi R = 2M_1$. The performance characteristics resemble those in Fig. 11.3. It is not, however, always possible to operate a motor at its maximum-power point, as this may be in conflict with the requirements specified for starting torque.

Various arrangements, Fig. 3.4, of ferrite magnets can be adopted. Commonly the magnets are fixed by adhesive to the inner wall of a steel tube thick enough to limit the yoke density to about 1.4 T. Manufacturing tolerances on ferrite are such that imperfect contact with the tube wall may reduce the available flux. The segment span is about $120°$. The flux depends on the physical dimensions and material of the magnet segments, on the gap length (typically 0.5 mm) and on the armature slotting. It is usual to fit magnets of axial length greater than the armature core length to increase the useful flux. The magnetic working point of ferrite may be affected by the sequence of assembly: in some cases the p.m. must be magnetized after the armature is in position, and its flux is impaired if the armature is withdrawn. In operation the flux is reduced by armature reaction but in small machines the effect is usually negligible.

The important parameters ($\Phi N$) and $R$ are interrelated. An increase in $N$ may demand a smaller conductor section in the armature winding with a consequent increase in $R$. Fewer slots make better use of the winding space, but introduce fluctuation in both torque and armature resistance between brushes. The armature design is worked out by optimizing $N$ and $R$ in relation to a core diameter $D$ and length $l$ with the assumption of a flux per pole of an available p.m. system. As small p.m. motors are often made in large production runs, the design has to provide for simplicity of manufacture, ease of assembly and economy of material.

EXAMPLE 13.1: Permanent-magnet 2-pole motor to operate from a 12 V secondary battery and to drive an automobile auxiliary (e.g. a windscreen wiper). Specified output through a 40/1 gear drive:

Run: 1.80 N-m at 50 r/min (5.2 rad/s) in 20°C ambient.
Start: 10 N-m at 20°C, 8 N-m at 70°C, 12 N-m at −10°C.

*Efficiencies* The gear-train efficiency is assumed to be 0.65 p.u. The motor power efficiency at rated running speed $\omega_1 = 5.2 \times 40 = 210$ rad/s is assumed to be 0.60 p.u., and the torque efficiency at start (when the core, windage and friction effects are small) as 0.96 p.u.

*Gross Torques* With the assumed efficiencies, the motor torque demands are

Run (20°C): $1.8/(40 \times 0.60 \times 0.65) = 0.115$ N-m
Start (20°C): $10/(40 \times 0.96 \times 0.65) = 10/25 = 0.40$ N-m
(70°C): $8/25 = 0.32$ N-m. (−10°C): $12/25 = 0.48$ N-m.

The starting torques for 20, 70 and −10°C are respectively in the ratio 1/0.8/1.2.

*Temperature Effects* The flux of a permanent-magnet ferrite varies with temperature, the coefficient being typically −0.002 per °C. The armature-circuit resistance $R$ includes the battery (0.01 $\Omega$ at 20°C), the armature winding, the brush drop (0.2−0.6 V with metal-carbon) and the buffer resistor. The combination is assumed to have an overall resistance-temperature coefficient of 0.004 per °C. The gross starting torque $M_{es}$ of the motor is proportion to $\Phi I = \Phi E/R$. The relative values of flux, resistance and starting torque at the three specified working temperatures are therefore

(20°C) $\Phi$, $R$, $M_{es}$; (70°C) $0.9\Phi$, $1.2R$, $0.75M_{es}$; (−10°C) $1.06\,\Phi$, $0.88\,R$,
$$1.20\,M_{es}$$

The torques are in the ratio 1/0.75/1.2. It is therefore necessary to raise the starting torque at 20°C to $0.40(0.8/0.75) = 0.43$ N-m in order to achieve the torque specified for 70°C. The following calculations are therefore based on this figure and evaluated for 20°C.

*Starting Condition* For zero speed, $E_r = 0$. Applying eq. (13.1) gives $M_{es} = (\Phi N)E/\pi R$, i.e. $0.43 = 3.80 \, (\Phi N)/R$. This yields the relation $R = 8.8 \, (\Phi N)$.

*Running Condition* Eq. (13.1) with $M_e = 0.115$ N-m, $E = 12$ V and $\omega_r = 210$ rad/s gives

$$0.115(\pi R) = (\Phi N) \, [12 - 67 \, (\Phi N)]$$

and with $R = 8.8 \, (\Phi N)$ this results in $(\Phi N) = 0.132$ Wb-t, $R = 1.2 \, \Omega$. The running torque is much less than the torque, $0.43/2 = 0.21$ N-m, for the maximum-power condition, and the efficiency will be substantially above 0.5 p.u.

*Magnet* It is now necessary to select magnets to provide the flux. A pair of ferrite magnets, of material with a remanent flux density $B_r = 0.38$ T and with the dimensions shown in (*a*) of Fig. 13.14, have the flux graphed in (*b*)

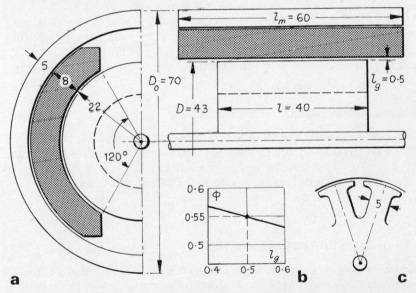

a       b       c

13.14 Permanent-magnet 12 V motor: Example 13.1 (full size)

when used with an *unslotted* rotor and a gap length $l_g$, the flux falling with increase of the gap length. The segment dimensions within a steel tube of wall thickness 5 mm determine the overall diameter $D_o = 70$ mm and the magnet length $l_m = 60$ mm.

*Useful Flux* With an airgap $l_g = 0.5$ mm the armature diameter is $D = 43$ mm. The core length has to be found by trial, leading to the choice of $l = 40$ mm. The ratio $l_m/l$ is 1.5, and the greater magnet length is estimated to raise the useful flux by a factor of 1.25. Armature slotting will, however, increase the effective airgap length by, say, 5%. The useful flux is therefore

$\Phi = 0.55 \times 1.25 \times 0.95 = 0.65$ mWb

from which, with $(\Phi N) = 0.132$ Wb-t, the number of armature turns required is $N = 0.132 \times 10^3/0.65 = 200$.

*Armature Winding* A 2-pole lap winding in 10 slots with 40 conductors per slot gives $N = 200$ turns. The mean length of turn is about 180 mm and the length of conductor in each armature path is $0.18 \times 100 = 18$ m. With copper wire of diameter 0.70 mm the resistance per path is $0.017 \times 18/0.365 = 0.84 \Omega$, so that the total armature resistance is $0.42 \Omega$. The brushes, connections and ballast resistor make up the total $R = 1.2 \Omega$.

*Slot* The conductor area per slot is $40 \times (\pi/4) \times 0.70^2 = 13.4$ mm$^2$. Estimating the slot space factor as about 0.38, the slot area is 40 mm$^2$. A suitable design with parallel-sided teeth is given in Fig. 13.14(*c*).

*Performance* For the running condition at 20°C, eq. (13.1) gives $E_r = (\Phi N)(\omega_1/\pi) = 0.13 \times 210/\pi = 8.7$ V, whence the current is $I = (E - E_r)/R = 2.75$ A. The gross electrical power is $12 \times 2.75 = 33$ W; the load power is provided by a torque of 1.8 N-m at 5.2 rad/s corresponding to 9.4 W. The overall efficiency is $9.4/33 = 0.29$ p.u. and the electrical efficiency is $E_r/E = 0.72$ p.u. The starting current is $E/R = 10$ A, developing

$$M_{es} = (\Phi N) E/\pi R = 0.42 \text{ N-m}$$

which is substantially that specified for 20°C.

*Armature Parameters* Circuit resistance, $R = 1.2 \Omega$; inductance, $L_a = 1.06$ mH from eq. (2.12); electrical time-constant, $\tau_e = 0.88$ ms; inertia, $J = 1.7 \times 10^{-3}$ kg-m$^2$; torque and e.m.f. constants, $k_t = k_e = 8.7/210 = 0.041$ V per rad/s or N-m per A; electromechanical time-constant, $\tau_{em} = RJ/k_t k_e = 1.2$ ms.

## 13.9 DESIGN: INDUSTRIAL MACHINES

Modern motors for industrial applications are built to the standards of IEC72, which has a coherent range of main structural dimensions with centre heights between 56 and 1000 mm. Some data are tabulated below.

Recommended Ratings and Dimensions of Rotating Machines:
IEC Publications 72 and 72A

| PREFERRED RATINGS (kW): | | | | | | | | | | |
|---|---|---|---|---|---|---|---|---|---|---|
| 0.06 | 0.37 | 2.2 | 15 | 45 | 132 | 220 | 335 | 450 | 600 | 800 |
| 0.09 | 0.55 | 3.7 | 18.5 | 55 | 150 | 250 | 355 | 475 | 630 | 850 |
| 0.12 | 0.75 | 5.5 | 22 | 75 | 160 | 280 | 375 | 500 | 670 | 900 |
| 0.18 | 1.1 | 7.5 | 30 | 90 | 185 | 300 | 400 | 530 | 710 | 950 |
| 0.25 | 1.5 | 11 | 37 | 110 | 200 | 315 | 425 | 560 | 750 | 1000 |

PREFERRED DIMENSIONS (mm):

*Foot-mounted Machines,* distance from shaft centre to mounting surface

56  63  71  80  90   100  112  132  160  180  200  225  250   280
315   355  400  450  500  630  710  800  900  1000

*Flange-mounted Machines,* various flange pitch-circle diameters over the range
55 to 1080

---

To select a motor for an industrial drive it is necessary to evaluate the duty
that the machine is called upon to perform. The standard duty cycles given in
Sect. 10.7 are utilized in determining the thermal rating, with specific
reference to ambient temperature (40°C or less, with higher temperatures
accommodated by derating), frequency of start and stop, and operating
conditions. Further conditions to be determined are: supply available;
restrictions on starting current; limitations on dimensions and mounting;
enclosure (where there are particular hazards such as explosive dust, flame-
proofing, corrosive chemicals, water or oil vapours, or gaseous atmospheres);
speed range; load characteristics (particularly inertia); acceleration and
retardation rates; stopping accuracy; mechanical load conditions likely to
impose sustained or shock stresses (axial or radial) on bearings; cyclic load
fluctuations; vibration and noise.

No meaning can be attached to the *power output* of a machine unless the
conditions in which that output is to be available are fully defined. The rating
is the set of numerical quantities (voltage, current, torque, speed . . .) that is
in accordance with the specified performance parameters.

Motors are designed in coherent ranges, comprising a number of
progressive centre heights selected from the table above. Each centre height
defines a *frame size,* and in each there is a range of core lengths, Fig. 13.15,
to cover variations in the required rating.

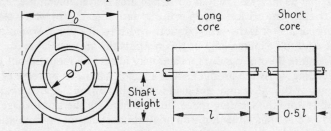

13.15 Main dimensions

**Power and loading**

Eqs. (2.7) and (2.8) show that the converted power is the product of torque
and speed. Although evaluated for a 2-pole machine, eq. (2.8) applies
generally. Consider a heteropolar machine with $2p$ poles, a gap flux $\Phi_g$ per
pole, an armature winding of $N_a$ turns connected in $2a$ parallel paths, and an

angular speed $\omega_r$ (or a rotational speed $n = \omega_r/2\pi$). The motional e.m.f. is given in eq. (4.8), and the converted power is

$$P = E_r I_a = (2p\Phi_g)(N_a/2a)(\omega_r/\pi)I_a$$

where $I_a$ is the armature terminal current. Let the armature have a diameter $D$ and a core length $l$. The total flux crossing the gap is $2p\Phi_g = \pi DlB$, where $B$ is the mean gap flux density, i.e. the specific magnetic loading. The current per armature path is $I = I_a/2a$, and the specific electric loading is $A = I(2N_a)/\pi D$, the linear current-sheet density around the armature periphery. Making the appropriate substitutions gives

$$P = \tfrac{1}{2}\pi D^2 lBA\,\omega_r = \pi^2 D^2 lBAn \tag{13.2}$$

### Design limitations

The d.c. machine has two basic limitations that result from commutation. One is the maximum allowable voltage between sectors under peak conditions of operation; the other is the inductive e.m.f. in the coils undergoing commutation. These are discussed in Sect. 5.7. For a given $D^2l$, reducing $D$ increases the voltage between sectors and the increased $l$ raises the inductive e.m.f. and involves more elaborate cooling of the axially extended armature and field systems. But if high acceleration and retardation rates are demanded, the diameter has to be minimized because the inertia is proportional to $D^4l$.

*Diameter/Length Ratio* Proportioning a machine requires both the magnetic and the electric circuits to be considered together within the limitations imposed by commutation. A typical practical limit for machines of armature diameter $D$ between 0.5 and 3.0 m is $l = 0.6(1 + D)$.

*Magnetic Loading* This is limited by magnetic-circuit saturation. The gap flux density $B$ can be increased only if sufficient area is available in the ferromagnetic paths. The critical region is the armature teeth, for which the significant parameter is the ratio of tooth-width to slot-pitch. Widening the teeth allows $B$ to be increased, but this restricts the area available for accommodating the slot conductors and may result in a reduction of the current loading $A$. As the product $BA$ is a major factor, an increase of one at the expense of the other may not lead to a higher output. A greater slot depth will restore the winding space, but will increase the slot leakage and the e.m.f. induced in coils undergoing commutation. If an optimum relation between $B$ and $A$ can be established — and it is independent of the diameter — the gap density of a range of machines will usually lie between 0.4 and 0.8 T.

*Electric Loading* For a given armature geometry, the slot area is proportional to $D^2$. For a given current density $J$ in the armature conductors, the current is proportional to $D^2$ and the electric loading to $D$. We may therefore write $P = kD^3 lBJ\omega_r$.

Compare two machines with the *same geometry*, the linear dimensions of one being $x$ times that of the other. With $B$ and $J$ the same for each, their conversion powers for the same speed are

$$P = k_1 D^3 l \quad \text{and} \quad P_x = k_1(xD)^3(xl) = k_1 x^4 P$$

On this basis, doubling the linear dimensions gives 16 times the conversion power, i.e. twice the power per unit mass. At the same time, the conductor area varies as $x^2$ and the length as $x$, so that the winding resistance varies as $1/x$, the $I^2 R$ loss as $x^4/x = x^3$, the cooling surface as $x^2$ and the $I^2 R$ loss per unit area as $x$. Thus the larger machine requires a more elaborate cooling system. We might therefore expect the electric loading to rise with the linear dimensions, the slope becoming less steep as heat dissipation becomes more difficult.

In practice, however, the slot depth is *not* proportional to $D$, for it is limited by slot leakage and the commutation inductive e.m.f. The slot depth may be of the order of 30 mm both for a high-speed machine with $D = 2.5$ m and for a standard armature with $D = 0.25$ m. Nevertheless, the specific electric loading can be increased with linear dimensions because the reduced tooth taper permits of wider slots (giving a lower slot permeance) without undue tooth flux density.

A rough indication of the range of $B$ and $A$ is given in Fig. 13.16. The

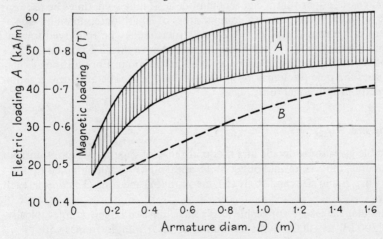

13.16 Typical specific loadings

value chosen for $A$ is influenced by speed, enclosure, insulation and cooling. The upper values are e.g. for machines with forced ventilation and Class F insulation.

*Pole Number* In a given design let the speed $n$, the main dimensions $D$ and $l$, and the loadings $B$ and $A$, be known. There is usually more than one pole number that would be suitable. Choice of the greater number of poles has the following effects:

There are more field and compole windings to be made and fitted.

The frequency $f = np$ of flux pulsation in the armature is increased, affecting the core loss (proportional roughly to $f^2$), especially in the teeth. Generally $f$ lies between 25 and 50 Hz, but may be more in small motors, typically high-speed series motors with low gap flux densities.

The smaller m.m.f. per field coil may lead to less field-conductor material. The shorter pole-pitch reduces the inactive overhang length of the armature coils.

The armature-reaction m.m.f. $F_a = N_a I_a / 4ap$ is reduced, which may obviate the need for compensating windings.

For a lap winding the number of armature coils is proportional to the number of poles, and so is the number of parallel paths, making necessary more equalizer connectors. For a wave winding the number of armature coils is unaffected.

There are more brush arms, a lower current per arm, and a shorter commutator.

There are more and narrower commutator sectors and the brush-arm spacing is reduced, increasing the possibility of flashover.

Commercial design is based on a relatively limited number of standards, in which for a given frame size the number of poles is optimized with respect to the range of designs that it can accommodate. It may sometimes be necessary to fit standard armatures with commutators of length determined by the current rather than the power. A 2-pole structure may be employed for small machines with $D$ up to about 100–150 mm. Above this, a 4-pole design is adopted until the pole-pitch $Y$ exceeds about 350 mm. For multipolar machines the pole-pitch may be about 400–450 mm. The armature reaction is directly proportional to $Y$.

### Frame size

The table gives some details of a modular range of general-purpose shunt (or compound) motors, ratings between 15 and 560 kW being obtained by the use of five frame sizes and short (S), medium (M) and long (L) cores in each frame.

Modular ranges are applicable also to machines of larger rating, typically from 200 kW at 1000 r/min to 2500 kW at 200 r/min, in frames with standard shaft-centre heights between 400 and 900 mm. Important design coefficients concern the torque/mass ratio (which affects economy of material) and the torque/inertia ratio (on which acceleration depends). With given specific magnetic and electric loadings and thermal stresses, the rated torque determines the rotor volume. The torque/inertia ratio for a given rotor volume is raised by an increase in torque and/or a reduction in inertia, the former by a rise in current loading and the latter by a smaller diameter. Both, however, impair the commutation. If the torque is raised by increasing the diameter, the commutation is less adversely affected. There is an optimum relation for a given frame. Fig. 13.17 shows that with a smaller diameter (a)

Typical Range of Shunt or Compound Motors

*Speed* Rated, 1500 r/min; maximum, 3000 r/min; control range, 100/1 by armature voltage, 3/1 by field weakening.

*Torque* Rated, $M_1$; maximum on full excitation and rated voltage (d.c. or r.m.s.) for 10 s, $1.6 M_1$; stall, $M_1$ for 30 s, $0.3 M_1$ for 180 s.

*Current* Rated, $I_1$; intermittent for $t$ sec., $50 I_1/t$.

| Frame size | | | 1 | 2 | 3 | 4 | 5 |
|---|---|---|---|---|---|---|---|
| Output power | (kW) | S | 15 | 32 | 64 | 125 | 290 |
| at 1500 r/min | | M | 19 | 40 | 80 | 155 | 400 |
| | | L | 22 | 50 | 100 | 215 | 560 |
| Maximum torque | (N-m) | S | 97 | 210 | 420 | 810 | 1900 |
| | | M | 124 | 260 | 520 | 1010 | 2600 |
| | | L | 143 | 325 | 650 | 1400 | 3640 |

there is more space for the field system and the mean gap density $B$ can be increased at the expense of the current loading. In (*b*) the field system is more constricted and the gap density is lower, but the electric loading can be raised. An optimum allocation of the diametral dimensions for the frame yields maximum torque.

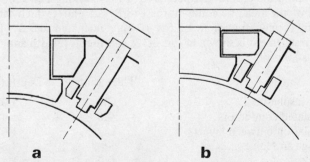

**a**          **b**

13.17 Effects of armature diameter

EXAMPLE 13.2: Industrial separately excited motor, 360 V, 4-pole, 100 kW maximum continuous rating, duty-cycle S1, 1400 r/min ($n$ = 23.3 r/s), screen ventilated, Class B insulation, shaft-centre height 400 mm. The machine is built into a standard frame size in which outputs are adjusted by variation of the active core length, and is designed for operation from a rectified a c. supply.

*Main Dimensions*   Eq. (13.2) relates the rotor dimensions $D$ and $l$, the speed $n$ and the specific loadings $B$ and $A$. Guessing the efficiency to be 0.94 p.u., the armature input power is $100/0.94 = 106.4$ kW and its current is $I_a = 106\,400/360 = 296$ A. Initial estimates of the specific loadings are $B = 0.5$ T and $A = 35\,000$ A/m. Then

$$D^2 l = 106\,400/(\pi^2 \times 0.5 \times 35\,000 \times 23.3) = 0.0264 \text{ m}^3$$

The standard rotor diameter for the specified frame is $D = 0.40$ m, whence the main dimensions are

$$D = 0.40 \text{ m} \quad l = 0.17 \text{ m} \quad Y = 0.32 \text{ m} \quad b = 0.21 \text{ m}$$

which gives a roughly square pole section. The frame could accommodate core lengths up to 0.25 m. The frequency $pn = 4 \times 23.3 = 47$ Hz is not excessive, and no radial ducts are needed. The active (iron) length of the core is $0.17 \times 0.9 = 0.153$ m.

*Armature Winding*   The input current is $I_a = 296$ A. The number of armature turns with $2a$ parallel paths is $N_a = a \cdot \pi DA/I_a = a \times 145$. A simplex wave winding ($a = 1$) has the closing rule given by eq. (4.4), from which $y_c = (C \pm 1)/p = 73$ or 74. With $y_c = 73$, $C = 147$ commutator sectors, $S = 49$ slots and $m = 6$ coil-sides per slot, the number of turns is $N_a = 147$ contained in 147 single-turn coils all of the same span. Thus the three top coil-sides of slot (1) are joined respectively to the three bottom coil-sides of slot (13). The winding pitches are given by $y_c = 2y_r = y_f + y_b$, so that $y_f = y_b = 73$ can be chosen.

*Slots*   The armature current per path is $I = 148$ A. Taking a current density $J = 6$ A/mm$^2$ the area of a conductor is $148/6 = 25$ mm$^2$, which can be provided by copper strap of area $2.5 \times 10$ mm$^2$ insulated to $3.2 \times 10.7$ mm$^2$. The slot arrangement is shown in Fig. 3.13, and the slot width and depth are made up as follows:

|                            | Width (mm) | Depth (mm) |
|----------------------------|------------|------------|
| Slot insulation            | 2.0        | 3.0        |
| Insulated conductors       | 9.6        | 21.4       |
| Insulation between layers  | –          | 1.0        |
| Wedge and lip              | –          | 3.5        |
| Slack                      | 0.4        | 1.1        |
| Total                      | 12.0       | 30.0       |

*Full-load Gap Flux*   This is obtained from eq. (4.9) with $p = 2, a = 1$, $N_a = 147$, $n = 23.3$ r/s, and $E_r$ estimated to be 345 V:

$$\Phi_g = 345/(4 \times 147 \times 23.3) = 25.2 \text{ mWb}$$

*Specific Loadings*   The actual values are now

$$B = 2p\Phi_g/\pi Dl = 0.47 \text{ T} \quad A = N_a I_a/\pi D = 34.5 \text{ k\bar{A}/m}$$

*Magnetic Circuit* The calculation is given in Example 3.2, Sect. 3.7. The full-load excitation required is 5.30 kA-t/pole.

*Field Winding* The field coil design is evaluated in Example 4.1, Sect. 4.3. Each coil has 2210 turns carrying 2.40 A supplied from a separate rectifier.

*Commutator* The commutator diameter $D_c$ must be less than the core diameter $D$ to allow for risers and connections to the armature winding. With a 5 mm width for a sector and its insulation and $C = 147$, then $D_c = 147 \times 5/\pi = 23.5$ mm. The peripheral speed is $u_c = 17.2$ m/s. A brush thickness (peripheral width) of $2\frac{1}{2}$ sector pitches, i.e. 12.5 mm, is suitable. There are four brush arms each carrying 148 A, and with a brush current density of 92 mA/mm$^2$ the brush contact area required per arm is 1600 mm$^2$. The total axial brush length per arm is $1600/12.5 = 128$ mm, satisfied by four brushes each 12.5 mm × 32 mm. The mean voltage between commutator sectors is 9.8 V.

*Compoles* As the load conditions are not severe, it is adequate to use the approximation in eq. (5.8) for the mean inductive e.m.f.:

$$e_{pc} = 4mzI (4l + 0.4\, l_o)/10^6 T_c$$

in which the number of coil-sides per slot is $m = 6$, the number of turns per coil is $z = 1$, the current per conductor is $I = 148$ A, the core length is $l = 0.17$ m and the estimated overhang length is $l_o = 0.34$ m. The effective commutation time, eq. (5.7), is

$$T_c = [\tfrac{1}{2} m - (a/p)]\, t_d + t_c$$

Here $a/p = 1/2$, $t_c = w_b/u_c = 12.5/17.2 = 0.73$ ms, $t_d = w_c/u_c = 5/17.2 = 0.29$ ms. Then $T_c = 1.45$ ms and $e_{pc} = 2.0$ V (mean). The peak inductive e.m.f. is estimated to be 2.5 V. The corresponding peak flux density of the commutating field is $B_{cm} = e_{pcm}/lu = 2.5(0.17 \times 29.3) = 0.50$ T. To stabilize the compole flux, the gap length is made $l_{gc} = 7$ mm, whence the airgap m.m.f. required is $800\,000 \times 0.50 \times 7 \times 10^{-3} = 2.80$ kA-t. The armature m.m.f. in the commutating zone is $N_a I_a/4ap = 5.44$ kA-t/pole. The total compole m.m.f. is therefore $5.44 + 2.80 = 8.24$, say 8.5 kA-t/pole to allow for the ferromagnetic path of the compole magnetic circuit. The number of compole turns per pole is thus $N_c = 8.5 \times 10^3/296 = 28$, formed of copper strap 2.5 mm × 45 mm.

The width of the commutating zone is $w_z = T_c u = 42$ mm. If the circumferential width of the compole face is 45 mm, it will both cover the zone and approximate to two armature slot-pitches. The compole flux is then $\Phi_c = B_{cm} l_c w_z/k_c = 1.9$ mWb with $k_c = 1.2$.

$I^2R$ *Loss* The loss components are calculated for a temperature of 75°C, for which the copper resistivity is 0.021 Ω per m length and per mm$^2$ cross-section.

Armature: The mean length of an armature turn is 1.02 m. The resistance per path is $0.021 \times (147/2) \times 1.02/25 = 0.063$ Ω. Hence $r_a = 0.0315$ Ω and the

loss is $\qquad$ $296^2 \times 0.0315 = 2760$ W
Brush Contact: 1 V per brush $\quad$ $296 \times 2.0 \qquad = 600$ W
Compoles: the mean length of a turn is 0.50 m, the 28 turns/pole being
provided by two coils, the inner of 12 and the outer of 16 turns. The
resistance per pole is $0.021 \times 28 \times 0.50/112.5 = 0.0026$ Ω. For four
compoles in series, $r_{cp} = 0.0104$ Ω, and the loss is
$\qquad$ $296^2 \times 0.0104 = 910$ W
Shunt Field: $\qquad$ $250 \times 2.40 \qquad = 650$ W

*Core Loss* The loss is calculated for the armature core and teeth separately.
Core: Volume 0.0089 m³, mass 70 kg, specific loss (at $B_c = 1.18$ T)
2.1 W/kg, giving the loss $\qquad$ $2.1 \times 70 \qquad = 150$ W
Teeth: Volume 0.0029 m³, mass 23 kg, specific loss (at $B_{t\frac{1}{3}} = 1.86$ T)
6.5 W/kg, giving the loss $\qquad$ $6.5 \times 23 \qquad = 150$ W
Total: 150 + 150 = 300 W + 20% for rectifier ripple = 360 W

*Mechanical Loss* The brush friction loss is estimated to be 200 W, and the
windage and friction to be 500 W, totalling 700 W

*Efficiency* The efficiency by loss summation is now obtainable.

| Losses: | | | Power: | | |
|---|---|---|---|---|---|
| | Armature $I^2 R$ | 2760 | | Output | 100 000 |
| | Brush contact | 600 | | Loss | 5 980 |
| | Compole $I^2 R$ | 910 | | | |
| | Shunt field | 650 | | Input (W) | 105 980 |
| | Core and teeth | 360 | | | |
| | Mechanical | 700 | | Design efficiency: 100/106 = 0.94 p.u. | |
| | Total (W) | 5980 | | Declared efficiency: 0.94 − 0.01 = 0.93 p.u. | |

Losses in the rectifier equipments are separately accounted.

*Temperature-rise* The method of Sect. 10.6 is applied.
Armature: Loss, 2760 + 360 = 3120 W. Peripheral speed $u = 29.3$ m/s. Axial
length of core and overhang, 0.39 m; diameter, 0.40 m; area,
$0.40 \times \pi \times 0.39 = 0.49$ m²; inner surface of back overhang,
$\pi \times 0.35 \times 0.10 = 0.11$ m²; one rotor end surface, $(\pi/4)0.4^2 = 0.12$ m².
Total cooling surface, (0.49 + 0.11 + 0.12) = 0.72 m². Temperature-rise

$$\theta_m = (3120/0.72) [0.03/(1 + 2.93)] = 44°C$$

Commutator: Loss, 200 + 600 = 800 W. Peripheral speed, $u_c = 17.2$ m/s.
Active cylindrical area, $0.235 \times \pi \times 0.165 = 0.12$ m²; half area of riser
annulus, 0.02 m². Total cooling surface, (0.12 + 0.02) = 0.14 m².
Temperature-rise

$$\theta_m = (800/0.14) [0.025/(1 + 1.72)] = 53°C$$

Compole: Loss per pole, 230 W. Exposed cooling surface, 0.085 m².

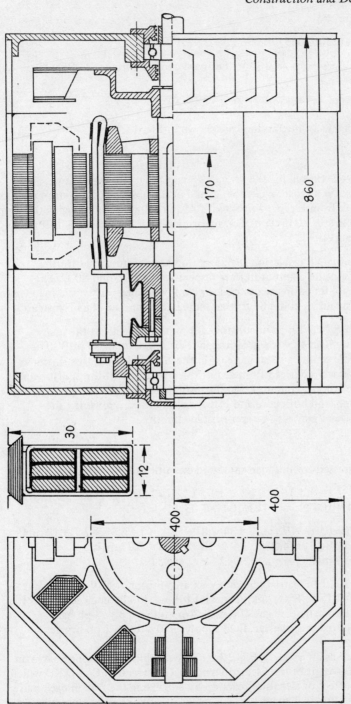

13.18 Separately excited 360 V 100 kW 1400 r/min motor: Example 13.2 (dimensions in mm)

Temperature-rise

$$\theta_m = (230/0.085) \ [0.075/(1 + 2.93)] = 52°C$$

Field Coil: Loss per pole, 130 W. Exposed cooling surface, 0.081 m². Temperature-rise

$$\theta_m = (130/0.081) \ [0.11/(1 + 2.93] = 45°C$$

The main details of the machine are shown in Fig. 13.18.

### Performance

*No-load speed.* On full load, $E_r = V_a - I_a(r_a + r_{cp}) - v = 346$ V and $n = 23.3$ r/s. On no load, $E_r \cong 358$ V, and in the absence of armature reaction the field flux from Fig. 3.14 rises to 1.045 p.u. The no-load speed is therefore $n \times (358/346) \ (1/1.045) \cong n$. The armature reaction produces a differential compounding effect.

*Rating* For Class B insulation the permitted temperature rises for the armature, field and commutator are respectively 80, 70 and 80°C. The machine could therefore be uprated, or employed on a duty-cycle involving overload. The rating could be further increased by use of Class F insulation.

*Rectified Supply* With an uncontrolled 3-ph bridge rectifier the major harmonics are relatively small, although their frequency may be 300 or 360 Hz. There results a small increase in the core and tooth loss. If the armature voltage is derived from a controlled rectifier for the adjustment of speed, the harmonic content is significantly increased, particularly for low-speed running, and both core and $I^2R$ losses are higher, lowering the efficiency. There may also be appreciable vibration.

### Parameters

These are assessed for full-load saturated conditions.

*Shunt Field* $L_f = 4(1.15 \times 25.2 \times 10^{-3}) \ 2210/2.40 = 107$ H; $r_f = 250/2.4 = 104 \ \Omega; \tau_f = 107/104 = 1.03$ s.

*Armature Circuit* $G = E_r/\omega_r I_f = 346/(146 \times 2.40) = 0.985; L_a \cong 12$ mH from eq. (2.12); $L_{cp} = 4(28 \times 1.9 \times 10^{-3})/296 = 0.72$ mH; $L_a + L_{cp} \cong 13$ mH; $r_a + r_{cp} = 0.042 \ \Omega; \tau_a = 0.31$ s.

EXAMPLE 13.3: Industrial separately excited motor, 480 V, 4-pole, 67.5/22.5 kW, 1500/500 r/min, constant torque, totally enclosed, Class F insulation, shaft-centre height 400 mm. Armature supply, 3-ph half-controlled thyristor converter; field supply, 370 V from 1-ph bridge rectifier.

The main details are given in Fig. 13.19. The machine has a split-pole main field system (patent of the Electro Dynamic Construction Co., U.K.) which reduces the effect of armature reaction by equalizing the flux in each half of a pole. It also provides space for four banks of aluminium cooling tubes.

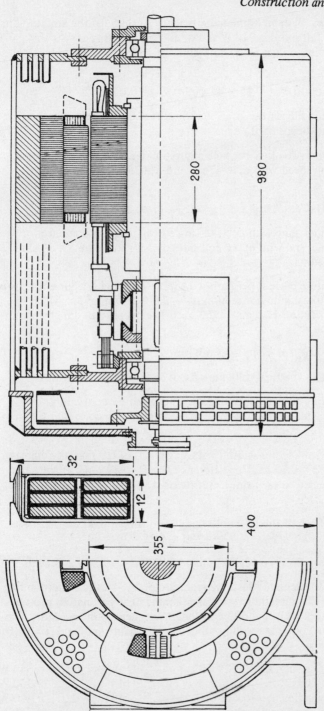

13.19 Separately excited 380 V 67.5/22.5 kW 1500/500 r/min totally enclosed motor: Example 13.3 (dimensions in mm)

*Main Dimensions* With an estimated efficiency of 0.95 for the armature, its input power at full speed is 71 kW and its current is 148 A at 480 V. Assuming specific loadings $B = 0.55$ T and $A = 15$ kA/m (limited by total enclosure), then

$$D^2 l = 71\ 000/(\pi^2 \times 0.55 \times 15\ 000 \times 25) = 0.035\ \text{m}^3$$

The diameter suitable for the given shaft height and split-pole construction is $D = 0.355$ m, whence $l = 0.28$ m.

*Armature Winding* The flux per pole approximates to $\Phi_g = B\pi Dl/2p = 43$ mWb. At the top speed with $E_r \cong 475$ V and $n = 25$ r/s, the number of armature turns is

$$N_a = (a/p)\,475 \times 10^3/(2 \times 25 \times 43) = (a/p)\,221$$

A simplex wave winding with $C = 111$ single-turn coils and commutator sectors is suitable, and with $m = 6$ coil-sides per slot there are $S = 37$ slots. Thus $N_a = 111$ turns.

*Slots* The armature current per path is $148/2 = 74$ A. With a current density $J = 2.6$ A/mm$^2$ the conductor section is $74/2.6 = 28.5$ mm$^2$, provided by copper strap of section 12.5 mm × 2.3 mm. The overall slot dimensions are 32 mm × 12 mm.

*Full-load Gap Flux* This is $\Phi_g = 43$ mWb/pole.

*Specific Loadings* The final values are $B = 0.55$ T, $A = 14.7$ kA/m.

*Magnetic Circuit* This is fully laminated. The armature core is built of 0.35 mm silicon-steel plates, the main and commutating poles of 1.0 mm low-loss steel plates.

*Commutator* With a sector width (including mica) of 6.5 mm, the diameter is $D_c = 111 \times 6.5/\pi = 0.23$ m. The difference $D - D_c = 125$ mm enables fan vanes to be attached to each third riser to assist in dissipating heat from the commutator surface.

*Field Winding* Each of the four coils has 1250 turns of 0.92 mm diameter copper wire. Excitation $F_f = N_f I_f = 1250 \times 1.55 = 1.94$ kA-t.

*Compole Winding* Strip-on-edge winding of 17 turns, 19 mm × 3.2 mm.

*$I^2R$ Loss* This is calculated for a temperature of 115°C, the corresponding copper resistivity being 0.024 Ω per m and mm$^2$. The calculation proceeds as in Example 13.2 with the following lengths of mean turn: armature winding, 1.25 m; compole winding, 0.66 m; main field winding, 1.10 m. The $I^2R$ losses are then

| | | |
|---|---|---|
| armature winding, $r_a = 0.029$ Ω | $148^2$ × 0.029 | = 640 W |
| brush contact | 148 × 2 | = 300 W |
| compole winding, $r_c = 0.018$ Ω | $148^2$ × 0.018 | = 400 W |
| main field winding, $r_f = 200$ Ω | $1.55^2$ × 200 | = 480 W |

The main field circuit is provided with an external rheostat for adjustment of the rated speed.

*Core Loss* The armature core below the teeth carries a flux density of 1.5 T, its mass is 85.5 kg, its specific loss (at 50 Hz) is 6.0 W/kg, giving
$$85.5 \times 6.0 = 510 \text{ W}$$
The mean tooth density is 1.60 T and a specific loss of 7.0 W/kg; the mass is 38.0 kg, whence $\quad 38.0 \times 7.0 = 270$ W.
The total core loss of 780 W, increased by 20% by reason of converter ripple, is then 930 W.

*Mechanical Loss* Estimates give the brush friction as 200 W, and windage and friction (including the fan) as 700 W, totalling 900 W.

*Efficiency* By loss summátion, the efficiency at top speed can now be assessed.

| Losses: | | | Power: | | |
|---|---|---|---|---|---|
| Armature $I^2 R$ | 640 | | Output | 67 500 |
| Brush contact | 300 | | Loss | 3650 |
| Compole $I^2 R$ | 400 | | | ——— |
| Field $I^2 R$ | 480 | | Input (W) | 71 150 |
| Core and teeth | 930 | | | |
| Mechanical | 900 | | Design efficiency: 67.5/71.15 = 0.95 p.u. |
| | ——— | | | |
| Total (W) | 3650 | | Declared efficiency: 0.95 − 0.01 = 0.94 p.u. |

At the bottom speed the frequency of flux pulsation in the armature core and teeth is 17 Hz, and the core loss is about 300 W. The brush friction is reduced to about 70 W and the friction and windage to about 300 W. The $I^2 R$ and brush-contact losses are unaffected. The total loss then sums to 2490 W. The output is 22.5 kW, the input is 25.0 kW, and the design efficiency is 0.90 p.u.

*Cooling* The motor is totally enclosed. The dissipating surface of the carcase is $(\pi \times 0.75 \times 0.98) + 2(\pi \times 0.75^2 /4) = 3.2$ m$^2$, which can dissipate about 12.5 W per m$^2$ per $^\circ$C temperature–rise above ambient − say 1 kW for a rise of 25$^\circ$C. As this is quite inadequate, the machine is provided with 36 aluminium cooling tubes, each 35 mm diameter and 800 mm long (a total cooling surface of 3.2 m$^2$) through which air is blown by a fan within the carcase but separated from the electrical part of the motor as in Fig. 13.19. Heat is circulated within the enclosure by the fanning action of the rotating armature, commutator and riser vanes. The critical condition occurs at the minimum rated speed, and the shaft-mounted fan must be designed to provide a sufficient flow of coolant at this speed.

EXAMPLE 13.4: Outline data are given in the following table and in Fig. 13.20 for two liberally rated industrial machines −
A.    Shunt generator, 350 kW (cont.), 440 V, 600 r/min, natural cooling
B.    Shunt motor, 110 kW (cont.), 500 V, 450 r/min, fan cooled, with a

series stabilizing winding to reduce the field-weakening effect of armature reaction and to provide a slightly drooping speed/load characteristic

The efficiencies and shunt-field currents have first to be estimated.

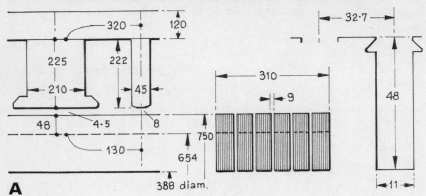

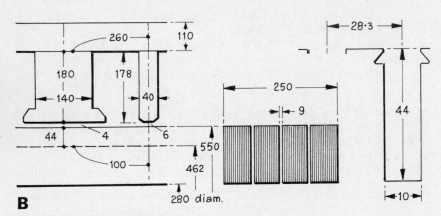

13.20 Main dimensions: Example 13.4

| Rating | | A | B |
|---|---|---|---|
| Output | kW | 350 | 110 |
| Voltage | V | 440 | 500 |
| Speed | r/min | 600 | 450 |
| Efficiency | p.u. | 0.95 | 0.92 |
| Input | kW | 368 | 120 |
| Terminal current | A | 795 | 240 |
| Shunt field current | A | 5 | 2 |
| Armature current | A | 800 | 238 |
| Number of poles | — | 6 | 6 |

## Main Dimensions

| | | | |
|---|---|---|---|
| Specific magnetic loading | T | 0.64 | 0.60 |
| Specific electric loading | kA/m | 32.6 | 34.2 |
| Armature diameter | m | 0.75 | 0.55 |
| Core length (gross) | m | 0.31 | 0.25 |
| Peripheral speed | m/s | 23.6 | 13.0 |
| Armature frequency | Hz | 30 | 22.5 |
| Pole-pitch | m | 0.393 | 0.288 |
| Pole-arc | m | 0.275 | 0.202 |
| Gap flux per pole | mWb | 78.4 | 43.0 |

## Armature Winding

| | | | |
|---|---|---|---|
| Type | — | lap | wave |
| Number of 1-turn coils | — | 288 | 244 |
| Number of slots/conductor per slot | — | 72/8 | 61/8 |
| Current per conductor | A | 133 | 119 |
| Current density | A/mm$^2$ | 4.35 | 4.96 |
| Conductor size | mm$^2$ | 18 × 1.7 | 16 × 1.5 |
| Length of turn | m | 1.78 | 1.40 |
| Resistance between brushes (hot) | mΩ | 9.7 | 73 |
| M.M.F. per pole | A-t | 6400 | 4950 |

## Magnetic Circuit

| | | | |
|---|---|---|---|
| Core: number of ducts/width | mm | 5/9 | 3/9 |
| net length | m | 0.265 | 0.223 |
| iron length | m | 0.236 | 0.198 |
| path area | m$^2$ | 0.032 | 0.018 |
| flux density | T | 1.22 | 1.19 |
| path length | m | 0.13 | 0.10 |
| Teeth: number per pole-arc | — | 8.4 | 7.1 |
| effective area (at 1/3) | m$^2$ | 0.0375 | 0.0215 |
| flux density | T | 2.09 | 2.00 |
| path length | m | 0.048 | 0.044 |
| Pole: leakage factor | — | 1.2 | 1.2 |
| effective flux | mWb | 94 | 51.6 |
| area | m$^2$ | 0.060 | 0.033 |
| flux density | T | 1.67 | 1.56 |
| path length | m | 0.225 | 0.180 |
| Yoke: area | m$^2$ | 0.036 | 0.020 |
| flux | mWb | 47 | 25.8 |
| flux density | T | 1.30 | 1.29 |
| path length | m | 0.32 | 0.26 |
| Airgap: length | mm | 4.5 | 4.0 |
| opening/gap coefficient | — | 1.13 × 1.05 | 1.13 × 1.04 |
| effective gap length | mm | 5.31 | 4.70 |
| flux density | T | 0.920 | 0.851 |

## Excitation per Pole

| | | | |
|---|---|---|---|
| Core | A-t | 50 | 30 |
| Teeth | A-t | 1250 | 1100 |
| Pole | A-t | 890 | 360 |
| Yoke | A-t | 130 | 100 |
| Gap | A-t | 3950 | 3200 |
| Armature reaction (15%) | A-t | 960 | 740 |
| Total f.l. m.m.f./pole | A-t | 7230 | 5530 |
| Shunt field m.m.f. | A-t | 7230 | 4800 |
| Series field m.m.f. | A-t | 0 | 730 |

## Shunt Field Coil

| | | | |
|---|---|---|---|
| Voltage | V | 64 | 83.3 |
| Current | A | 5.30 | 2.17 |
| Number of turns | — | 1360 | 2240 |
| Conductor diameter | mm | 1.90 | 1.20 |
| Conductor area | mm$^2$ | 2.83 | 1.13 |
| Mean length of turn | m | 1.20 | 0.94 |
| Resistance (hot) | Ω | 12.1 | 38.3 |
| Cooling surface | m$^2$ | 3.06 | 1.80 |

## Series Field Coil

| | | | |
|---|---|---|---|
| Current | A | — | 238 |
| Number of turns | — | — | 3 |
| Conductor area | mm$^2$ | — | 110 |
| Mean length of turn | m | — | 0.95 |
| Resistance (hot) | mΩ | — | 0.55 |

## Commutator

| | | | |
|---|---|---|---|
| Diameter | m | 0.55 | 0.40 |
| Peripheral speed | m/s | 17.3 | 9.4 |
| Number of sectors | — | 288 | 244 |
| Sector pitch | mm | 6.00 | 5.15 |
| Volts between sectors (mean/max) | V | 9.2/12.9 | 12.3/17.5 |

## Brushgear

| | | | |
|---|---|---|---|
| Current per brush-arm | A | 267 | 79 |
| Number of brushes per arm | — | 8 | 4 |
| Contact area per brush | mm$^2$ | 15.9 × 31.7 | 12.7 × 31.7 |
| Contact area per arm | mm$^2$ | 4040 | 1610 |
| Current density | mA/mm$^2$ | 67 | 49 |
| Brush pressure | mN/mm$^2$ | 12 | 14 |

## Compole

| | | | |
|---|---|---|---|
| Axial length | m | 0.16 | 0.14 |
| Circumferential width | m | 0.045 | 0.040 |

| Flux density | T | 0.35 | 0.35 |
|---|---|---|---|
| Airgap | mm | 8 | 6 |
| Airgap coefficient | — | 1.10 | 1.10 |
| Net m.m.f. | A-t | 2440 | 1850 |
| Armature m.m.f. | A-t | 6400 | 4950 |
| Compole m.m.f. | A-t | 8840 | 6800 |
| Current | A | 800 | 238 |
| Conductor area | mm$^2$ | 350 | 110 |
| Number of turns | — | 11 | 28 |
| Mean length of turn | m | 0.50 | 0.42 |
| Resistance (hot) | mΩ | 0.33 | 2.24 |

## Core Loss

| Frequency | Hz | 30 | 22.5 |
|---|---|---|---|
| Armature core: flux density | T | 1.22 | 1.19 |
|   specific loss | W/kg | 2.4 | 1.8 |
|   mass | kg | 410 | 172 |
|   loss | W | 980 | 310 |
| Teeth: flux density | T | 2.09 | 2.09 |
|   specific loss | W/kg | 6.8 | 5.0 |
|   mass | kg | 120 | 63 |
|   loss | W | 820 | 320 |
| Pole-face pulsation loss | W | 720 | 250 |
| Total core loss | kW | 2.52 | 0.88 |

## $I^2R$ Loss

| Armature | kW | 6.21 | 4.14 |
|---|---|---|---|
| Shunt field | kW | 2.33 | 1.09 |
| Series field | kW | 0 | 0.19 |
| Compole field | kW | 1.27 | 0.76 |
| Total $I^2R$ loss | kW | 9.81 | 6.18 |

## Other Losses

| Brush contact | kW | 1.60 | 0.48 |
|---|---|---|---|
| Brush friction | kW | 0.54 | 0.27 |
| Bearing friction and windage | kW | 2.80 | 1.60 |
| Stray load | kW | 1.66 | 0.65 |
| Total other loss | kW | 6.60 | 3.00 |

## Performance (f.l.)

| Rotational e.m.f. | V | 451 | 477 |
|---|---|---|---|
| Output | kW | 350 | 110 |
| Loss summation | kW | 18.9 | 10.1 |
| Input | kW | 368.9 | 120.1 |
| Efficiency | p.u. | 0.95 | 0.915 |

# References

1. Say, M. G.: *Introduction to the unified theory of electromagnetic machines* (Pitman, 1971); *Alternating current machines* (Pitman, 1976).
2. Nasar, S. A.: Electromagnetic theory of electrical machines, *Proc. IEE*, **111**, 1123 (1964).
3. Lynch, R.: Development of samarium-cobalt permanent-magnet d.c. servomotors, *IEE Conf.* **136**, 5 (1976); Gould, J.E.: Permanent magnets, *Proc. IEE*, **125** (No. 11R), 1137 (1978).
4. Binns, J. K., Jabbar, M. A. and Bernard, W. R.: Computation of the magnetic field of permanent magnets in iron cores, *Proc. IEE*, **122**, 1377 (1975); Reichert, K. and Freundl, H.: Computer-aided design of permanent-magnet systems, *Brown Boveri Rev.*, **59**, 469 (1972).
5. Macfadyen, W. K., Simpson, R. R. S., Slater, R. D., Teape, J. W. and Wood, W. S.: Representation of magnetization curves by exponential series, *Proc. IEE*, **120**, 902 (1973); **121**, 1019 (1974).
6. Carter, F. W.: Magnetic field of the dynamo-electric machine, *JIEE*, **64**, 1115 (1926).
7. Campbell, P.: Principles of a permanent-magnet axial-field d.c. machine, *Proc. IEE*, **121**, 1489 (1974).
8. Green, C. W. and Paul, R. J. A.: Application of d.c. linear machines as short-stroke and static actuators, *Proc. IEE*, **116**, 599 (1969); Performance of d.c. linear machines based on an assessment of flux distribution, *Ibid*, **118**, 1413 (1971).
9. Hague, B.: *Electromagnetic problems in electrical engineering* (Dover, 1955).
10. Carpenter, C. J.: Application of the method of images to machine and winding fields, *Proc. IEE*, **107A**, 467 (1960).
11. Mukherjee, K. C. and Neville, S.: Magnetic permeance of identical double slotting, *Proc. IEE*, **118**, 1257 (1971).
12. *Proceedings of the international conference on stepping motors and systems* (University of Leeds, 1974 and 1976).
13. Grimshaw, K. P.: Electromagnetic principles of commutation in d.c. machines, *Int. Jnl. Elec. Eng. Educn.*, **2**, 59 (1964).
14. Taylor, P. L.: Improving commutation in d.c. machines by the use of flux traps, *Proc. IEE*, **117**, 1269 (1970).
15. Erdelyi, E. A.: Nonlinear magnetic field analysis of d.c. machines, *Trans. IEEE*, PAS-88, 1546 (1970).

16. Bates, J. J. and Stanway, J.: Development of a 300 kW.3000 r/min d.c. machine using thyristor-assisted commutation, *Proc. IEE*, **123**, 76 (1976).

17. Jones, C. V.: Analysis of commutation for unified machine theory, *Proc. IEE*, **108C**, 1 (1958).

18. Gibbs, W. J.: *Electric machine analysis using matrices* (Pitman, 1962).

19. Barney, G. C.: Method of rapidly reversing large currents in highly inductive loads, *Proc. IEE*, **113**, 2031 (1966).

20. Mellitt, R. and Rashid, M. H.: Analysis of d.c. chopper circuits by computer-based piecewise linear technique, *Proc. IEE*, **121**, 173 (1974).

21. Dubey, G. K. and Shepherd, W.: Analysis of a d.c. series motor controlled by power pulses, *Proc. IEE*, **122**, 1397 (1975); Transient analysis of a d.c. motor controlled by power pulses, *Ibid*, **124**, 229 (1977).

22. Mazda, F. F.: *Thyristor control* (Newnes-Butterworths, 1973); Ramshaw, R. S.: *Power electronics* (Chapman and Hall, 1973); *Power engineering using thyristors* (Mullard, 1970).

23. Holmes, P. G.: Speed control of a d.c. shunt motor using silicon controlled rectifiers, *Int. Jnl. Elec. Eng. Educn.*, **2**, 555 (1965); Merrett, J.: Thyristor control of a d.c. motor from a 1-ph supply, *Mullard Tech. Commn.*, No. 80 (1966).

24. *Electricity Council Engineering Recommendations* G5/2 (1976).

25. Farrer, W. and Andrew F. F.: Fully-controlled regenerative bridges with half-controlled characteristics, *Proc. IEE*, **125**, 109 (1978).

26. Rudenberg, R.: *Transient performance of electric power systems* (McGraw-Hill, 1950); Shegekazu *et al.*: Advanced theory of the delay of the commutating-pole flux of a d.c. machine, *Proc. IEE*, **125**, 857 (1978).

27. Gallagher, P. J., Barrett, A. and Shepherd, W.: Analysis of 1-ph rectified thyristor-controlled load with integral-cycle triggering, *Proc. IEE*, **117**, 409 (1970).

28. Franklin, P. W.: Theory of a d.c. motor controlled by power pulses, *Trans. IEEE* PAS-91, 242 (1972); Parimelalagan, R. and Rajagopalan, V.: Steady-state investigation of a chopper-fed d.c. motor with separate excitation, *Trans. IEEE* IGA-7, 101 (1971).

29. McLellan, P. R.: Thyristor chopper using a bridge-connected capacitor for commutation, *Proc. IEE*, **122**, 514 (1975).

30. Mangan, M. F. and Griffith, J. T.: Motors and controllers for electric cars, *Elec. Cncl. Research Cntr.*, Rpt. M238 (1970).

31. Tustin, A. and Bates, J. J.: Temperature-rises in electrical machines on variable load and with variable speed, *Proc. IEE*, **103A**, 483 (1956); Temperature-rises in electrical machines as related to the properties of thermal networks, *Ibid*, **103A**, 471 (1956); Tustin, A., Nettell, D. F. and Solt, R.: Performance and heating curves for motors on short-run duties, *Ibid*. **103A**, 493 (1956).

32. *Electrical performance of rotating electrical machines*, BS 2613 (1970).

33.    Hanitsch, R. and Meyna, A.: Buerstenloser Gleichstrommotor mit digitaler Ansteuerung, *ETZ-A*, **4**, 204 (1976).

34.    Lawrenson, P. J. and Kingham, I. E.: Resonance effects in stepping motors, *Proc. IEE*, **124**, 445 (1977); Viscously coupled inertial damping of stepping motors, *Ibid*, **122**, 1137 (1975).

35.    Hughes, A. and Lawrenson, P. J.: Electromagnetic damping in stepping motors, *Proc. IEE*, **122**, 819 (1975).

36.    Hammad, A. E. and Mathur, R. M.: Dynamic performance of permanent-magnet stepping motors, *Proc. IEE*, **124**, 651 (1977).

37.    Ellis, P. J.: Analysis and control of the permanent-magnet stepping motor, *Radio and Electronic Eng.*, **41**, 302 (1971).

38.    Pickup, I. E. D. and Tipping, D.: Method for predicting the dynamic response of a variable-reluctance stepping motor, *Proc. IEE*, **120**, 757 (1973); Prediction of pull-in rate and settling time characteristics of a variable-reluctance stepping motor, *Ibid*, **123**, 213 (1976).

39.    Pasek, Z.: Novy zpusob urcem zakladnich dynamickych parametru steinosmerneho motoru, *Elektrotech. Obzor.*, **51**, 109 (1962); Lord, W. and Hwang, J. H.: D.C. servomotors, modelling and parameter determination, *IEEE Pubn.* No. 73, CHO 763-31A, 69 (1973).

40.    Schieber, D.: Unipolar induction braking of thin metal sheets, *Proc. IEE*, **119**, 1499 (1972); Force on a moving conductor due to a magnetic pole array, *Ibid*, **120**, 1519 (1973); Braking torque on a rotating sheet in a stationary magnetic field, *Ibid*, **121**, 117 (1974).

41.    Venkataratnam, K. and Ramachandra Raju, D.: Analysis of eddy-current brakes with nonmagnetic rotors, *Proc. IEE*, **124**, 67 (1977).

42.    Singh, A.: Theory of eddy-current brake with thick rotating disc, *Proc. IEE*, **124**, 373 (1977).

43.    Carpenter, C. J.: Finite-element network models and their application to eddy-current problems, *Proc. IEE*, **122**, 455 (1975).

44.    Bloxham, D. A. and Wright, M. T.: Eddy-current coupling as an industrial variable-speed drive, *Proc. IEE*, **119**, 1149 (1972).

45.    Lawrence, P. J. and Ralph, M. C.: General three-dimensional solution of eddy-current and Laplacian fields in cylindrical structures, *Proc. IEE*, **117**, 467 (1970).

46.    Davies, E. J.: Experimental and theoretical study of eddy-current couplings and brakes, *Trans. IEEE*, PAS-82, 401 (1963); General theory of eddy-current couplings and brakes, *Proc. IEE*, **113**, 825 (1966); Davies, E. J., James, B. and Wright, M. T.: Experimental verification of the generalised theory of eddy-current couplings, *Proc. IEE*, **122**, 67 (1975).

47.    Blake, L. R.: Conduction and induction pumps for liquid metals, *Proc. IEE*, **104A**, 49 (1957).

48.    Boldea, I. and Babescu, M.: Multilayer theory of d.c. linear brake with solid-iron secondary, *Proc. IEE*, **123**, 220 (1976).

49.    Watt, D. A.: Design of electromagnetic pumps for liquid metals, *Proc. IEE*, **106A**, 199 (1959).

50.   Maddock, B. J. and James, G. B.: Protection and stabilisation of large superconducting coils, *Proc. IEE,* **115**, 543 (1968).
51.   *Small electrical machines*. (IEE Conf. Pubn. 136 (1976)).

# List of Symbols

All basic quantities and expressions are in SI units. Unit symbols, and most quantity symbols, correspond to those listed in *Symbols and Abbreviations for Electrical and Electronic Engineering* (Institution of Electrical Engineers, 1979). For voltage and current, constant values are indicated by capitals, time-varying values by small letters.

## CAPITALS

| | | | |
|---|---|---|---|
| $A$ | amplifier gain | $L$ | inductance |
| | angular current density | $M$ | torque |
| | linear current density | $N$ | number of turns |
| | specific electric loading | $P$ | active power |
| $B$ | flux density | | power |
| | specific magnetic loading | $Q$ | flow rate |
| $C$ | capacitance | | reactive power |
| | number of coils or commutator | $R$ | radius |
| | sectors | | resistance |
| $D$ | diameter | $S$ | apparent power |
| $E$ | electromotive force | | number of slots |
| $F$ | magnetomotive force | | reluctance |
| $G$ | mass | | surface area |
| | motional inductance | $T$ | thermodynamic temperature |
| $H$ | magnetic field strength | | total time |
| $I$ | current | $V$ | voltage |
| $J$ | current density | $W$ | energy |
| | moment of inertia | $Y$ | pole pitch |
| $K$ | coefficient | $Z$ | number of armature conductors |

## SMALL LETTERS

| | | | |
|---|---|---|---|
| $a$ | acceleration | $c_p$ | specific heat capacity |
| | number of pairs of paths | $d$ | damping coefficient |
| | sectional area | | diameter |
| $b$ | breadth | | plate thickness |
| | pole arc or width | $e$ | electromotive force |
| | pressure ratio | $f$ | force |
| $c$ | cooling coefficient | | frequency |
| | leakage path dimension | $g$ | conductance |
| | number of periods | | |

370

| $h$ | height | $q$ | $\partial/\partial x$ |
| | pressure head | $r$ | radius |
| $i$ | current | | resistance |
| $k$ | coefficient | $s$ | cross-section |
| | constant | | slip |
| $l$ | length | $s$ | Laplace operator |
| $m$ | mass | $t$ | time |
| | mean radius | $u$ | peripheral speed |
| | coil-sides per slot | | translational speed |
| $n$ | harmonic order | $v$ | brush drop |
| | rotational frequency | | voltage |
| | speed (r/s) | $w$ | energy |
| $p$ | pole pairs | | width |
| | power | $x$ | exponent |
| | power loss | | length variable |
| | pressure | | ratio |
| $p$ | operator $d/dt$ | | reactance |
| $q$ | coil perimeter | | turns-ratio |
| | number of pulses | $y$ | pitch of slot or winding |
| | quantity | $z$ | impedance |
| | short-pitching | | conductors per coil-side |

## GREEK LETTERS

| $\Lambda$ | permeance | $\theta$ | temperature-rise |
| $\Phi$ | magnetic flux | | time-angle $\omega t$ |
| $\Psi$ | magnetic linkage | $\kappa$ | thermal conductivity |
| $\alpha$ | angular acceleration | $\lambda$ | permeance coefficient |
| | current ratio | | surface emissivity |
| | delay angle | | torque angle |
| | resistance ratio | $\mu$ | absolute permeability $\mu_r \mu_0$ |
| | resistance-temperature | | friction coefficient |
| | coefficient | $\mu_0$ | magnetic space constant |
| $\beta$ | angular pole-arc | $\mu_r$ | relative permeability |
| $\delta$ | angle | $\rho$ | resistivity |
| | density | $\sigma$ | conductivity |
| $\epsilon$ | chording angle | $\tau$ | time-constant |
| | radiation constant | $\phi$ | phase angle |
| $\eta$ | efficiency | $\omega$ | angular frequency |
| $\theta$ | angular displacement | $\omega_r$ | rotational frequency (rad/s) |
| | temperature | | |

SUBSCRIPTS

| | | | | |
|---|---|---|---|---|
| *a* | absorption | *m* | magnet |
| | acceleration | | maximum |
| | alternating | | mechanical |
| | armature | | motor |
| | phase A | | peak |
| *b* | back | | test value |
| | booster | *n* | compensating |
| | bottom | | neutralizing |
| | braking | | harmonic order |
| | brush | *o* | opening |
| | phase B | | operating |
| *c* | capacitor | | outer |
| | channel | | output |
| | coercive | | overhang |
| | commutator | ' | overlap |
| | conduction | *oc* | open-circuit |
| | core | *p* | pole |
| | critical | | potential |
| | phase C | | pulsational |
| | ratio | *q* | quadrature-axis |
| *cp* | compole | *r* | motional |
| *d* | damping | | radiation |
| | delay | | relative |
| | direct | | remanent |
| | direct-axis | | resistive |
| | diverter | | resultant |
| | duct | | rotational |
| | eddy | *s* | saturation |
| *e* | electric | | series |
| | e.m.f. | | shaft |
| | external | | slot |
| *em* | electromagnetic | | space |
| *f* | field | | stalling |
| | fluid | | storage |
| | friction | | stray |
| | front | | surface |
| *g* | gap | *sc* | short-circuit |
| | generator | *t* | time |
| *h* | hysteresis | | tooth |
| *i* | input | | top |
| | insulation | | torque |
| | iron | | turn |
| *l* | leakage | $t\frac{1}{3}$ | one-third tooth |
| | line | *v* | convection |
| | load | *w* | wedge |

| | | | |
|---|---|---|---|
| $x$ | excitation | 1 | final value |
| $y$ | yoke | | input |
| $z$ | zone | | primary |
| | | | rated |
| | | | source |
| 0 | free-space | | stator |
| | initial value | 2 | rotor |
| | no-load | | secondary |

# Index